Problems and Solutions in
Mathematical Olympiad

Secondary 2

Related Titles

*Problems and Solutions in Mathematical Olympiad
Secondary 1*
by Zun Shan
translated by Yi-Yang She
ISBN: 978-981-12-8720-6
ISBN: 978-981-12-8742-8 (pbk)

*Problems and Solutions in Mathematical Olympiad
Secondary 3*
by Jun Ge
translated by Huan-Xin Xie
ISBN: 978-981-12-2982-4
ISBN: 978-981-12-3141-4 (pbk)

*Problems and Solutions in Mathematical Olympiad
High School 1*
by Bin Xiong and Zhi-Gang Feng
translated by Tian-You Zhou
ISBN: 978-981-122-985-5
ISBN: 978-981-123-142-1 (pbk)

*Problems and Solutions in Mathematical Olympiad
High School 2*
by Shi-Xiong Liu
translated by Jiu Ding
ISBN: 978-981-122-988-6
ISBN: 978-981-123-143-8 (pbk)

*Problems and Solutions in Mathematical Olympiad
High School 3*
by Hong-Bing Yu
translated by Fang-Fang Lang and Yi-Chao Ye
ISBN: 978-981-122-991-6
ISBN: 978-981-123-144-5 (pbk)

Problems and Solutions in
Mathematical Olympiad

Secondary 2

Editors-in-Chief
Zun Shan *Nanjing Normal University, China*
Bin Xiong *East China Normal University, China*

Original Author
Xiong-Hui Zhao *Hunan Province Institute of Educational Science, China*

English Translator
Tian-You Zhou *Shanghai High School, China*

Copy Editors
Ming Ni *East China Normal University Press, China*
Ling-Zhi Kong *East China Normal University Press, China*
Lei Rui *East China Normal University Press, China*

**East China Normal
University Press**

World Scientific

Published by

East China Normal University Press
3663 North Zhongshan Road
Shanghai 200062
China

and

World Scientific Publishing Co. Pte. Ltd.

5 Toh Tuck Link, Singapore 596224

USA office: 27 Warren Street, Suite 401-402, Hackensack, NJ 07601

UK office: 57 Shelton Street, Covent Garden, London WC2H 9HE

Library of Congress Cataloging-in-Publication Data
Names: Zhao, Xiong-Hui, author. | Zhou, Tian-You (Translator), translator.
Title: Problems and solutions in Mathematical Olympiad : secondary 2 /
    Xiong-Hui Zhao, Hunan Province Institute of Educational Science, China,
    Tian-You Zhou, Shanghai High School, China.
Description: Singapore ; Hackensack : World Scientific, 2024.
Identifiers: LCCN 2024000594 | ISBN 9789811287237 (hardcover) |
    ISBN 9789811287435 (paperback) | ISBN 9789811287244 (ebook) |
    ISBN 9789811287251 (ebook other)
Subjects: LCSH: Mathematics--Problems, exercises, etc. | Mathematics--Competitions. |
    International Mathematical Olympiad.
Classification: LCC QA43 .Z485 2024 | DDC 510.76--dc23/eng/20240314
LC record available at https://lccn.loc.gov/2024000594

British Library Cataloguing-in-Publication Data
A catalogue record for this book is available from the British Library.

For any available supplementary material, please visit
https://www.worldscientific.com/worldscibooks/10.1142/13702#t=suppl

Desk Editors: Nambirajan Karuppiah/Tan Rok Ting

Typeset by Stallion Press
Email: enquiries@stallionpress.com

Editorial Board

Hong-Bing Yu
Ph.D. in Mathematics
Honorary Doctoral Supervisor
Professor, Suzhou University

Zun Shan
Ph.D. in Mathematics
Honorary Doctoral Supervisor
Chinese Team Leader, The 30th and 31st IMOs
Professor, Nanjing Normal University

Shun-Qing Hang
Mathematics Grand Grade Teacher
Senior Coach, The Chinese Mathematical Olympiad

Xiong-Hui Zhao
Ph.D. in Education
Deputy Director, Hunan Province Institute
of Educational Science

Ming Ni
Master Planner, Mathematics Olympiad Book Series
Director, Teaching Materials Branch
East China Normal University Press

Jun Ge
Ph.D. in Education
Senior Coach, The Chinese Mathematical Olympiad
Principal, High School Affiliated to Nanjing
Normal University, China

Bin Xiong
Honorary Doctoral Supervisor
Director, Shanghai Key Laboratory of Pure
Mathematics and Mathematical Practice
Professor, East China Normal University
Member, The Chinese Mathematical
Olympiad Committee

Preface

It is said that in many countries, especially the United States, children are afraid of mathematics and regard it as an "unpopular subject." But in China, the situation is very different. Many children love mathematics, and their math scores are also very good. Indeed, mathematics is a subject that the Chinese are good at. If you see a few Chinese students in elementary and middle schools in the United States, then the top few in the class of mathematics are none other than them.

At the early stage of counting numbers, Chinese children already show their advantages.

Chinese people can express integers from 1 to 10 with one hand, whereas those in other countries would have to use two.

The Chinese have long had the concept of digits, and they use the most convenient decimal system (many countries still have the remnants of base 12 and base 60 systems).

Chinese characters are all single syllables, which are easy to recite. For example, the multiplication table can be quickly mastered by students, and even the slow learners know the concept of "three times seven equals twenty one." However, for foreigners, as soon as they study multiplication, their heads get bigger. Believe it or not, you could try and memorize the multiplication table in English and then recite it; it is actually much harder to do so in English.

It takes the Chinese one or two minutes to memorize $\pi = 3.14159\cdots$ to the fifth decimal place. However, in order to recite these digits, the Russians wrote a poem. The first sentence contains three words, the second sentence contains one, and so on. To recite π, recite poetry first. In our opinion, as conveyed by *Problems and Solutions in Mathematical Olympiad*

Secondary 3, this is just simply asking for trouble, but they treat it as a magical way of memorization.

Application problems for the four arithmetic operations and their arithmetic solutions are also a major feature of Chinese mathematics. Since ancient times, the Chinese have compiled a lot of application questions which have contact or close relations with reality and daily life. Their solutions are simple and elegant, as well as smart and diverse, which helps increase students' interest in learning and enlighten students. For example: "There are one hundred monks and one hundred buns. One big monk eats three buns and three little monks eat one bun. How many big monks and how many little monks are there?"

Most foreigners can only solve equations, but Chinese have a variety of arithmetic solutions. As an example, one can turn each big monk into 9 little monks, and 100 buns indicate that there are 300 little monks, which contain 200 added little monks. As each big monk becomes a little monk, 8 more little monks are created, so $200/8 = 25$ is the number of big monks, and naturally, there are 75 little monks. Another way to solve the problem is to group a big monk and three little monks together, and so each person eats a bun on average, which is exactly equal to the overall average. Thus, the big monks and the little monks are not more and less after being organized this way; that is, the number of big monks is $100/(3 + 1) = 25$.

The Chinese are good at calculating, especially mental arithmetic. In ancient times, some people used their fingers to calculate (the so-called "counting by pinching fingers"). At the same time, China has long had computing devices, such as counting chips and abaci. The latter can be said to be the prototype of computers.

In the introductory stage of mathematics – the study of arithmetic, our country had obvious advantages, so mathematics is often the subject that our smart children love.

Geometric reasoning was not well developed in ancient China (but there were many books on the calculation of geometric figures in our country), and it was slightly inferior to that of the Greeks. However, the Chinese are good at learning from others. At present, the geometric level of middle school students in our country is far ahead of the rest of the world. Once, a foreign education delegation came to a junior high school class in our country. They thought that the geometric content taught was too in-depth for students to comprehend, but after attending the class, they had to admit that the content was not only understood by Chinese students but also well mastered.

The achievements of mathematics education in our country are remarkable. In international mathematics competitions, Chinese contestants have won numerous medals, which is the most powerful proof. Ever since our country officially sent a team to participate in the International Mathematical Olympiad in 1986, the Chinese team has won 14 team championships, which can be described as quite impressive. Professor Shiing-Shen Chern, a famous contemporary mathematician, once admired this in particular. He said, "One thing to celebrate this year is that China won the first place in the international math competition ... Last year it was also the first place." (Shiing-Shen Chern's speech, How to Build China into a Mathematical Power, at Cheng Kung University in Taiwan in October 1990.)

Professor Chern also predicted: "China will become a mathematical power in the 21st century."

It is certainly not an easy task to become a mathematical power. It cannot be achieved overnight. It requires unremitting efforts. The purpose of this series of books is as follows: (1) to further popularize the knowledge of mathematics, to make mathematics be loved by more young people, and to help them achieve good results; (2) to enable students who love mathematics to get better development and learn more knowledge and methods through the series of books.

"The important things in the world must be done in detail." We hope and believe that the publication of this series of books will play a role in making our country a mathematical power. This series was first published in 2000. According to the requirements of the curriculum reform, each volume is revised to different degrees.

A well-known mathematician, academician of the Chinese Academy of Sciences, and former chairman of the Chinese Mathematical Olympiad, Professor Yuan Wang, served as a consultant for this series of books and wrote inscriptions for young math enthusiasts. We express our heartfelt thanks. We would also like to thank East China Normal University Press, and in particular Mr. Ming Ni and Mr. Ling-zhi Kong. Without them, this series of books would not have been possible.

Zun Shan and Bin Xiong
May 2018

Contents

Solutions

Chapter 1

Linear Equations with Absolute Values

We know that the absolute value of a positive number is itself, and the absolute value of a negative number is its opposite, while the absolute value of 0 is 0. In other words,

$$|x| = \begin{cases} x, & x \geq 0, \\ -x, & x < 0. \end{cases}$$

Consequently,

$|x| = |-x|, |xy| = |x||y|$;

If $|x + y| = a$, then $a \geq 0$;

If $|x + y| = |x - y|$, then $x = 0$ or $y = 0$.

On the number axis, if a point A represents the number a, then $|a|$ is the distance from A to the origin, and $|a - b|$ is the distance between the points representing the numbers a and b. For example, $|9 + 7| = |9 - (-7)|$, which is the distance between the points representing 9 and -7.

Equations that have unknowns inside the absolute value symbol are called equations with absolute values.

To solve an equation with absolute values, the key is to remove the absolute value symbol based on its definition. In order to remove the absolute value symbol, we sometimes need to divide the range of a variable into parts for further analysis.

Example 1. The sum of all roots of the equation $|2x - 4| = 5$ is ( ).

(A) -0.5 (B) 4.5   (C) 5   (D) 4

1

Solution   From $|2x-4| = 5$, we obtain $2x-4 = 5$ or $2x-4 = -5$. Solving the equations separately, we have $x = 4.5$ or $x = -0.5$. Hence, the sum of all roots is $4.5 + (-0.5) = 4$, and the answer is (D).

Remark: If the absolute value of an expression is 5, then there are two possibilities: either it equals 5 or it equals -5.

Example 2. Solve the following equations:

(1) $|-5x - 7| = -6$;
(2) $|4x + 3| = 2x + 9$.

Solution   (1) Since the absolute value of a number is always non-negative, we have $|-5x - 7| \geq 0$, regardless of the value of x, so it is impossible that $|-5x - 7| = -6$, and the equation has no solution.

(2) The original equation implies

$$4x + 3 = \pm(2x + 9) \text{ and } 2x + 9 \geq 0.$$

If $4x + 3 = 2x + 9$, then $x = 3$.
   If $4x + 3 = -(2x + 9)$, then $x = -2$.
   When $x = 3$, we have $2x + 9 = 15 > 0$, and when $x = -2$, we have $2x + 9 = 5 > 0$, so both are solutions of the original equation. Therefore, the solutions of the original equation are $x = 3$ and $x = -2$.

Remark: For an equation $|ax + b| = cx + d$, we can convert it to

$$ax + b = \pm(cx + d),$$

where $cx + d \geq 0$. A solution to the equation $ax + b = \pm(cx + d)$ is also a solution to the original equation only if $cx + d \geq 0$; otherwise, it is an extraneous solution and needs to be abandoned.

Example 3. Solve the following equations:

(1) $|x - 1| + |x - 4| = 3$;
(2) $|3x| + |x - 2| = 4$.

Solution   (1) On the number axis, $|x-1|$ equals the distance between the points x and 1, while $|x-4|$ equals the distance between the points x and 4. Thus, $|x - 1| + |x - 4|$ is the sum of the distances from x to 1 and 4, which is greater than or equal to 3. When x lies between 1 and 4 (including the endpoints), the sum of the two distances is exactly 3. Therefore, we always have $|x - 1| + |x - 4| \geq 3$, where the equality is attained if and only if $1 \leq x \leq 4$, and the solution to the original equation is $1 \leq x \leq 4$.

(2) If $x < 0$, then the equation becomes $-3x - x + 2 = 4$, which yields $x = -\frac{1}{2}$.

If $0 \le x \le 2$, then the equation becomes $3x - x + 2 = 4$, which yields $x = 1$.

If $x > 2$, then the equation becomes $3x + x - 2 = 4$, which yields $x = \frac{3}{2}$. However, $x = \frac{3}{2}$ is not a solution to the original equation and is abandoned.

Therefore, the original equation has two solutions, namely $x = \frac{1}{2}$ and $x = 1$.

Remark: Linear equations with absolute values can have more than one solution. Problem (1) can also be solved through piecewise analysis.

Example 4. Suppose $0 < x < 10$ and $|x - 3| = a$, where a is an integer. Find all possible values of a.

Solution If $0 < x < 3$, then the equation becomes $3 - x = a$, or equivalently $x = 3 - a$. In this case, a can be 1 or 2.

If $3 \le x < 10$, then the equation becomes $x - 3 = a$, or equivalently $x = 3 + a$. In this case, a can be $0, 1, 2, 3, 4, 5$, or 6.

In combination, the value of a can be $0, 1, 2, 3, 4, 5$, or 6.

Example 5. Solve the equation: $|||x| - 2| - 1| = 3$.

Solution By the definition of absolute values, we have

$$||x| - 2| - 1 = \pm 3.$$

Thus, $||x| - 2| = 4$ or $||x| - 2| = -2$, where the latter has no solution.

Again, by definition, $||x| - 2| = 4$ implies $|x| - 2 = \pm 4$, so $|x| = 6$ or $|x| = -2$, and the latter has no solution.

Hence, we have $|x| = 6$, and $x = \pm 6$, and the solutions to the original equation are $x = 6$ and $x = -6$.

Remark: Equations with iterations of absolute values, such as $|||ax - b| - c| - d| = e$, can be solved by successively removing the absolute value symbols from outside to inside and finally converting them into linear equations with no absolute values.

Example 6. Solve the equation: $|x - |3x + 2|| = 4$.

Solution It follows by the definition of absolute values that

$$x - |3x + 2| = \pm 4,$$

so $|3x + 2| = x - 4$ or $|3x + 2| = x + 4$.

(1) $|3x + 2| = x = 4$:

Then, we have $3x + 2 = \pm(x - 4)$ and $x - 4 \geq 0$.

Solving $3x + 2 = x - 4$, we get $x = -3$, and solving $3x + 2 = -(x - 4)$, we get $x = \frac{1}{2}$. If $x = -3$, then $x - 4 = -7 < 0$, and if $x = \frac{1}{2}$, then $x - 4 = -\frac{7}{2} < 0$. So, both are extraneous solutions and should be abandoned.

(2) $|3x + 2| = x + 4$:

Then, $3x + 2 = \pm(x + 4)$ and $x + 4 \geq 0$.

Solving $3x + 2 = x + 4$, we get $x = 1$, and solving $3x + 2 = -(x + 4)$, we get $x = -\frac{3}{2}$.

If $x = 1$, then $x + 4 = 5 > 0$, and if $x = -\frac{3}{2}$, then $x + 4 = \frac{5}{2} > 0$. So, both are solutions to the original equation. Therefore, the solutions to the original equation are $x = 1$ and $x = -\frac{3}{2}$.

Example 7. Suppose the equation (where x is the unknown) $||x-2|-1| = a$ has three integer solutions, find the value of a.

Solution   It follows that $a \geq 0$ and $|x - 2| = 1 \pm a$.

If $a > 1$, then necessarily $|x - 2| = 1 + a$, and there are at most two solutions.

If $0 \leq a \leq 1$, then $|x - 2| = 1 \pm a$ yields

$$x = 3 + a, \ 3 - a, \ 1 - a, \ 1 + a.$$

Since there are three integer solutions x, it follows that a is an integer, and $a = 0$ or 1. If $a = 0$, then the original equation has only two solutions, so $a = 1$.

In this case, the solutions of the original equation $x = 0, 2, 4$.

Example 8. Suppose the equation (where x is the unknown) $|\frac{3}{5}|x| - 100| = a$ has four solutions, and if we arrange the four solutions in increasing order, the difference between consecutive terms is a constant. Find the value of a.

Solution   It follows that $\frac{3}{5}|x| - 100 = \pm a$, and $a \geq 0$.

If $a = 0$, then the equation has only two solutions, which does not satisfy the conditions.

If $a > 0$, then $|x| = \frac{5}{3}(100 \pm a)$, and the solutions are

$$-\frac{5}{3}(100 + a), -\frac{5}{3}(100 - a), \frac{5}{3}(100 + a), \frac{5}{3}(100 - a).$$

If $a = 100$, then $-\frac{5}{3}(100-a) = \frac{5}{3}(100-a)$, so there are only three solutions.

If $0 < a < 100$, then the solutions (in increasing order) are

$$-\frac{5}{3}(100 + a), -\frac{5}{3}(100 - a), \frac{5}{3}(100 - a), \frac{5}{3}(100 + a).$$

The differences between consecutive terms are equal, so we have $\frac{10}{3}a = \frac{10}{3} \times 100 - \frac{10}{3}a$, and $a = 50$.

If $a > 100$, then the equation $|x| = \frac{5}{3}(100 \pm a)$ has only two solutions, which does not satisfy the conditions.

Therefore, $a = 50$.

Have a Taste

The Nine Chapters on the Mathematical Art is an ancient mathematics book in China. As a mathematics textbook for China and its neighboring countries, it has experienced prosperity for 2000 years. The author of the book did not leave a surname or the year of the book. However, from the 246 problems collected in the book and their answers, it can be traced back to the pre-Qin dynasty era.

The Nine Chapters on the Mathematical Art is divided into nine chapters, namely "Fang Tian" (方田), "Su Mi" (粟米), "Shuai Fen" (衰分), "Shao Guang" (少广), "Shang Gong" (商功), "Jun Shu" (均输), Surplus and Loss (盈不足), Equation (方程), and Pythagorean (勾股). From the titles of those chapters, it may be difficult to guess their contents. If you want to know the problems and solutions, you need to take time to taste them.

Exercises

(I) Multiple-choice questions:

1. Which of the following statements is true? (   )

   (A) The solutions to the equation $a|x| = a$ are $x = \pm 1$.
   (B) The solutions to the equation $|a|x = |a|$ are $x = \pm 1$.
   (C) The solutions to the equation $|a|x = a$ are $x = \pm 1$.
   (D) The solutions to the equation $(|a|+1)|x| = |a|+1$ are $x = \pm 1$.

2. Suppose $|3990x + 1995| = 1995$, then the value of x is ( ).

   (A) 0 or 1   (B) -1 or 0   (C) 1 or -1 (D) -1 or -2

3. If $|2010x + 2010| = 20 \times 2010$, then x equals ( ).

   (A) 20 or -21 (B) -20 or 21   (C) -19 or 21   (D) 19 or -21

4. Suppose the equation (where x is the unknown) $|x + 1| + |x - 1| = a$ has at least one solution, then the value range of a is ( ).

   (A) $a \geq 0$ (B) $a > 0$ (C) $a \geq 1$ (D) $a \geq 2$

5. Suppose $||x + 2| - |y + 2|| = ||y + 2| - |x - y|| = ||x - y| - |x + 2||$, then $x + y = $ ( ).

   (A) -4 (B) -2 (C) 0   (D) 4

(II) Fill in the blanks:

6. If $|x| = x + 2$, then $19x^{94} + 3x + 27 = $ _____.

7. If $|x| + |2x| + |3x| + |4x| = 5x$, then $x = $ _____.

8. If $|5 - 4x| = 8$, then $x = $ _____.

9. When $x = $ _____, it holds that $|x - |x|| = -2x$.

10. If the equation $\frac{a}{1997}|x| - x - 1997 - 0$ has only negative solutions, then the value range of a is _____.

11. The equation $|| \cdots ||x - 10| - 9| - \cdots - 2| - 1| = 0$ has _____ different solutions.

(III) Solve the following equations:

12. $|5x + 6| = 6x - 5$.

13. $||x| - 4| = 5$.

14. $|5x - 3| = 2 + 3x$.

15. $|||x - 2| + 1| + 3| = 12$.

16. $||4x + 8| - 3x| = 5$.

17. $|x - 2| + |x - 3| = 1$.

(IV) Answer the following question:

18. Suppose the equation (where x is the unknown) $||x - 1| - 2| + a = 0$ has exactly three different integer solutions, find the value of a.

Chapter 2

Linear Inequalities with Absolute Values

There are several basic properties of inequalities with absolute values, such as

$$|a + b| \leq |a| + |b|, \quad |a - b| \leq |a| + |b|,$$

$$|a + b| \geq |a| - |b|, \quad |a - b| \geq |a| - |b|.$$

If $|a| < |b|$, then $a^2 < b^2$.

For an inequality $|x| > a$, where $a > 0$, we may obtain $x > a$ or $x < -a$. For $|x| \leq a$, where $a > 0$, we obtain $-a \leq x \leq a$.

To solve an inequality with absolute values, we usually first remove the absolute value symbol and then solve the resulting inequality.

Example 1. If $a > b > c$, then ( ).

(A) $|a|c > b|c|$ (B) $|ab| > |bc|$
(C) $|ab| \geq |ac|$ (D) $a|c| \geq b|c|$
(E) $a|c| > b|c|$

Solution We eliminate the false answers by giving counterexamples. Let $a = 5, b = 2, c = -6$, and we see that (A), (B), and (C) are all wrong. (E) is wrong because c can be zero, in which case $a|c| = b|c|$, and in all other cases, we have $a|c| > b|c|$. Therefore, the answer is (D).

Example 2. If a and b satisfy the inequality $||a| - (a + b)| < |a - |a + b||$, then which of the following statements is true? ( )

(A) $a > 0$ and $b > 0$ (B) $a < 0$ and $b > 0$
(C) $a > 0$ and $b < 0$ (D) $a < 0$ and $b < 0$

Solution   We square both sides of the given inequality and get

$$|a|^2 - 2|a| \cdot (a+b) + (a+b)^2 < a^2 - 2a \cdot |a+b| + |a+b|^2.$$

Simplify the above inequality, and we have

$$|a| \cdot (a+b) > a \cdot |a+b|.$$

Apparently $a \neq 0$, so it follows that $a + b > \frac{a}{|a|}|a+b|$.

In this case, $\frac{a}{|a|}$ can only be -1, so $a < 0$. Thus, $a + b > -|a+b|$, which implies $a + b > 0$, and $b > -a > 0$. Therefore, (B) is true.

Example 3. Suppose $|a| < |c|, |b| < 2|a|, b = \frac{a+c}{2}$, and $S_1 = |\frac{a-b}{c}|, S_2 = |\frac{b-c}{a}|, S_3 = |\frac{a-c}{b}|$. Compare the numbers $S_1, S_2,$ and S_3.

Solution   Plugging $b = \frac{a+c}{2}$ into S_1 and S_2, we have

$$S_1 = \left| \frac{a - \frac{a+c}{2}}{c} \right| = \left| \frac{a-c}{2c} \right|.$$

$$S_2 = \left| \frac{\frac{a+c}{2} - c}{a} \right| = \left| \frac{a-c}{2a} \right|.$$

Since $|a| < |c|$, it follows that $S_1 < S_2$.

Also, since $b < 2|a|$, we have $S_2 < S_3$. Therefore, $S_1 < S_2 < S_3$.

Example 4. Find the number of integer solutions to the inequality $|x - 2006| + |x| \leq 9999$.

Solution   If $x \geq 2006$, then the inequality becomes $x - 2006 + x \leq 9999$, so that $x \leq 6002$, where $x \geq 2006$ is satisfied simultaneously.

If $0 \leq x < 2006$, then the inequality becomes $2006 - x + x \leq 9999$, which apparently holds for all x in this range.

If $x < 0$, then the inequality becomes $2006 - x - x \leq 9999$, so that $x \geq -3996$.

In combination, the solution set to the original inequality is $-3996 \leq x \leq 6002$, in which there are $6002 - (-3996) + 1 = 9999$ integer solutions.

Remark: After removing the absolute value symbol through piecewise analysis, we need to examine whether the solution belongs to the corresponding interval.

Example 5. Solve the inequality: $|x + 2| > |x - 3|$.

Solution   The solutions to equations $x + 2 = 0$ and $x - 3 = 0$ are $x = -2$ and $x = 3$, respectively. Thus, the inequality can be divided into the following three systems of inequalities:

(1) $\begin{cases} x \geq 3, \\ x + 2 > x - 3; \end{cases}$

(2) $\begin{cases} -2 \leq x < 3, \\ x + 2 > 3 - x; \end{cases}$

(3) $\begin{cases} x < -2, \\ -(x + 2) > 3 - x. \end{cases}$

Solving (1), we have $x \geq 3$. Solving (2), we have $\frac{1}{2} < x < 3$. The system (3) has no solution.

Therefore, the solution set to the original inequality is $x > \frac{1}{2}$.

Remark: On observing the number axis, one can also assert that the distance from the point x to -2 is greater than its distance to 3, which also implies $x > \frac{1}{2}$.

Example 6. Solve the inequality: $|x - 4| - |2x - 3| \leq 1$.

Solution   The zeros of $|x-4|$ and $|2x-3|$ are $x = 4$ and $x = \frac{3}{2}$, respectively. Hence, we may divide the range of x into three pieces: $x \leq \frac{3}{2}, \frac{3}{2} < x \leq 4$, and $x > 4$.

(1) If $x \leq \frac{3}{2}$, then the inequality becomes

$$-x + 4 + 2x - 3 \leq 1,$$

so that $x \leq 0$. In this case, the solution is $x \leq 0$.

(2) If $\frac{3}{2} < x \leq 4$, the inequality becomes

$$-x + 4 - 2x + 3 \leq 1,$$

so that $x \geq 2$. In this case, the solution is $2 \leq x \leq 4$.

(3) If $x > 4$, then the inequality becomes

$$x - 4 - 2x + 3 \leq 1,$$

so that $x \geq -2$. In this case, the solution is $x > 4$.

In summary, the solution set to the original inequality is $x \leq 0$ or $x \geq 2$.

Example 7. Solve the inequality (where x is the unknown):

$$|ax - 1| > ax - 1.$$

Solution　Since $|ax - 1| > ax - 1$, it follows that $ax - 1 < 0$, so that $ax < 1$.

　　If $a > 0$, then the solution is $x < \frac{1}{a}$.

　　If $a < 0$, then the solution is $x > \frac{1}{a}$.

　　If $a = 0$, then x can be any real number.

Example 8. Suppose x, y satisfy the inequality $|x| + |y| \leq 1$, find the maximum of $x^2 - xy + y^2$.

Solution　We have $x^2 - xy + y^2 = \frac{1}{4}(x + y)^2 + \frac{3}{4}(x - y)^2$.

　　Since $|x| + |y| \leq 1$, it follows that

$$|x + y| \leq |x| + |y| \leq 1,$$

which also implies $(x + y)^2 \leq 1$. Similarly, $(x - y)^2 \leq 1$. Hence,

$$x^2 - xy + y^2 \leq \frac{1}{4} + \frac{3}{4} = 1.$$

The equality can be attained when one of x, y is 1 and the other is 0. Therefore, the maximum of $x^2 - xy + y^2$ is 1.

Reading

On the 10th anniversary of the death of the British mathematician and inventor of the telescope, T. Harriot (1560–1621) in 1631, his posthumous book *Artis analyticae Praxis* was published in order to commemorate his great achievements in mathematical research. In this book, he created a number of mathematical symbols, including the "greater than" and "less than" symbols, which were used for the first time. $a > b$ indicates that the quantity of a is greater than the quantity of b, and $a < b$ indicates that the quantity of a is less than the quantity of b. These symbols were not widely used until the early 18th century.

　　The first to use the "$\geq$" and "$\leq$" symbols was the French mathematician P. Bougeur (1698–1758), who used these symbols in 1734. The absolute value symbol "$|\ \ |$" was first introduced in 1841 by the German mathematician K. Weierstrass (1815–1897).

Exercises

(I) Multiple-choice questions:

1. If x, y are rational numbers, and $x < y$, then ( ).

    (A) $|x| < |y|$ (B) $|x| > |y|$
    (C) $|x| = |y|$ (D) All three above are possible

2. If a, b are rational numbers, which of the following is always true? ( ).

    (A) $a + b > a - |b|$ (B) $|a + b| > |a| + |b|$
    (C) $a - |a| \leq 0$ (D) $|a| > \frac{1}{b}$

3. If $a < 0, p > q > 0$, then ( ).

    (A) $|pa| > |qa|$ (B) $|pa| < |qa|$
    (C) $\left|\frac{a}{p}\right| > \left|\frac{a}{q}\right|$ (D) $\left|\frac{p}{a}\right| < \left|\frac{q}{a}\right|$

4. If $|a - c| < |b|$, then ( ).

    (A) $a - c < b$ (B) $a > c - b$
    (C) $|a| > |b| - |c|$ (D) $|a| < |b| + |c|$

5. If real numbers m, n satisfy $n < m < 2n$, then it follows that ( ).

    (A) $2(|m| + |n|) > 5|mn|$ (B) $1 < \left|\frac{m}{n}\right| < 2$
    (C) $|n - m| < |m - 2n|$ (D) $2(|m| - |n|) < |mn|$

6. If a, b, c are negative numbers that are greater than -1, then ( ).

    (A) $a^2 - b^2 - c^2 < 0$ (B) $a + b - c > 0$
    (C) $abc > -1$ (D) $|abc| > 1$

(II) Answer the following questions:

7. Suppose a is an integer, and the solution of the equation (in x) $a^2 x - 20 = 0$ is a prime number, and the following inequality is satisfied: $|ax - 7| > a^2$. Find the value of a.

8. Suppose the inequality $|x + 1| + |x - 3| \leq a$ has at least one solution, find the value range of a.

(III) Solve the following inequalities:

9. $|3x + 5| \leq 10$.

10. $|x + 2| > \frac{3x + 14}{5}$.

11. $|x + 3| - |2x - 1| < 2$.

12. $|x + 2| + |x - 3| > 2$.

Chapter 3

Polynomial Factorization (I)

Polynomial factorization means writing a polynomial as the product of several polynomials (factors). Usually, we require the factors to be irreducible, which are also called prime polynomials.

Common methods of factorization include common factors, grouping, the cross-method, and applying known identities.

When factorizing a polynomial, if all the summands are divisible by a common factor, we first take out this common factor. A common factor can be a number, a monomial, or a polynomial. If the summands do not have a common factor, we then try the cross-method or known identities. If these methods do not work, we may try grouping or other methods. In factorization, we need to factorize until no factor can be further factorized. The result is a product of polynomials, and duplicating factors are written as powers.

In this chapter, we discuss factorization based on the methods above as well as applying factorization to solve some problems.

Example 1. Factorize

$$(x-y)^{2n+1} + 2(y-x)^{2n}(y-z) - (x-z)(x-y)^{2n},$$

where n is a positive integer.

Solution Note that $(y-x)^{2n} = (x-y)^{2n}$, so

$$(x-y)^{2n+1} + 2(y-x)^{2n}(y-z) - (x-z)(x-y)^{2n}$$
$$= (x-y)^{2n}[(x-y) + 2(y-z) - (x-z)]$$
$$= (x-y)^{2n}(y-z).$$

Remark: The common factor method is based on the distributive law of multiplication, and it takes out the common factor of all terms in a polynomial. We need to pay attention to the signs of the terms. Sometimes, expressions that do not seem the same become indeed equal as common factors through appropriate transformation. For example, $(x-y)^{2n} = (y-x)^{2n}$, $(x-y)^{2n+1} = -(y-x)^{2n+1}$, where n is a positive integer.

Example 2. Suppose $x^4 + x^3 + x^2 + x + 1 = 0$, find the value of

$$1 + x + x^2 + \cdots + x^{2014}.$$

Solution   We first factorize the polynomial and plug in the known condition:

$$1 + x + x^2 + \cdots + x^{2014}$$
$$= (1 + x + x^2 + x^3 + x^4) + (x^5 + x^6 + x^7 + x^8 + x^9) + \cdots$$
$$\quad + (x^{2010} + x^{2011} + x^{2012} + x^{2013} + x^{2014})$$
$$= (1 + x + x^2 + x^3 + x^4)(1 + x^5 + x^{10} + \cdots + x^{2010})$$
$$= 0.$$

Remark: When there are ellipses in the sum, we need to examine how many terms there are in the summation and how to divide them into groups to ensure correctness.

Example 3. Factorize:

(1) $x^3(x - 2y) + y^3(2x - y)$;
(2) $x(x + 1)(x + 2)(x + 3) + 1$;
(3) $(x + y - 2xy)(x + y - 2) + (1 - xy)^2$.

Solution

(1) $x^3(x - 2y) + y^3(2x - y)$
$$= (x^4 - y^4) - 2xy(x^2 - y^2)$$
$$= (x^2 - y^2)(x^2 + y^2) - 2xy(x^2 - y^2)$$
$$= (x^2 - y^2)(x^2 + y^2 - 2xy)$$
$$= (x + y)(x - y)^3.$$
(2) $x(x + 1)(x + 2)(x + 3) + 1$
$$= [x(x + 3)][(x + 1)(x + 2)] + 1$$
$$= (x^2 + 3x)(x^2 + 3x + 2) + 1$$
$$= (x^2 + 3x)^2 + 2(x^2 + 3x) + 1$$
$$= (x^2 + 3x + 1)^2.$$

(3) $(x + y - 2xy)(x + y - 2) + (1 - xy)^2$
$$= (x - y)^2 - 2xy(x + y) - 2(x + y) + 4xy + 1 - 2xy + x^2 y^2$$
$$= [(x + y)^2 - 2(x + y) + 1] - 2xy(x + y - 1) + x^2 y^2$$
$$= (x + y - 1)^2 - 2(x + y - 1)xy + (xy)^2$$
$$= (x + y - 1 - xy)^2$$
$$= (xy - x - y + 1)^2$$
$$= (x - 1)^2 (y - 1)^2.$$

Remark: In this problem, we have repeatedly used fundamental identities. Such identities include:
$$a^2 - b^2 = (a - b)(a + b),$$
$$a^2 \pm 2ab + b^2 = (a \pm b)^2,$$
$$a^3 \pm b^3 = (a \pm b)(a^2 \mp ab + b^2),$$
$$a^3 \pm 3a^2 b + 3ab^2 \pm b^3 = (a \pm b)^3,$$
$$a^3 + b^3 + c^3 - 3abc = (a + b + c)(a^2 + b^2 + c^2 - ab - bc - ca).$$

The following identities are also used in factorization:

When n is a positive odd number,
$$a^n + b^n = (a + b)(a^{n-1} - a^{n-2}b + a^{n-3}b^2 - \cdots - ab^{n-2} + b^{n-1}).$$

When n is a positive integer,
$$a^n - b^n = (a - b)(a^{n-1} + a^{n-2}b + a^{n-3}b^2 + \cdots + ab^{n-2} + b^{n-1}).$$

In particular, if p, q are positive integers that are greater than 1, and q is odd, then
$$2^{pq} + 1 = (2^p)^q + 1 = (2^p + 1)[(2^p)^{q-1} - (2^p)^{q-2} + \cdots - 2^p + 1].$$

Factorization through identities can help in solving some interesting problems, such as Examples 4–6, as follows.

Example 4. It is known that $2^{48} - 1$ is divisible by two integers between 60 and 70. Find the two integers.

Solution   We have $2^{48} - 1$
$$= (2^{24})^2 - 1$$
$$= (2^{24} + 1)(2^{24} - 1)$$
$$= (2^{24} + 1)(2^{12} + 1)(2^{12} - 1)$$
$$= (2^{24} + 1)(2^{12} + 1)(2^6 + 1)(2^6 - 1).$$

It is easy to find that $2^6 + 1 = 65, 2^6 - 1 = 63$, while $2^{12} + 1 > 70$, $2^{24} + 1 > 70$, so the two integers are 65 and 63.

Example 5. Find the minimum value of the expression $2x^2 + 4xy + 5y^2 - 4x + 2y - 5$.

Solution   The given expression equals

$$(x^2 + 4xy + 4y^2) + (x^2 - 4x + 4) + (y^2 + 2y + 1) - 10$$
$$= (x + 2y)^2 + (x - 2)^2 + (y + 1)^2 - 10.$$

Since $(x + 2y)^2 \geq 0, (x - 2)^2 \geq 0, (y + 1)^2 \geq 0$, and when $x = 2, y = -1$, all three squares are equal to 0, we conclude that the minimum value of the given expression is -10.

Example 6. As shown in Figure 3.1, there is a number on each face of the cube, and the sums of numbers on each pair of opposite faces are equal. Suppose the number opposite 18 is a prime number a, the number opposite 14 is a prime number b, and the number opposite 35 is a prime number c. Find the value of the expression $a^2 + b^2 + c^2 - ab - bc - ca$.

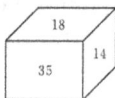

Fig. 3.1

Solution   It follows that $a + 18 = b + 14 = c + 35$.
      This implies $a - b = -4, b - c = 21, c - a = -17$.
      We have $a^2 + b^2 + c^2 - ab - bc - ca$

$$= \frac{1}{2}(a^2 - 2ab + b^2 + b^2 - 2bc + c^2 + c^2 - 2ac + a^2)$$

$$= \frac{1}{2}[(a - b)^2 + (b - c)^2 + (c - a)^2]$$

$$= \frac{1}{2}[(-4)^2 + 21^2 + (-17)^2]$$

$$= 373.$$

Therefore, the value of the given expression is 373.

Remark: In this problem, we have not used the condition that a, b, c are prime numbers. In fact, this condition combined with $18 + a = 14 + b = 35 + c$ implies that c has a different parity from that of a and b. Since $a \neq b$, and there is only one even prime number, the only possibility is $c = 2$, then $a = 19, b = 23$.

Using the method followed in this problem, when the condition $a = 1990x + 1989, b = 1990x + 1990, c = 1990x + 1991$ is given, we may still find the value of $a^2 + b^2 + c^2 - ab - bc - ca$.

Example 7. Factorize using the cross-method:

(1) $x^2 - 1007x - 2018$;
(2) $6x^2 + x - 35$;
(3) $x^2 + x + 6y^2 + 3y + 5xy$.

Solution

(1) $x^2 - 1007x - 2018 = (x - 1009)(x + 2)$.

$$1 \qquad -1009$$
$$1 \qquad 2$$
$$\overline{1 \times 2 + 1 \times (-1009) = -1007}$$

(2) $6x^2 + x - 35 = (2x + 5)(3x - 7)$.

$$2 \qquad 5$$
$$3 \qquad -7$$
$$\overline{2 \times (-7) + 3 \times 5 = 1}$$

(3) $x^2 + x + 6y^2 + 3y + 5xy$

$$= x^2 + (5y + 1)x + (6y^2 + 3y)$$
$$= x^2 + (5y + 1)x + 3y(2y + 1)$$
$$= (x + 3y)(x + 2y + 1).$$

$$1 \qquad 3y$$
$$1 \qquad 2y + 1$$
$$\overline{2y + 1 + 3y = 5y + 1}$$

Remark: In (3), we view the expression as a degree 2 polynomial in x, and y is treated as a constant, and then we apply the cross-method. This is also called the "pivot method." When a polynomial has multiple variables, we choose one variable to be the pivot and treat other variables as constants. Then, we rewrite the polynomial in descending power in the pivot to factorize. When applying the cross-method, if the quadratic term has a negative coefficient, we may first take out the negative sign. For example, we may write $-6x^2 + 12 - x = -(6x^2 + x - 12)$, and then factorize $6x^2 + x - 12$.

Example 8. Compute

$$\frac{(2 \times 5 + 2)(4 \times 7 + 2)(6 \times 9 + 2)(8 \times 11 + 2) \cdots (2016 \times 2019 + 2)}{(1 \times 4 + 2)(3 \times 6 + 2)(5 \times 8 + 2)(7 \times 10 + 2) \cdots (2015 \times 2018 + 2)}.$$

Solution The expression inside every pair of parentheses can be written as $n(n+3)+2$. Indeed, $n(n+3)+2 = n^2 + 3n + 2 = (n+1)(n+2)$. Hence, the given expression equals

$$\frac{(3 \times 4)(5 \times 6) \cdots (2017 \times 2018)}{(2 \times 3)(4 \times 5) \cdots (2016 \times 2017)}$$

$$= \frac{2018}{2} = 1009.$$

Remark: For the evaluation of complex calculation problems similar to Example 8, it is necessary to observe the characteristics of numbers, find out the rules, express numbers in letters, decompose the general case, and identify the method of simplification.

Example 9. Suppose the quadratic polynomial (in x) $x^2 - mx - 8$ (where m is an integer) can be factorized into two linear factors with integer coefficients. Find all possible values of m.

Solution It follows that we may factorize -8 as the product of two integers, whose sum is $-m$. Since -8 can be written as

$$(-1) \times 8, (-2) \times 4, (-4) \times 2, (-8) \times 1.$$

The possible values of $-m$ are $(-1)+8, (-2)+4, (-4)+2, (-8)+1$. Therefore, the possible values of m are $-7, -2, 2$, and 7.

Remark: If integer coefficients are not required, then the above method is not applicable. If the problem changed so that $x^2 - 8x - m$ can be factorized with integer coefficients, then -8 can be written as the sum of two integers (for example, $-50 + 42$ or $-9 + 1$), whose product is $-m$. In this case, there are infinitely many possible values.

Example 10. On the blackboard is a polynomial with missing coefficients:

$$x^3 + \square\, x^2 + \square\, x + \square.$$

Now, two people play a game. The first player chooses a blank and fills in a nonzero integer (which can be positive or negative). Then, the second player chooses one of the remaining blanks and fills in an integer. Finally, the first player writes an integer in the remaining blank.

Prove that, regardless of how the second player acts, the first player can make the polynomial factorizable into three monic linear factors with integer coefficients.

Proof: The first player may choose the blank in front of x and write integer -1, so that the polynomial is now $x^3 + \square\, x^2 - x + \square$.

Next, if the second player writes a number a in one blank, then the first player writes $-a$ in the remaining blank. Consequently, the polynomial becomes

$$x^3 + ax^2 - x - a \text{ or } x^3 - ax^2 - x + a.$$

Note that $x^3 + ax^2 - x - a = x(x^2 - 1) + a(x^2 - 1) = (x + a)(x + 1)(x - 1)$, and $x^3 - ax^2 - x + a = x(x^2 - 1) - a(x^2 - 1) = (x - a)(x + 1)(x - 1)$, we conclude that the first player can always reach the goal.

Reading

Thales (c. 624–c. 546 BC) was a Greek mathematician who was regarded as a versatile genius. Many stories about him have been told by people over the years.

Once, Thales was engaged in a small business where every day he had to cross a river with a mule carrying salt. One day, the mule slipped in the river. As it struggled to stand up, it found the bag much lighter since the salt had dissolved in the water. The mule remembered this experience, and in the following three days, it deliberately fell in the river and let the salt dissolve so that it could carry less burden. Thales observed the mule's behavior. So, on the fourth day, he let the mule carry a bag of sponges. The mule still fell in the water, but this time it found the bag much heavier. After that, the mule did not dare fall into the water again.

The story might not be true, but Thales did have great wisdom.

Exercises

(I) Multiple-choice questions:

1. If $x^2 + kx + \frac{1}{9}$ is the square of a polynomial, then k equals (   ).

   (A) -3 (B) 3 (C) $\frac{1}{3}$ or $-\frac{1}{3}$ (D) $\frac{2}{3}$ or $-\frac{2}{3}$

2. Suppose $x^2 + ax - 6$ can be written as the product of two linear factors with coefficients, and $a < 0$, then a equals (   ).

   (A) -2 or 3 (B) -7 or -4 (C) -1 or -5

   (D) any negative rational number

3. If n is an odd number, then $\frac{1}{4}(n^2 - 1)$ (   ).

   (A) must be odd

   (B) must be even

   (C) may be odd or even

   (D) may be an integer or a fraction (whose denominator is not 1)

4. Suppose n is a natural number, and four students calculate the value of $n^3 - n$, whose answers are listed as follows. The correct answer can be (   ).

   (A) 388944   (B) 388945   (C) 388954   (D) 388948

5. Let x be a nonzero natural number, and $y = x^4 + 2x^3 + 2x^2 + 2x + 1$, then (   ).

   (A) y must be a perfect square

   (B) there are finitely many x such that y is a perfect square

   (C) y must not be a perfect square

   (D) there are infinitely many x such that y is a perfect square

6. Suppose $x = 1\underbrace{00\cdots00}_{n\text{ copies of }0}1\underbrace{00\cdots0}_{(n+1)\text{ copies of }0}50$, then (   ).

   (A) x is a perfect square   (B) $x - 25$ is a perfect square

   (C) $x - 50$ is a perfect square   (D) $x + 50$ is a perfect square

7. Let a, b be positive numbers such that $12345 = (111 + a)(111 - b)$, then (   ).

   (A) $a > b$ (B) $a < b$ (C) $a = b$ (D) all three are possible

8. Numbers a, b satisfy $a + b - ab = 1$. It is known that a is not an integer. It follows that (   ).

   (A) b is not an integer   (B) b must be a positive integer

   (C) b is a negative integer   (D) b is an even integer

(II) Factorize the following polynomials:

9. $2018x^2 - (2018^2 - 1)x - 2018$.

10. $(a - b)a^6 + (b - a)b^6$.

11. $(x + y)(x + y + 2xy) + (xy + 1)(xy - 1)$.

12. $(x + y)(x - y) + 4(y - 1)$.

13. $x^2y - y^2z + z^2x - x^2z + y^2x + z^2y - 2xyz$.

14. $x^2 - (m^2 + n^2)x + mn(m^2 - n^2)$.

15. $(ax - by)^3 + (by - cz)^3 - (ax - cz)^3$.

(III) Compute the following:

16. $\dfrac{1^2 - 2^2 + 3^2 - 4^2 + \cdots + 2017^2 - 2018^2}{1 + 2 + 3 + 4 + \cdots + 2017 + 2018}$.

17. $\dfrac{20012000^2}{20011999^2 + 20012001^2 - 2}$.

18. $\dfrac{2019^3 - 2 \times 2019^2 - 2017}{2019^3 + 2019^2 - 2020}$.

19. $\dfrac{45.1^3 - 13.9^3}{31.2} + 45.1 \times 13.9$.

(IV) Answer the following questions:

20. Find all integers n between 1 and 100 such that $x^2 + x - n$ can be written as the product of two linear factors with integer coefficients.

21. Prove that $81^7 - 27^9 - 9^{13}$ is divisible by 45.

22. Prove that the product of four consecutive integers plus 1 is a perfect square.

23. Prove that if an integer m is equal to the sum of two perfect squares, then $2m$ is also equal to the sum of two perfect squares.

24. It is known that

$$11 - 2 = 3^2,$$

$$1111 - 22 = 33^2,$$

$$111111 - 222 = 333^2,$$

$$11111111 - 2222 = 3333^2.$$

Based on the above equalities, what general result can be inferred? Prove your result.

Chapter 4

Polynomial Factorization (II)

Factorization is a kind of identical transformation of polynomials, so it is an important tool when dealing with problems such as simplification, solving equations, evaluating expressions, and deriving proofs.

There are many ways of factorization. In addition to the basic methods learned in the previous chapter, other important methods are also commonly used, such as adding and splitting terms, changing variables, the method of undetermined coefficients, and the factor theorem.

1. Adding and Splitting Terms

When factorizing, we often need to transform the polynomial appropriately so that it can be factorized by grouping. Adding terms and splitting terms are two important transformation techniques. Adding terms means adding an additional term, as well as its opposite, to the polynomial (so that the value of the expression is unchanged). For example, if we add $4a^2 + (-4a^2)$ to $a^4 + 4$, then we get $a^4 + 4 = (a^4 + 4a^2 + 4) - 4a^2 = (a^2 + 2)^2 - (2a)^2$. Then, we can proceed to factorize. Splitting terms means writing a term in the polynomial as the sum of two or more terms, and then factorizing by grouping.

Example 1. Factorize the following polynomials:

(1) $x^4 + x^2y^2 + y^4$;
(2) $x^3 + 9x^2 + 26x + 24$;
(3) $x^4 + 2019x^2 + 2018x + 2019$;
(4) $a^3 + b^3 + c^3 - 3abc$.

Solution

(1) $x^4 + x^2y^2 + y^4$

$$= x^4 + 2x^2y^2 + y^4 - x^2y^2$$
$$= (x^2 + y^2)^2 - (xy)^2$$
$$= (x^2 + y^2 + xy)(x^2 + y^2 - xy).$$

(2) $x^3 + 9x^2 + 26x + 24$

$$= (x^3 + 7x^2 + 12x) + (2x^2 + 14x + 24)$$
$$= x(x^2 + 7x + 12) + 2(x^2 + 7x + 12)$$
$$= (x + 2)(x^2 + 7x + 12)$$
$$= (x + 2)(x + 3)(x + 4).$$

(3) $x^4 + 2019x^2 + 2018x + 2019$

$$= (x^4 + x^3 + x^2) - (x^3 + x^2 + x) + (2019x^2 + 2019x + 2019)$$
$$= x^2(x^2 + x + 1) - x(x^2 + x + 1) + 2019(x^2 + x + 1)$$
$$= (x^2 + x + 1)(x^2 - x + 2019).$$

(4) $a^3 + b^3 + c^3 - 3abc$

$$= a^3 + 3a^2b + 3ab^2 + b^3 + c^3 - 3a^2b - 3ab^2 - 3abc$$
$$= (a + b)^3 + c^3 - 3ab(a + b + c)$$
$$= (a + b + c)[(a + b)^2 - (a + b)c + c^2 - 3ab]$$
$$= (a + b + c)(a^2 + b^2 + c^2 - ab - bc - ca).$$

Remark: The aim of adding and splitting terms is to regroup them to find common factors or to fit them into recognizable patterns, such as perfect squares, difference of squares, sum/difference of cubes, and perfect cubes. Therefore, we need to do some trial and error based on the specific characteristics of the polynomial.

In (2), we can also group as $(x^3 + 2x^2) + (7x^2 + 14x) + (12x + 24)$, $(x^3 + 3x^2) + (6x^2 + 18x) + (8x + 24)$, or $(x^3 + 4x^2) + (5x^2 + 20x) + (6x + 24)$. Thus, it can be seen that there may be different ways of adding and splitting terms.

Example 2. Let a be a natural number. Is $a^4 - 3a^2 + 9$ a prime number or a composite number? Prove your result.

Analysis A prime number is divisible only by 1 and itself, while a composite number has other divisors. In order to determine whether $a^4 - 3a^2 + 9$ is prime or composite, we may try to factorize this polynomial.

Solution

$$a^4 - 3a^2 + 9 = (a^4 + 6a^2 + 9) - 9a^2$$
$$= (a^2 + 3)^2 - (3a)^2$$
$$= (a^2 + 3a + 3)(a^2 - 3a + 3).$$

If $a = 0$, then $a^4 - 3a^2 + 9 = 9$, which is composite.
If $a = 1$, then $a^4 - 3a^2 + 9 = 7$, which is prime.
If $a = 2$, then $a^4 - 3a^2 + 9 = 13$, which is prime.
If $a > 2$, then $a^2 + 3a + 3 > 1$, and $a^2 - 3a + 3 = (a-2)(a-1) + 1 > 1$.
So in this case, $a^4 - 3a^2 + 9$ is the product of two positive integers greater than 1, which means it must be a composite number.

Therefore, if $a = 1, 2$, then $a^4 - 3a^2 + 9$ is a prime number, and in the other cases, $a^4 - 3a^2 + 9$ is a composite number.

Remark: According to the national standard, natural numbers include and positive integers. In this problem, we may first take $a = 0, 1, 2, 3, \ldots$ and find the values of $a^4 - 3a^2 + 9$, then guess the answer, and try to find the proof for the result. It is noteworthy that factorizing $a^4 - 3a^2 + 9$ does not immediately imply that it is composite, and we need to determine whether the values of the factors are greater than 1.

2. Changing Variables

Some complicated polynomials may be simplified if we treat some of their parts as a whole and use a new alphabet to represent them. Such changes not only simplify the original expression but also emphasize the characteristics of the expression. Thus, we may first change a variable, then factorize the polynomial with the new variable, and finally plug in the original variable. This is the application of changing variables in polynomial factorization.

Example 3. Factorize the following polynomials:

(1) $(x^2 + x + 1)(x^2 + x + 2) - 12$;
(2) $(y + 1)^4 + (y + 3)^4 - 272$.

Analysis In (1), the expressions in the parentheses share the same quadratic and linear terms. If we let $y = x^2 + x + 1$, then the expression is converted into a simpler form, which is easier to factorize. In (2), we have a sum of two quartic polynomials, which is quite complicated to calculate directly. In this problem, we may take the arithmetic mean for a change of variable. Let $u = \frac{y+1+y+3}{2} = y + 2$, so that $y + 1 = u - 1, y + 3 = u + 1$. Then, some terms can be canceled, which simplifies our calculation.

Solution

(1) Let $x^2 + x + 1 = y$, then

$$(x^2 + x + 1)(x^2 + x + 2) - 12$$
$$= y(y + 1) - 12$$
$$= (y - 3)(y + 4)$$
$$= (x^2 + x - 2)(x^2 + x + 5)$$
$$= (x - 1)(x + 2)(x^2 + x + 5).$$

(2) Let $y + 2 = u$, then

$$(y + 1)^4 + (y + 3)^4 - 272$$
$$= (u - 1)^4 + (u + 1)^4 - 272$$
$$= (u^2 - 2u + 1)^2 + (u^2 + 2u + 1)^2 - 272$$
$$= [(u^2 + 1) - 2u]^2 + [(u^2 + 1) + 2u]^2 - 272$$
$$= 2(u^2 + 1)^2 + 8u^2 - 272$$
$$= 2u^4 + 12u^2 - 270$$
$$= 2(u^2 - 9)(u^2 + 15)$$
$$= 2(u - 3)(u + 3)(u^2 + 15)$$
$$= 2(y - 1)(y + 5)(y^2 + 4y + 19).$$

Example 4. Factorize the following polynomials:

(1) $(a - b)^4 + (a + b)^4 + (a^2 - b^2)^2$;
(2) $(x - 2xy + y)(x + y - 2) + (1 - xy)^2$.

Solution

(1) Let $a - b = x, a + b = y$, then

$$(a - b)^4 + (a + b)^4 + (a^2 - b^2)^2$$
$$= x^4 + y^4 + x^2 y^2$$
$$= (x^2 + y^2)^2 - x^2 y^2$$
$$= (x^2 + y^2 - xy)(x^2 + y^2 + xy)$$
$$= (2a^2 + 2b^2 - a^2 + b^2)(2a^2 + 2b^2 + a^2 - b^2)$$
$$= (a^2 + 3b^2)(3a^2 + b^2).$$

(2) Let $u = x + y, v = xy$, then

$$(x - 2xy + y)(x + y - 2) + (1 - xy)^2$$
$$= (u - 2v)(u - 2) + (1 - v)^2$$
$$= u^2 - 2u - 2uv + 4v + 1 - 2v + v^2$$
$$= (u - v)^2 - 2(u - v) + 1$$
$$= (u - v - 1)^2$$
$$= (x + y - xy - 1)^2$$
$$= (x - 1)^2 (1 - y)^2.$$

Remark: Changing variables is like building a bridge to cross a river and then removing the bridge after crossing it. The aim of changing variables is to highlight the characteristic of the expression and simplify the calculation. Without changing variables, one can also factorize directly.

3. The Method of Undetermined Coefficients

Some polynomials may not be factorized conveniently, but some characteristics, such as the highest degree terms and specific coefficients, may show the pattern of the result of factorization. For example, a cubic polynomial in one variable can be factorized either as the product of a linear factor and a quadratic factor or as the product of three linear factors. Thus, we may first assume the given polynomial equals the product of several factors, then write the equations that the coefficients need to satisfy, and finally solve for these coefficients. This is called the method of undetermined coefficients.

When using the method of undetermined coefficients, the following important property of polynomials is used:

If $a_n x^n + a_{n-1} x^{n-1} + \cdots + a_1 x + a_0 \equiv b_n x^n + b_{n-1} x^{n-1} + \cdots + b_1 x + b_0$, then $a_n = b_n, a_{n-1} = b_{n-1}, \ldots, a_1 = b_1, a_0 = b_0$. In other words, the coefficients of corresponding terms in an identity must be equal. Here, the symbol "$\equiv$" means that the two expressions are equal for all values of x.

Example 5. If the polynomial $x^2 - (a+5)x + 5a - 1$ can be written as the product of two linear factors $(x+b)$ and $(x+c)$, where b, c are integers, find the value of a.

Solution   We have

$$x^2 - (a+5)x + 5a - 1 = (x+b)(x+c) = x^2 + (b+c)x + bc.$$

It follows that

$$\begin{cases} b + c = -a - 5, & \text{①} \\ bc = 5a - 1. & \text{②} \end{cases}$$

With ① $\times$ 5 + ②, we have $5(b+c) + bc = -26$. Equivalently, $bc + 5(b+c) + 25 = -1$, or $(b+5)(c+5) = -1$.

Since b, c are integers, necessarily $\begin{cases} b = -4, \\ c = -6 \end{cases}$ or $\begin{cases} b = -6, \\ c = -4. \end{cases}$

Therefore, $a = -b - c - 5 = 5$.

Example 6. Factorize $6x^2 + xy - 2y^2 + 2x - 8y - 8$.

Analysis   The given expression is a quadratic polynomial in two variables, and the quadratic term $6x^2 + xy - 2y^2 = (3x + 2y)(2x - y)$. So, it can be inferred that the result of factorization must be $(3x + 2y + a)(2x - y + b)$, where a, b are undetermined constants. Thus, we may use the method of undetermined coefficients.

Solution   Suppose the given polynomial equals $(3x + 2y + a)(2x - y + b)$, then by expanding we have

$$6x^2 + xy - 2y^2 + 2x - 8y - 8$$
$$= 6x^2 + xy - 2y^2 + (2a + 3b)x + (-a + 2b)y + ab.$$

Then, comparing the coefficients on both sides, we have

$$\begin{cases} 2a + 3b = 2, & \text{①} \\ -a + 2b = -8, & \text{②} \\ ab = -8. & \text{③} \end{cases}$$

Using ① and ②, we solve that $a = 4, b = -2$, and is also satisfied.

Therefore, the original polynomial equals $(3x + 2y + 4)(2x - y + 2)$.

Remark: Comparing the coefficients gives three equations, while there are only two unknowns, so we should choose two to solve for the unknowns (preferably simpler ones) and then check whether the third equation is satisfied. If the third equation is not satisfied, then the solution is not valid, and the given polynomial cannot be factorized into the assumed form.

In this problem, another method to determine the coefficients is to take specific values of x, y and obtain equations for a, b.

4. The Factor Theorem

For a polynomial $a_n x^n + a_{n-1} x^{n-1} + \cdots + a_1 x + a_0$, if $x - a$ is one of its factors, then the value of the polynomial for $x = a$ must be 0. The converse, which is also true, is stated in the following theorem.

The Factor Theorem. If the value of the polynomial $a_n x^n + a_{n-1} x^{n-1} + \cdots + a_1 x + a_0$ is 0 for $x = a$, then $x - a$ is a factor of the polynomial.

For a polynomial $a_n x^n + a_{n-1} x^{n-1} + \cdots + a_1 x + a_0$ whose coefficients are all integers, if its value is 0 for $x = \frac{q}{p}$ (where p, q are coprime integers), then $x - \frac{q}{p}$ is a factor of the polynomial. In this case, it follows that a_n is divisible by p, and a_0 is divisible by q.

For polynomials with integer coefficients and $a_n = 1$, if $x - q$ is a factor, then q is a divisor of its constant term.

With the theorem above, when factorizing a polynomial with integer coefficients, we may factorize the coefficient of the highest term and the constant term to form some fractions, examine whether one or some of these fractions give a factor of the polynomial, and then do polynomial division.

Example 7. Suppose that $x^2 + x - 6$ is a factor of the polynomial $2x^4 + x^3 - ax^2 + bx + a + b - 1$, determine the values of a, b.

Solution Since $x^2 + x - 6 = (x + 3)(x - 2)$, it follows that both $x + 3$ and $x - 2$ divide $2x^4 + x^3 - ax^2 + bx + a + b - 1$. Thus, the value of the polynomial for $x = -3$ and $x = 2$ must be 0, which means
$$2 \times (-3)^4 + (-3)^3 - a \times (-3)^2 + b \times (-3) + a + b - 1 = 0,$$
$$2 \times 2^4 + 2^3 - a \times 2^2 + b \times 2 + a + b - 1 = 0.$$
By simplification,
$$\begin{cases} -8a - 2b + 134 = 0, \\ -a + b + 13 = 0. \end{cases}$$
Solving the equations, we get $a = 16, b = 3$.

Example 8. Factorize $x^4 + 2x^3 - 9x^2 - 2x + 8$.

Analysis   In this problem, we may use the method of undetermined coefficients or adding and splitting terms, while the factor theorem is also applicable. Since the leading coefficient is 1, and the constant term is 8, where the divisors of 8 are $\pm 1, \pm 2, \pm 4, \pm 8$, it follows that if $x - a$ is a factor of the polynomial, then a is necessarily one of the eight numbers. Plugging in the numbers gives us all the linear factors.

Solution   Since all the divisors of 8 are $\pm 1, \pm 2, \pm 4, \pm 8$, we plug in these numbers and calculate the value of the polynomial to find that the value is 0 for $x = 1, -1, 2, -4$. This implies that $x - 1, x + 1, x - 2, x + 4$ are all factors of the polynomial. On the other hand, since the polynomial has a degree of 4 and a leading coefficient of 1, we conclude that

$$x^4 + 2x^3 - 9x^2 - 2x + 8 = (x - 1)(x + 1)(x - 2)(x + 4).$$

Remark: If all the coefficients of the polynomial sum to 0, then $x - 1$ is a factor. If the sum of the coefficients of odd power terms is equal to the sum of the coefficients of even power terms, then $x + 1$ is a factor. Thus, the polynomial in this problem has factors $x + 1, x - 1$.

Exercises

(I) Multiple-choice questions:

1. If m, n are both integers greater than 1, then $m^4 + 4n^4$ must be ( ).

    (A) odd   (B) even   (C) prime   (D) composite

2. If $m = 2006^2 + 2006^2 \times 2007^2 + 2007^2$, then m ( ).

    (A) is an odd perfect square (B) is an even perfect square
    (C) is an even non-square (D) is a prime number

3. If $3x^3 - kx^2 + 4$ is divided by $3x - 1$ and the remainder is 3, then k equals ( ).

    (A) 3   (B) 4   (C) 9   (D) 10

4. Which of the following is a factor of $2x^3 + x^2 - 13x + 6$? ( ).

    (A) $x + 2$ (B) $x - 3$ (C) $2x - 1$ (D) $2x + 1$

(II) Fill in the blanks:

5. If the polynomial $x^3 + ax^2 + bx + 15$ has a factor $x^2 - 2$, then $a^2b^2 = $ _____ .

6. If $4x - 3$ is a factor of $4x^2 + 5x + a$, then the value of a is _____ .

7. If the polynomial (in x, y) $x^2 + kxy + 4y^2 - 3x - 6y + 2$ can be factorized into two linear factors with rational coefficients, then the value of k is _____ .

8. If the polynomial (in x, y) $6x^2 + 17xy + my^2 + 22x + 31y + 20$ can be factorized into two linear factors, then the value of m is _____ .

(III) Factorize the following polynomials:

9. $a^4 + 64b^4$.

10. $x^3 + 2x^2 - 5x - 6$.

11. $x^3 + 3x^2 + 3x + 2$.

12. $(x^2 - x - 3)(x^2 - x - 5) - 3$.

13. $3x^2 + 5xy - 2y^2 + x + 9y - 4$.

14. $x^3 + 6x^2 + 11x + 6$.

15. $x^4 - 2018x^2 + 2019x - 2018$.

(IV) Factorize using the given method:

16. Factorize with change of variable: $(x^2 + 5x + 6)(x^2 + 7x + 6) - 3x^2$.

17. Factorize with the method of undetermined coefficients: $x^2 - xy - 2y^2 - x + 5y - 2$.

18. Factorize with the factor theorem: $x^3 - 3x^2 - 13x + 15$.

(V) Answer the following questions:

19. Prove that $2x + 3$ is a factor of the polynomial $2x^4 - 5x^3 - 10x^2 + 15x + 18$.

20. Prove that $8x^2 - 2xy - 3y^2$ can be written as the difference of the squares of two polynomials with integer coefficients.

21. Prove that regardless of the values of x, y, the value of $3x^2 - 8xy + 9y^2 - 4x + 6y + 13$ is always positive.

22. Suppose that $12a^2 + 7b^2 + 5c^2 \le 12a|b| - 4b|c| - 16c - 16$, find the values of a, b, c.

Chapter 5

Calculation of Rational Fractions

A rational fraction is a fraction whose numerator and denominator are both polynomials. The denominator must contain unknowns, while the numerator need not contain unknowns. If the numerator equals 0 and the denominator is nonzero, the value of the rational fraction is 0. If the denominator equals 0, its value is undefined.

The properties of rational fractions show us how to perform the calculation. We have $\frac{A}{B} = \frac{A \times M}{B \times M}$, $\frac{A}{B} = \frac{A \div M}{B \div M}$, where M is a nonzero polynomial.

The calculation of rational fractions is important in algebraic transformation, and its rules include:

$$\frac{a}{c} \pm \frac{b}{c} = \frac{a \pm b}{c},$$

$$\frac{a}{b} \pm \frac{c}{d} = \frac{ad \pm bc}{bd},$$

$$\frac{a}{b} \cdot \frac{c}{d} = \frac{ac}{bd}, \quad \frac{a}{b} \div \frac{c}{d} = \frac{ad}{bc},$$

$$\left(\frac{a}{b}\right)^n = \frac{a^n}{b^n} \ (n \text{ is a positive integer}).$$

Adding and subtracting rational fractions with different denominators usually involves reduction to a common denominator by multiplying both the numerator and the denominator by the same object. Sometimes, we can avoid complicated calculations through certain techniques, such as step-by-step reduction, splitting terms, reduction by groups, and changing variables.

The concept and arithmetic of rational fractions are very similar to those of common fractions, so it is important to emphasize the analogy between problems of rational fractions and problems of common fractions.

Example 1. Suppose the value of $\dfrac{4}{3+\frac{2}{1+\frac{1}{x}}}$ is defined. Find the value range of x.

Solution The condition means that all the denominators in the expression are nonzero, so

$$x \neq 0, \quad 1+\frac{1}{x} \neq 0, \quad 3+\frac{2}{1+\frac{1}{x}} \neq 0.$$

Hence, $x \neq 0, x \neq -1, x \neq -\frac{3}{5}$, and the value of $\dfrac{4}{3+\frac{2}{1+\frac{1}{x}}}$ is defined if and only if x does not equal any of $0, -1, -\frac{3}{5}$.

Example 2. Suppose $a = \dfrac{x}{x^2+x+1}, x \neq 0$. Express $\dfrac{x^2}{x^4+x^2+1}$ in terms of a.

Solution It follows that $\frac{x^2+x+1}{x} = \frac{1}{a}$, so $x + \frac{1}{x} = \frac{1}{a} - 1$. Also,

$$\frac{x^4 + x^2 + 1}{x^2} = x^2 + \frac{1}{x^2} + 1$$

$$= \left(x + \frac{1}{x} \right)^2 - 1$$

$$= \left(\frac{1}{a} - 1 \right)^2 - 1$$

$$= \frac{1 - 2a}{a^2}.$$

Thus, $\frac{x^2}{x^4+x^2+1} = \frac{a^2}{1-2a}$.

Remark: When the denominator is a polynomial and the numerator is a monomial whose degree is lower than the denominator, it is common to write its reciprocal.

Example 3. Suppose $x \neq 0, \pm 1$, and we are given two numbers, x and 1, which can be used arbitrarily many times. We are allowed to do addition and subtraction, take reciprocals, and add parentheses to the numbers. Try to give a six-step expression (where parentheses do not count) that produces x^2.

Solution We easily have $\frac{1}{x} - \frac{1}{x+1} = \frac{1}{x(x+1)}$, so $x(x+1) = \frac{1}{\frac{1}{x} - \frac{1}{x+1}}$.

Thus, $x^2 = x(x+1) - x = \frac{1}{\frac{1}{x} - \frac{1}{x+1}} - x$.

Remark: We may obtain $1 + x, 1 - x\frac{1}{x}$ with the restricted arithmetic and further attempt to reach the answer. Another expression is $\frac{1}{\frac{1}{x-1} - \frac{1}{x}} + x$.

Example 4. Simplify

$$\frac{x-c}{(x-a)(x-b)} + \frac{b-c}{(a-b)(x-b)} + \frac{b-c}{(b-a)(x-a)}.$$

Solution The given expression equals

$$\frac{(x-c)}{(x-a)(x-b)} + \frac{b-c}{a-b}\left(\frac{1}{x-b} - \frac{1}{x-a}\right)$$

$$= \frac{(x-c)}{(x-a)(x-b)} - \frac{b-c}{(x-b)(x-a)}$$

$$= \frac{x-b}{(x-a)(x-b)} = \frac{1}{x-a}.$$

Remark: When adding several rational fractions, we first observe the characteristics of the summands and their relations. We do not necessarily add from left to right or write the common denominator of all summands. Instead, we treat the expressions with flexibility and simplify the calculation with methods such as step-by-step reduction.

Example 5. Simplify

$$\frac{x_2}{x_1(x_1 + x_2)} + \frac{x_3}{(x_1 + x_2)(x_1 + x_2 + x_3)}$$

$$+ \cdots + \frac{x_n}{(x_1 + x_2 + \cdots + x_{n-1})(x_1 + x_2 + \cdots + x_n)}.$$

Solution The given expression equals

$$\frac{1}{x_1} - \frac{1}{x_1 + x_2} + \frac{1}{x_1 + x_2} - \frac{1}{x_1 + x_2 + x_3}$$

$$+ \cdots + \frac{1}{x_1 + x_2 + \cdots + x_{n-1}} - \frac{1}{x_1 + x_2 + \cdots - x_n}$$

$$= \frac{1}{x_1} - \frac{1}{x_1 + x_2 + \cdots + x_n} = \frac{x_2 + x_3 + \cdots + x_n}{x_1(x_1 + x_2 + \cdots + x_n)}.$$

Example 6. Simplify

$$\frac{1}{1+x} + \frac{2}{2+x^2} + \frac{4}{1+x^4} + \cdots + \frac{2^n}{1+x^{2^n}}.$$

Analysis Observing the characteristics of the summands, we may add $\frac{1}{1-x}$ and $-\frac{1}{1-x}$ to the expression and do step-by-step reduction.

Solution The given expression equals

$$\frac{1}{1-x} + \frac{1}{1+x} + \frac{2}{2+x^2} + \frac{4}{1+x^4} + \cdots + \frac{2^n}{1+x^{2^n}} - \frac{1}{1-x}$$

$$= \frac{2}{1-x^2} + \frac{2}{2+x^2} + \frac{4}{1+x^4} + \cdots + \frac{2^n}{1+x^{2^n}} - \frac{1}{1-x}$$

$$= \cdots$$

$$= \frac{2^n}{1-x^{2^n}} + \frac{2^n}{1+x^{2^n}} - \frac{1}{1-x}$$

$$= \frac{2^{n+1}}{1-x^{2^{n+1}}} - \frac{1}{1-x}.$$

Example 7. Simplify

$$\frac{2a-b-c}{a^2-ab-ac+bc} + \frac{2b-c-a}{b^2-bc-ab+ac} + \frac{2c-a-b}{c^2-ac-bc+ab}.$$

Solution

$$\frac{2a-b-c}{a^2-ab-ac+bc} + \frac{2b-c-a}{b^2-bc-ab+ac} + \frac{2c-a-b}{c^2-ac-bc+ab}$$

$$= \frac{(a-b)+(a-c)}{(a-b)(a-c)} + \frac{(b-c)+(b-a)}{(b-c)(b-a)} + \frac{(c-a)+(c-b)}{(c-a)(c-b)}$$

$$= \frac{1}{a-c} + \frac{1}{a-b} + \frac{1}{b-a} + \frac{1}{b-c} + \frac{1}{c-b} + \frac{1}{c-a}$$

$$= 0.$$

Remark: It is quite complicated to directly reduce to a common denominator in this problem. By splitting terms with the identity $\frac{x+y}{xy} = \frac{1}{x} + \frac{1}{y}$, we have significantly reduced the amount of calculation.

Example 8. Simplify

$$\frac{b-c}{(a-b)(a-c)} + \frac{c-a}{(b-c)(b-a)} + \frac{a-b}{(c-a)(c-b)} + \frac{2}{b-a} - \frac{2}{c-a}.$$

Analysis By observation, we have $\frac{b-c}{(a-b)(a-c)} = \frac{1}{a-b} - \frac{1}{a-c}$, $\frac{c-a}{(b-c)(b-a)} = \frac{1}{b-c} - \frac{1}{b-a}$, $\frac{a-b}{(c-a)(c-b)} = \frac{1}{c-a} - \frac{1}{c-b}$. Such splitting makes it easier to reach the result.

Solution

$$\frac{b-c}{(a-b)(a-c)} + \frac{c-a}{(b-c)(b-a)} + \frac{a-b}{(c-a)(c-b)} + \frac{2}{b-a} - \frac{2}{c-a}$$

$$= \frac{1}{a-b} - \frac{1}{a-c} + \frac{1}{b-c} - \frac{1}{b-a} + \frac{1}{c-a} - \frac{1}{c-b} + \frac{2}{b-a} - \frac{2}{c-a}$$

$$= \frac{2}{b-c}.$$

Example 9. Simplify

$$\frac{3x^2 + 9x + 7}{x+1} - \frac{2x^2 + 4x + 3}{x-1} - \frac{x^3 + x + 1}{x^2 - 1}.$$

Analysis In the three summands, the numerators all have a higher degree than the denominators. So, we may do polynomial division with remainder and split every summand into a polynomial and a simpler fraction in order to simplify the calculation.

Solution

$$\frac{3x^2 + 9x + 7}{x+1} - \frac{2x^2 + 4x + 3}{x-1} - \frac{x^3 + x + 1}{x^2 - 1}$$

$$= (3x + 6) + \frac{1}{x+1} - (2x + 6) - \frac{3}{x-1} - x - \frac{2x+1}{x^2 - 1}$$

$$= \frac{1}{x+1} - \frac{3}{x-1} - \frac{2x+1}{x^2 - 1}$$

$$= \frac{-2x - 4}{x^2 - 1} - \frac{2x+1}{x^2 - 1}$$

$$= \frac{-4x - 5}{x^2 - 1}.$$

Reading

In ancient Greece, there was a fast-running god whose name was Achilles. Someone asked him to race with a turtle, and let the turtle start 1000 m ahead of Achilles. Suppose Achilles ran at 10 times the speed of the turtle. After the beginning of the race, when Achilles ran 1000 m, the turtle was 100 m ahead of him. After Achilles ran the next 100 m, the turtle was still 10 m ahead of him. Thus, some people concluded that Achilles could continue to approach the turtle, but he could never catch up with the turtle. Do you agree with this claim?

Exercises

(I) Fill in the blanks:

1. Suppose $a \neq 0$, and the value of $\frac{x-a}{\frac{1}{x}-a}$ is defined, then the value range of x is _____ .

2. The sum of all natural numbers x such that the value of $\frac{x^2+11}{x+1}$ is an integer is _____ .

3. If $\frac{a^{-1}+b}{a+b^{-1}} = m$, then $\frac{a^{-2}+b^2}{a^2+b^{-2}}$ equals _____ .

4. If $\frac{x-6y}{4x+3y} = -\frac{16}{17}$, then $\frac{x}{y} =$ _____ .

5. If $\frac{a}{b} = 20$, $\frac{b}{c} = 10$, then $\frac{a+b}{b+c} =$ _____ .

6. The minimum value of $\frac{6x^2+12x+10}{x^2+2x+2}$ is _____ .

7. Suppose $y_1 = 2x, y_2 = \frac{2}{y_1}, y_3 = \frac{2}{y_2}, \cdots, y_{2019} = \frac{2}{y_{2018}}, y_{2020} = \frac{2}{y_{2019}}$, then $y_1 \cdot y_{2020} =$ _____ .

(II) Simplify the following expressions:

8. $\frac{a^3 - a^2 b - ab^2 + b^3}{a^2 - 2|ab| + b^2}$.

9. $\left(\frac{\frac{1}{a}}{\frac{1}{a}-\frac{1}{b}} - \frac{\frac{1}{b}}{\frac{1}{a}+\frac{1}{b}}\right) \left(\frac{1}{a} - \frac{1}{b}\right) \cdot \frac{1}{\frac{1}{a^2}+\frac{1}{b^2}}$.

10. $\frac{1}{x-1} - \frac{1}{x+1} - \frac{2}{x^2+1} - \frac{4}{x^4+1} - \frac{8}{x^8+1}$.

11. $\frac{1}{x-1} + \frac{1}{(x-1)(x-2)} + \frac{1}{(x-2)(x-3)} + \cdots + \frac{1}{(x-99)(x-100)}$.

12. $\frac{(1+ax)^2-(a+x)^2}{(1+bx)^2-(b+x)^2} \div \frac{(1+ay)^2-(a+y)^2}{(1+by)^2-(b+y)^2}$.

13. $\frac{x^3-1}{x^3+2x^2+2x+1} + \frac{x^3+1}{x^3-2x^2+2x-1} - \frac{2(x^2+1)}{x^2-1}$.

(III) Answer the following questions:

14. Suppose that the value of $\frac{ax+7}{bx+11}$ is a constant for all x such that it is defined. Find the condition that a, b need to satisfy.

15. The expression $\frac{a}{b} - \frac{b}{a}$ can be written as the product of two factors whose sum is $\frac{a}{b} + \frac{b}{a}$. Find the two factors.

16. If $a_1 = x, a_{n+1} = 1 - \frac{1}{a_n} (n = 1, 2, 3, \ldots)$. Find a_{2006}.

17. Suppose n is a given positive integer, and there is exactly one integer k such that the inequality $\frac{8}{15} < \frac{n}{n+k} < \frac{7}{13}$ is satisfied. Find the maximum of n.

Chapter 6

Partial Fractions

A proper rational fraction is a rational fraction whose denominator has a higher degree than the numerator. For example, $\frac{2x^2}{6x^3 + 9x - 7}$ is a proper rational fraction. A non-proper rational fraction can be written as the sum of a polynomial and a proper rational fraction through polynomial division with the remainder.

Sometimes, we also need to write a proper rational fraction as the sum of several simpler proper rational fractions. For example, $\frac{2x}{x^2-4} = \frac{1}{x-2} + \frac{1}{x+2}$. This is called conversion to partial fractions.

When converting a proper rational fraction to a partial fraction, we first factorize the denominator, then set undetermined coefficients based on the degrees of the factors, and finally solve for the coefficients.

If one of the factors is $(x-a)^n$, then the corresponding partial fraction is

$$\frac{A_n}{(x-a)^n} + \frac{A_{n-1}}{(x-a)^{n-1}} + \cdots + \frac{A_1}{x-a},$$

where $A_1, A_2, \ldots, A_n$ are undetermined constants.

Similarly, if one of the factors is $(x^2 + px + q)^2$, then the corresponding partial fraction is

$$\frac{Ax + B}{x^2 + px + q} + \frac{Cx + D}{(x^2 + px + q)^2},$$

where A, B, C, D are undetermined constants.

Example 1. Suppose $\frac{5x+11}{2x^2+7x-4} = \frac{A}{2x-1} + \frac{B}{x+4}$, where A, B are constants. Find the value of $2020A - 2019B$.

Solution   We reduce the right-hand side to a common denominator for summation so that

$$\frac{5x+11}{2x^2+7x-4} = \frac{(A+2B)x+(4A-B)}{2x^2+7x-4}.$$

Thus, $\begin{cases} A+2B = 5, \\ 4A-B = 11. \end{cases}$ Solving the equations, we have $A = 3, B = 1$, so $2020A - 2019B = 4041$.

Example 2. Simplify

$$\frac{1}{x^2-7x+10} + \frac{1}{x^2-x-2} + \frac{1}{x^2+5x+4} + \frac{1}{x^2+11x+28}.$$

Solution   We convert each part to a partial fraction:

$$\frac{1}{x^2-7x+10} = \frac{1}{(x-5)(x-2)} = \frac{1}{3}\left(\frac{1}{x-5} - \frac{1}{x-2}\right),$$

$$\frac{1}{x^2-x-2} = \frac{1}{(x-2)(x+1)} = \frac{1}{3}\left(\frac{1}{x-2} - \frac{1}{x+1}\right),$$

$$\frac{1}{x^2+5x+4} = \frac{1}{(x+1)(x+4)} = \frac{1}{3}\left(\frac{1}{x+1} - \frac{1}{x+4}\right),$$

$$\frac{1}{x^2+11x+28} = \frac{1}{(x+4)(x+7)} = \frac{1}{3}\left(\frac{1}{x+4} - \frac{1}{x+7}\right).$$

Hence, the original expression equals

$$\frac{1}{3}\left(\frac{1}{x-5} - \frac{1}{x-2} + \frac{1}{x-2} - \frac{1}{x+1} + \frac{1}{x+1} - \frac{1}{x+4} + \frac{1}{x+4} - \frac{1}{x+7}\right)$$

$$= \frac{1}{3}\left(\frac{1}{x-5} - \frac{1}{x+7}\right) = \frac{4}{(x-5)(x+7)}.$$

Remark: Sometimes, converting to partial fractions can make things much easier to calculate than directly reducing to a common denominator.

Example 3. Suppose that $\frac{x^2+2x+3}{(x+2)^3} = \frac{A}{x+2} + \frac{B}{(x+2)^2} + \frac{C}{(x+2)^3}$. Find the value of A^3B^2C.

Solution   Removing the denominator, we have

$$x^2 + 2x + 3 = Ax^2 + (4A+B)x + (4A+2B+C).$$

Comparing the coefficients, it follows that

$$\begin{cases} A = 1, \\ 4A+B = 2, \\ 4A+2B+C = 3. \end{cases}$$

Therefore, $A = 1, B = -2, C = 3$, so $A^3B^2C = 12$.

Example 4. Convert $\frac{13x+14}{2x^3-13x^2-7x}$ to a partial fraction.

Solution   Since $2x^3 - 13x^2 - 7x = x(x-7)(2x+1)$, we may assume

$$\frac{13x+14}{2x^3-13x^2-7x} = \frac{A}{x} + \frac{B}{x-7} + \frac{C}{2x+1},$$

Where multiplying both sides by $x(x-7)(2x+1)$, we have

$$13x + 14 = A(x-7)(2x+1) + Bx(2x+1) + Cx(x-7),$$

or, equivalently,

$$13x + 14 = (2A + 2B + C)x^2 + (-13A + B - 7C)x - 7A.$$

Comparing coefficients, we obtain

$$\begin{cases} 2A + 2B + C = 0, \\ -13A + B - 7C = 13, \\ -7A = 14. \end{cases}$$

Thus, $A = -2, B = 1, C = 2$, and

$$\frac{13x+14}{2x^3-13x^2-7x} = -\frac{2}{x} + \frac{1}{x-7} + \frac{2}{2x+1}.$$

Remark: In the result following the conversion, all summands need to be in the simplest form.

Example 5. Convert $\frac{x^3+16}{(x-2)^4}$ to a partial fraction.

Solution   Let $x - 2 = t$, then $x = t + 2$, $t \neq 0$. Thus,

$$x^3 + 16 = (2+t)^3 + 16 = t^3 + 6t^2 + 12t + 24,$$

so we have

$$\frac{x^3+16}{(x-2)^4} = \frac{t^3+6t^2+12t+24}{t^4} = \frac{1}{t} + \frac{6}{t^2} + \frac{12}{t^3} + \frac{24}{t^4}.$$

Equivalently,

$$\frac{x^3+16}{(x-2)^4} = \frac{1}{x-2} + \frac{6}{(x-2)^2} + \frac{12}{(x-2)^3} + \frac{24}{(x-2)^4}.$$

Remark: The denominator here is $(x+a)^n$, but we have not used the method of undetermined coefficients. Instead, we have solved it by changing variables. This happens when the denominator has a relatively high degree and the numerator is relatively simple, in which case changing variables can be easier than using undetermined coefficients.

Example 6. Convert $\frac{x+4}{x^3+2x-3}$ to a partial fraction.

Solution

$$x^3 + 2x - 3 = (x^3 - 1) + (2x - 2)$$
$$= (x - 1)(x^2 + x + 1) + 2(x - 1)$$
$$= (x - 1)(x^2 + x + 3).$$

Assume $\frac{x+4}{x^3+2x-3} = \frac{A}{x-1} + \frac{Bx+C}{x^2+x+3}$, and we have

$$x + 4 = A(x^2 + x + 3) + (Bx + C)(x - 1)$$
$$= (A + B)x^2 + (A - B + C)x + 3A - C.$$

Comparing coefficients, we get

$$\begin{cases} A + B = 0, \\ A - B + C = 1, \\ 3A - C = 4. \end{cases}$$

Finally, solve the equations, and we have $A = 1, B = -1, C = -1$. Therefore,

$$\frac{x + 4}{x^3 + 2x - 3} = \frac{1}{x - 1} - \frac{x + 1}{x^2 + x + 3}.$$

Exercises

(I) Fill in the blanks:

1. If $\frac{1}{x^2+3n} = \frac{A}{n} + \frac{B}{n+3}$, then $A =$ _____, $B =$ _____.

2. If $\frac{3x-4}{x^2+2-3x} = \frac{a}{x-1} + \frac{b}{x-2}$, then $ab =$ _____.

3. If $\frac{x^2-1}{x^2-5x+6} = M + \frac{a}{x-2} + \frac{b}{x-3}$, where M, a, b are constants, then $M + a + b =$ _____.

4. Suppose $\frac{ax-1}{(x-1)(x^2+1)} = \frac{b}{x+m} + \frac{cx+5}{x^2+n}$ is an identity, where $\frac{ax-1}{(x-1)(x^2+1)}$ and $\frac{cx+5}{x^2+n}$ are both in the simplest form, and a, b, c, m, n are all constants. Then, $c =$ _____, $b =$ _____, $a =$ _____.

5. If $\frac{A-17x}{7x^2+41x-6} = \frac{B}{x+6} + \frac{4}{7x-1}$, then $A =$ _____, $B =$ _____.

(II) Answer the following questions:

6. Suppose $\frac{3x^2-7x+2}{(x-1)(x+1)} = 3 + \frac{A}{x-1} + \frac{B}{x+1}$, where A, B are constants, find the value of $4A - 2B$.

7. Suppose $\frac{2x^2+x-11}{x^3-x^2} = \frac{A}{x} + \frac{B}{x^2} + \frac{C}{x-1}$, where A, B, C are constants. Find the value of $A^2 + B^2 + C^2$.

8. Suppose that in the identity $\frac{Mx+N}{x^2+x-2} = \frac{2}{x+a} - \frac{c}{x+b}$, $\frac{Mx+N}{x^2+x-2}$ is in the simplest form, where M, N, a, b, c are constants such that $a > b, a + b = c$. Find the value of N.

(III) Convert to partial fractions:

9. $\frac{x^2+2}{(x-1)^3}$.

10. $\frac{2x^2+x+2}{x^3+1}$.

11. $\frac{x^2+1}{(x+1)^2(x+2)}$.

12. $\frac{12x^2+20x-29}{(2x-1)(4x^2-4x-15)}$.

Chapter 7

Polynomial Equations and Fractional Equations with Unknown Constants

Some equations may have alphabets that represent constants (coefficients) rather than unknowns. These are called equations with unknown constants. When solving such equations, we need to discuss different cases based on the values of the unknown constants.

For example, in the equation $ax = b$, where x is the unknown, if $a \neq 0$ is not certain, then we need to discuss the following cases:

(1) If $a \neq 0$, then $ax = b$ is a linear equation with a unique solution $x = \frac{b}{a}$.
(2) If $a = 0$ and $b \neq 0$, then the equation has no solution.
(3) If $a = b = 0$, then the equation is $0x = 0$, which has infinitely many solutions.

In this chapter, we are mainly concerned with equations with unknown constants that can be converted into linear equations.

A fractional equation not only involves fractions but also needs to have unknowns in the denominator.

The primary idea of solving a fractional equation is to convert it into a polynomial equation, solve the polynomial equation, and finally check the denominator for extraneous solutions.

There are multiple ways of conversion to polynomial equations. One is to directly multiply both sides by the least common denominator, and another is to apply a change of variable. Since converting to polynomial equations may produce extraneous solutions, we always need to test the validity of the solutions.

Example 1. Solve the following equation for x:

$$(a - 1)(a - 3)x + (x + 2) = a.$$

Solution   The equation is equivalent to

$$(a - 2)^2 x = a - 2.$$

(1) If $a \neq 2$, then $(a - 2)^2 \neq 0$, so the solution is $x = \frac{1}{a-2}$.
(2) If $a = 2$, then the equation becomes $0x = 0$, so x can be any number.

Example 2. Suppose a, b are constants, and the equation (for x) $\frac{2kx+a}{3} = 2 + \frac{x-bk}{6}$ always has a solution $x = 1$, regardless of the value of k. Find the value of a, b.

Solution   From the conditions, we may plug in $x = 1$, so

$$\frac{2k + a}{3} = 2 + \frac{1 - bk}{6}$$

is an identity. Hence,

$$4k + 2a = 13 - bk,$$

for all k. Choosing $k = 0$, we get $a = 6.5$, and choosing $k = -1$, with the same value of a, we have $b = -4$.

Example 3. Suppose the solutions of the equation (for x) $x + \frac{b}{x} = a + \frac{b}{a}$ are $x_1 = a$, $x_2 = \frac{b}{a}$. Find the solution(s) of the equation $x - \frac{2}{x-1} = a - \frac{2}{a-1}$.

Solution   We observe that the equation to solve has similarities to the known equation. Hence, we rewrite the equation as

$$x - 1 + \frac{-2}{x - 1} = a - 1 + \frac{-2}{a - 1}.$$

Let $x - 1 = y, a - 1 = m, -2 = n$, then the equation becomes

$$y + \frac{n}{y} = m + \frac{n}{m}.$$

From the given result, we have $y_1 = m, y_2 = \frac{n}{m}$, so $x_1 - 1 = a - 1, x_2 - 1 = \frac{-2}{a-1}$.

Therefore, the solutions of the equation are

$$x_1 = a, \quad x_2 = \frac{-2}{a - 1} + 1 = \frac{a - 3}{a - 1}.$$

Example 4. Suppose the equation (where x is the unknown) $\frac{1}{x-1} - \frac{a}{2-x} = \frac{2(a+1)}{x^2-3x+2}$ has no solution. Find the value of a.

Solution Removing the denominator, we have

$$(x-2) + a(x-1) = 2(a+1).$$

or, equivalently,

$$(a+1)x = 3a + 4. \qquad \qquad ①$$

If $a = -1$, then (①) has no solution, and neither does the original equation.

If $a \neq -1$, then there must be an extraneous solution, either $x = 1$ or $x = 2$.

If $x = 1$ is the extraneous solution, then $\frac{3a+4}{a+1} = 1$, so $a = -\frac{3}{2}$.

If $x = 2$ is the extraneous solution, then $\frac{3a+4}{a+1} = 2$, so $a = -2$.

Therefore, if the equation has no solution, then the value of a is one of $-1, -\frac{3}{2}, -2$.

Example 5. Solve the equation

$$\frac{x+2015}{x+2014} + \frac{x+2017}{x+2016} = \frac{x+2018}{x+2017} + \frac{x+2014}{x+2013}.$$

Solution The equation can be simplified as

$$\frac{1}{x+2014} + \frac{1}{x+2016} = \frac{1}{x+2017} + \frac{1}{x+2013}.$$

Or, equivalently, $\frac{1}{x+2016} - \frac{1}{x+2017} = \frac{1}{x+2013} - \frac{1}{x+2014}$. Hence,

$$\frac{1}{(x+2016)(x+2017)} = \frac{1}{(x+2013)(x+2014)},$$

so $(x+2016)(x+2017) = (x+2013)(x+2014)$.

Expanding both sides yields

$$6x = 2013 \times 2014 - 2016 \times 2017,$$

which means $x = -2015$.

One can check that $x = -2015$ is the solution to the original equation.

Remark: When solving relatively complicated fractional equations, we should carefully inspect the equation and simplify based on its characteristics, rather than removing the denominator hastily.

Example 6. Solve the equation

$$\frac{4x^3 + 10x^2 + 16x + 1}{2x^2 + 5x + 7} = \frac{6x^3 + 10x^2 + 5x - 1}{3x^2 + 5x + 1}.$$

Analysis Both sides of the equation are quite complicated, and direct calculation seems cumbersome. However, we observe that the ratios of the highest terms of the numerator and denominator are the same on both sides, as $\frac{4x^3}{2x^2} = \frac{6x^3}{3x^2} = 2x$. So, we may simplify the equation first.

Solution The equation can be written as

$$\frac{2x(2x^2 + 5x + 7) + 2x + 1}{2x^2 + 5x + 7} = \frac{2x(3x^2 + 5x + 1) + 3x - 1}{3x^2 + 5x + 1},$$

which is equivalent to $\frac{2x+1}{2x^2+5x+7} = \frac{3x-1}{3x^2+5x+1}$.

If $2x + 1 = 0$ or $3x - 1 = 0$, it is easy to check that the corresponding values of x are not solutions of the equation. Hence, we have

$$\frac{2x^2 + 5x + 7}{2x + 1} = \frac{3x^2 + 5x + 1}{3x - 1},$$

which can be reduced to $\frac{4x+7}{2x+1} = \frac{6x+1}{3x-1}$. Further reduction gives $\frac{5}{2x+1} = \frac{3}{3x-1}$, whose solution is $x = \frac{8}{9}$. We check that $x = \frac{8}{9}$ is the solution of the original equation.

Remark: In this problem, we have successively simplified the equation to find the solution. One should carefully inspect whether the simplified equation has the same solution(s) as the original equation in order to avoid missing solutions.

Example 7. Solve the system of equations:

$$\begin{cases} \dfrac{10}{x + y} + \dfrac{3}{x - y} = -5, \\[2mm] \dfrac{15}{x + y} - \dfrac{2}{x - y} = -1. \end{cases}$$

Analysis Directly removing denominators would produce a system of quadratic equations with two unknowns, which is out of scope. However, the characteristic of the equations shows that we may try the method of changing variables.

Solution   Let $\frac{1}{x+y} = u$, $\frac{1}{x-y} = v$, then the system becomes

$$\begin{cases} 10u + 3v = -5, \\ 15u - 2v = -1. \end{cases}$$

Solving the system of equations, we obtain $\begin{cases} u = -\frac{1}{5} \\ v = -1 \end{cases}$. Hence, $\begin{cases} x + y = -5 \\ x - y = -1 \end{cases}$.

Therefore, $\begin{cases} x = -3 \\ y = -2 \end{cases}$, and one can check that it is the solution of the original equations.

Example 8. A reservoir has three inlets, A, B, and C. If A and B are simultaneously open for one hour, then $\frac{1}{2}$ of the reservoir is filled with water. If B and C are simultaneously open for one hour, then $\frac{2}{3}$ of the reservoir is filled with water. If A and C are simultaneously open for 1 h and 12 min, then the whole reservoir is filled with water. If all three inlets are open, how long does it take to fill $\frac{1}{3}$ of the reservoir?

Solution   Suppose it takes a min to fill the reservoir if only A is open, and define b and c similarly. Also, suppose it takes x min to fill $\frac{1}{3}$ of the reservoir if all three are open. Then,

$$\frac{1}{a} + \frac{1}{b} = \frac{1}{2} \div 60, \qquad \text{①}$$

$$\frac{1}{b} + \frac{1}{c} = \frac{2}{3} \div 60, \qquad \text{②}$$

$$\frac{1}{c} + \frac{1}{a} = \frac{1}{72}, \qquad \text{③}$$

$$\frac{1}{a} + \frac{1}{b} + \frac{1}{c} = \frac{1}{3x}. \qquad \text{④}$$

Note that ① + ② + ③ implies $\frac{1}{a} + \frac{1}{b} + \frac{1}{c} = \frac{1}{60}$. Plugging into ④, we obtain $x = 20$.

Therefore, it takes 20 min to fill $\frac{1}{3}$ of the reservoir if all three inlets are open.

Reading

Proverbs give people encouragement, advice, enlightenment, and wisdom. Thinking of some proverbs when solving problems may bring us a different feeling.

(1) When understanding the meaning of a problem, remember:
   If you know yourself and your enemy, you will not be defeated in a hundred battles.
   Wise people think twice, while fools act recklessly.
(2) When looking for problem-solving ideas, remember:
   Dripping water penetrates the stone and succeeds naturally.
   Do not be discouraged by the first defeat, and continue working hard.
   All roads lead to Rome.
   The wise adapt to changes, and fools are stubborn.
   The purpose of fishing is fish, not fishing.
(3) When answering, remember:
   Mind the weather before going out, and ask for directions before traveling far. (Do not act recklessly.)
   Nothing ventured, nothing gained.
(4) When reviewing the answers, remember:
   Gain new knowledge by reviewing the old.
   A second thought brings more insight.

It is recommended to collect more proverbs to help solve problems.

Exercises

(I) Multiple-choice questions:

1. Let a be any real number. How many of the following statements are true? ( ).

   (1) The solution of the equation $a^2 x = 0$ is $x = 0$.
   (2) The solution of the equation $ax - a = 0$ is $x = 1$.
   (3) The solution of the equation $ax + 1 = 0$ is $x = -\frac{1}{a}$.
   (4) The solution of the equation $|a| x = a$ is $x = \pm 1$.

   (A) 0   (B) 1   (C) 2   (D) 3

2. If the equation (where x is the unknown) $a(3x + 2) + b(3 - 2x) = 5x + 12$ has infinitely many solutions, then ( ).

   (A) $a = 3$, $b = -2$ (B) $a = -2$, $b = 3$
   (C) $a = 3$, $b = 2$ (D) $a = -3$, $b = -2$

3. If the equation (where x is the unknown) $\frac{a-2}{2-x} = \frac{a-2}{x-2}$ has infinitely many solutions, then a can take ( ) different values.

   (A) 0   (B) 1   (C) 2   (D) infinitely many

4. If the solution to the equation $\frac{x+1}{x+2} - \frac{x}{x-1} = \frac{1}{x^2+x-2}$ is positive, then a should satisfy (   ).

(A) $a > -1$ (B) $a \neq -3$
(C) $a < -1$ and $a \neq -3$ (D) $a < 2$ and $a \neq -3$

(II) Solve the following equations:

5. $b(b^2 + ax) - a^2(x + 2b) = b^3 - 2a^3 (a \neq b, a \neq 0)$.

6. $\frac{z}{2z-4} - \frac{z+2}{3z+3} = \frac{z^2}{6z^2-6z-12}$.

7. $\frac{1-ax}{mx} + \frac{1+bx}{nx} = \frac{ab}{mn}$, where $ab + an \neq bm$.

8. $\frac{x+1}{a+b} + \frac{x-1}{a-b} = \frac{2a}{a^2-b^2}$.

9. $\frac{1}{(x+1)(x+2)} + \frac{1}{(x+2)(x+3)} + \cdots + \frac{1}{(x+99)(x+100)} + \frac{1}{x+100} = \frac{1999}{2000}$.

10. $\frac{x+6}{x+5} + \frac{x+10}{x+9} = \frac{x+7}{x+6} + \frac{x+9}{x+8}$.

(III) Answer the following questions:

11. Suppose the solution of the equation (where x is the unknown) $\frac{1}{4}\left(x + \frac{a}{3}\right) - \frac{1-5x}{8} = 1$ has the same absolute value as the solution of the equation $4x = 3\left[x - 2\left(x - \frac{a}{3}\right)\right]$. Find the solutions of both equations.

12. If the equation (for x) $\frac{2}{x-2} + \frac{mx}{x^2-4} = \frac{3}{x+2}$ has an extraneous solution, find the value of m.

13. Suppose there is a fraction (not necessarily in the simplest form) that is equal to $\frac{2}{3}$ by reduction. If we add a number a to both the numerator and the denominator, the fraction will be equal to $\frac{8}{11}$ after reduction. On the other hand, if we subtract $a + 1$ from both the numerator and the denominator (of the original fraction), the fraction will be equal to $\frac{5}{9}$ after reduction. Find the original fraction.

(IV) Solve the systems of equations:

14. $\begin{cases} \frac{2}{3x+2y} + \frac{5}{3x-2y} = 3, \\ \frac{9}{3x+2y} - \frac{2}{3x-2y} = \frac{5}{4}. \end{cases}$

15. $\begin{cases} \frac{xy}{x+y} = 1, \\ \frac{yz}{z+y} = 2, \\ zx = 3(z+x). \end{cases}$

(V) Application problems:

16. Suppose that container A contains 40 kg of salt water, whose salinity is 20%, and container B contains 60 kg of salt water, whose salinity is 4%. (1 kg of salt water with salinity a% contains $\frac{a}{100}$ kg of dissolved salt.) Now, we move some salt water in A into container C and move some salt water in B into container D. Next, we pour the salt water in D into A, and pour the salt water in C into B. Finally, we observe that the salt water in A and B have the same salinity.

    Assume that the amount we moved from B to D is 6 times the amount we moved from A to C (in terms of weight), then how much salt water have we moved from A to C?

17. Three people, A, B, and C, are working on a project. If A works alone, the time consumed is a times the time consumed if B and C work together. If B works alone, the time consumed is a times the time consumed if A and C work together. If C works alone, what would be the ratio of time consumption relative to A and B working together?

18. A toy car has moved 12 m, during which the front wheel has turned 6 more rounds than the rear wheel. If the circumference of the front wheel is increased by $\frac{1}{4}$, and the circumference of the rear wheel is increased by $\frac{1}{5}$, then the front wheel will turn 4 more rounds than the rear wheel (for the same distance). Find the original circumferences of the front and rear wheels.

Chapter 8

Real Numbers

Rational numbers and irrational numbers constitute real numbers.

The decimal representation of a rational number is either finite or periodic. Rational numbers can be written as $\frac{q}{p}$, where p, q are coprime integers and $p \neq 0$. Conversely, if a number can be written as $\frac{q}{p}$, where p, q are coprime integers and $p \neq 0$, then it is a rational number. There are infinitely many rational numbers between any two different rational numbers, and the sum, difference, product, and quotient (where the divisor is nonzero) of any two rational numbers is still a rational number. Thus, we say that rational numbers are closed under addition, subtraction, multiplication, and division (for nonzero divisors).

For any two different rational numbers a, b with $a < b$, the number $\frac{a+b}{2}$ is a rational number that lies between a and b. Repeat this process, and we can prove that there are infinitely many rational numbers between a and b.

Irrational numbers are those whose decimal representations are infinite and non-periodic, which cannot be written as $\frac{q}{p}$ (p, q are coprime integers and $p \neq 0$). Irrational numbers are not closed under elementary arithmetic. For example, $\sqrt{7}$ is irrational, but $\sqrt{7} \times \sqrt{7}$ and $\sqrt{7} - \sqrt{7}$ are not irrational.

For a rational number a and an irrational number b, the numbers $a + b$ and $a - b$ are both irrational, and if $a \neq 0$, then ab, $\frac{a}{b}$, and $\frac{b}{a}$ are also irrational. However, if $a = 0$, then ab and $\frac{a}{b}$ are rational.

There are infinitely many real numbers, and there is no maximum or minimum in real numbers.

Real numbers are in one-to-one correspondence with the points on the number axis. Within real numbers, all real numbers have odd-degree roots, while only nonnegative real numbers have even-degree roots.

The absolute value of a real number is a non-negative real number, and the absolute values of opposite real numbers are the same, so $|a| = |-a|$.

Nonnegative real numbers include positive real numbers and 0. If a is a real number, then $|a|, a^2, \sqrt{a}$ (if $a \geq 0$) are all nonnegative real numbers. Nonnegative real numbers have many properties. The following are a few examples: the sum of finitely many nonnegative real numbers is still nonnegative; if the sum of finitely many nonnegative real numbers is 0, then they must all be equal to 0; the product of finitely many nonnegative real numbers is still nonnegative; and 0 is the minimal nonnegative real number, and there is no maximum.

Example 1. Compute the following:

(1) $\sqrt{(-5)^2} - \sqrt[4]{(-34)^4} + |-3|^2 - \sqrt[3]{(-4)^3}$;

(2) $\sqrt{(\underbrace{99\cdots9}_{1999\,\text{copies}})^2 + 1\underbrace{99\cdots9}_{1999\,\text{copies}}}$;

(3) $\left(\frac{7}{3}\right)^{1007} \cdot \sqrt{\frac{3^{2014}+15^{2014}}{7^{2014}+35^{2014}}}$;

(4) $\sqrt[2018]{\frac{2^{2018}+6^{2018}+8^{2018}}{3^{2018}+9^{2018}+12^{2018}}}$.

Solution   (1) It equals $5 - 34 + 9 + 4 = -16$.

(2) It equals

$$\sqrt{\left(\underbrace{99\cdots9}_{1999\,\text{copies}}\right)^2 + 2 \times \underbrace{99\cdots9}_{1999\,\text{copies}} + 1}$$

$$= \sqrt{\left(\underbrace{99\cdots9}_{1999\,\text{copies}} + 1\right)^2} = \sqrt{(10^{1999})^2} = 10^{1999}.$$

(3) It equals

$$\left(\frac{7}{3}\right)^{1007} \cdot \sqrt{\frac{3^{2014}(1 + 5^{2014})}{7^{2014}(1 + 5^{2014})}}$$

$$= \left(\frac{7}{3}\right)^{1007} \cdot \sqrt{\left(\frac{3}{7}\right)^{2014}} = \left(\frac{7}{3}\right)^{1007} \cdot \left(\frac{3}{7}\right)^{1007} = 1.$$

(4) It equals

$$\sqrt[2018]{\frac{2^{2018} + 2^{2018} \cdot 3^{2018} + 2^{2018} \cdot 4^{2018}}{3^{2018} + 3^{2018} \cdot 3^{2018} + 3^{2018} \cdot 4^{2018}}}$$

$$= \sqrt[2018]{\frac{2^{2018}(1 + 3^{2018} + 4^{2018})}{3^{2018}(1 + 3^{2018} + 4^{2018})}} = \sqrt[2018]{\left(\frac{2}{3}\right)^{2018}} = \frac{2}{3}.$$

Example 2. For an irrational number m, we call the greatest integer that does not exceed m the integer part of m, and the difference between m and its integer part is called the fractional part of m. Let $x = \sqrt{2} + 1$, a be the fractional part of x, and b be the fractional part of $-x$. Find the value of $a^3 + b^3 + 3ab$.

Solution Since $2 < \sqrt{2} + 1 < 3$, the integer part of x is 2, and $a = (\sqrt{2} + 1) - 2 = \sqrt{2} - 1$.

On the other hand, $-3 < -\sqrt{2} - 1 < -2$, so the integer part of $-x$ is -3, and

$$b = -\sqrt{2} - 1 - (-3) = 2 - \sqrt{2}.$$

Hence, $a + b = 1$, and

$$a^3 + b^3 + 3ab = (a + b)(a^2 - ab + b^2) + 3ab$$
$$= a^2 + 2ab + b^2$$
$$= (a + b)^2 = 1.$$

Example 3. Suppose that $\sqrt{3}$ lies between $\frac{x+3}{x}$ and $\frac{x+4}{x+1}$, where x is a positive integer. Find x.

Solution Note that

$$\frac{x+3}{x} = 1 + \frac{3}{x}, \quad \frac{x+4}{x+1} = 1 + \frac{3}{x+1}.$$

Since x is a positive integer, it follows that $\frac{x+4}{x+1} < \sqrt{3} < \frac{x+3}{x}$.

By $\frac{x+4}{x+1} < \sqrt{3}$, we have $x > \frac{1+3\sqrt{3}}{2}$.

By $\sqrt{3} < \frac{x+3}{x}$, we have $x < \frac{3}{\sqrt{3}-1} = \frac{3+3\sqrt{3}}{2}$.

Since $\sqrt{3} \approx 1.732$, we have the estimation $3.09 < x < 4.09$. Therefore, $x = 4$.

Example 4. If two real numbers a, b satisfy that $a^2 + b$ and $a + b^2$ are both rational, then (a, b) is called a harmonic pair.

(1) Try to find a harmonic pair (a, b), where both a and b are irrational.

(2) Prove that if (a, b) is a harmonic pair, and $a + b$ is a rational number not equal to 1, then a, b are both rational.

(3) Prove that if (a, b) is a harmonic pair, and $\frac{a}{b}$ is rational, then a, b are both rational.

Solution   (1) Let $a = \sqrt{2} + \frac{1}{2}$, $b = \frac{1}{2} - \sqrt{2}$, then they are both irrational. Further, $a^2 + b = 2 + \frac{3}{4} = \frac{11}{4}$ and $a + b^2 = \frac{11}{4}$, which are both rational. Hence, (a, b) is a harmonic pair.

(2) Since
$$m = (a^2 + b) - (a + b^2) = (a^2 - b^2) - (a - b) = (a - b)(a + b - 1),$$
and by assumption, $(a^2 + b) - (a + b^2)$ is rational, while $a + b - 1$ is rational and nonzero, we derive that $a - b = \frac{m}{a+b-1}$ is also rational. Since $a = \frac{(a+b)+(a-b)}{2}$ and $b = \frac{(a+b)-(a-b)}{2}$, they are both rational.

(3) If $a + b^2 = 0$, then $b = -\frac{a}{b}$. Since $\frac{a}{b}$ is rational by assumption, b is also rational, and $a = (a + b^2) - b^2$ is rational.

If $a + b^2 \neq 0$, then consider
$$x = \frac{a^2 + b}{a + b^2} = \frac{\left(\frac{a}{b}\right)^2 + \frac{1}{b}}{\frac{a}{b} \cdot \frac{1}{b} + 1}.$$

Since $a^2 + b$ and $a + b^2$ are both rational, x is also rational, and we have
$$\left(\frac{a}{b} \cdot x - 1\right) \cdot \frac{1}{b} = \left(\frac{a}{b}\right)^2 - x.$$

Equivalently, $b = \frac{\frac{a}{b}x - 1}{(\frac{a}{b})^2 - x}$. Since $\frac{a}{b}, x$ are both rational, we conclude that b is rational, and $a = \frac{a}{b} \cdot b$ is also rational.

Example 5. Initially, we have three numbers: $89, 12, 3$, and we perform the following operations. Every time we choose two numbers a, b, we replace them with $\frac{a+b}{\sqrt{2}}$ and $\frac{a-b}{\sqrt{2}}$. Is it possible that after finitely many operations, the three numbers become $90, 14, 10$? Prove your result.

Solution   Suppose that at some stage the three numbers are a, b, c, and after an operation, they become $\frac{a+b}{\sqrt{2}}, \frac{a-b}{\sqrt{2}}, c$. Note that
$$\left(\frac{a + b}{\sqrt{2}}\right)^2 + \left(\frac{a - b}{\sqrt{2}}\right)^2 + c^2$$
$$= \frac{a^2 + 2ab + b^2}{2} + \frac{a^2 - 2ab + b^2}{2} + c^2$$
$$= a^2 + b^2 + c^2,$$
so the sum of the squares of the numbers is invariant.

However, we have $89^2 + 12^2 + 3^2 = 8074$, while

$$90^2 + 14^2 + 10^2 = 8396 > 8074.$$

Therefore, it is impossible to obtain $90, 14, 10$ after finitely many operations.

Example 6. Suppose x, y are real numbers such that $(x - y)^2$ and $\sqrt{5x - 3y - 16}$ are opposite real numbers. Find the value of $\sqrt{x^2 + y^2}$.

Solution It follows that

$$(x - y)^2 + \sqrt{5x - 3y - 16} = 0.$$

Since $(x - y)^2$ and $\sqrt{5x - 3y - 16}$ are both nonnegative numbers, they must be both equal to 0, which means

$$\begin{cases} x - y = 0, \\ 5x - 3y - 16 = 0. \end{cases}$$

Solving this system of equations, we have

$$\begin{cases} x = 8, \\ y = 8. \end{cases}$$

Therefore, $\sqrt{x^2 + y^2} = \sqrt{8^2 + 8^2} = 8\sqrt{2}$.

Example 7. Suppose that x, y are rational numbers, and

$$\left(\frac{1}{3} + \frac{\sqrt{3}}{2}\right)x + \left(\frac{1}{4} - \frac{\sqrt{3}}{12}\right)y - 2.25 - 1.45\sqrt{3} = 0.$$

Find the values of x, y.

Analysis If a, b are rational and m is irrational, then $a + bm = 0$ if and only if $a = b = 0$. This is the key idea of this problem.

Solution The given equation can be written as

$$\left(\frac{1}{3}x + \frac{1}{4}y - 2.25\right) + \left(\frac{1}{2}x - \frac{1}{12}y - 1.45\right)\sqrt{3} = 0.$$

Since x, y are rational, it follows that

$$\begin{cases} \dfrac{1}{3}x + \dfrac{1}{4}y - 2.25 = 0, \\ \dfrac{1}{2}x - \dfrac{1}{12}y - 1.45 = 0. \end{cases}$$

Solving the system of equations, we have $\begin{cases} x = 3.6, \\ y = 4.2. \end{cases}$

Example 8. Let $y = \frac{ax+b}{cx+d}$, where a, b, c, d are rational numbers and $cd \neq 0$. If x is irrational and $bc = ad$, prove that y is rational.

Solution Since $cd \neq 0$, necessarily $c \neq 0$ and $d \neq 0$. Thus, $bc = ad$ implies $a = \frac{bc}{d}$, and plugging into the expression of y, we have

$$y = \frac{\frac{bc}{d}x + b}{cx + d} = \frac{bcx + bd}{cdx + d^2} = \frac{b(cx + d)}{d(cx + d)} = \frac{b}{d}.$$

Therefore, y is rational.

Remark: To prove a number is rational, we usually express it as the result of elementary arithmetic of several rational numbers. To prove a number is irrational, we usually prove it by contradiction.

Additional Reading

The ancient Greek Pythagorean (c. 580 –c. 500 BC) School regarded integers as gods and believed that all phenomena in mathematics could be attributed to integers and their ratios. When Hippasus (in the 5th century BC) found that the diagonal of a unit square could not be expressed by a rational number, it caused great panic in the Pythagorean School. It is said that Hippasus was thrown into the sea for revealing his secret, sacrificing his life for truth.

It can be easily found that the diagonal of a unit square is $\sqrt{2}$. Why is it not a rational number? Assuming that $\sqrt{2}$ is rational, one can set $\sqrt{2} = \frac{p}{q}$, where p, q are coprime positive integers. Thus, $q = \sqrt{2}p$, and squaring both sides, we have $q^2 = 2p^2$. This means q is even, and let $q = 2t$. Then, since $q^2 = 2p^2$, we obtain $4t^2 = 2p^2$, and $2t^2 = p^2$. This means p is also even, and this is contradictory to the assumption that p, q are coprime.

Hence, $\sqrt{2}$ is not a rational number, and thus, it is an irrational number. Consider $\sqrt{2} = 1 + (\sqrt{2} - 1)$, where $\sqrt{2} - 1 = \frac{1}{\sqrt{2}+1}$, so it follows that

$$\sqrt{2} = 1 + \frac{(\sqrt{2} - 1)(\sqrt{2} + 1)}{\sqrt{2} + 1} = 1 + \frac{1}{2 + (\sqrt{2} - 1)} = 1 + \frac{1}{2 + \frac{1}{\sqrt{2}+1}}$$

$$= \cdots = 1 + \cfrac{1}{2 + \cfrac{1}{2 + \cfrac{1}{2 + \cfrac{1}{\sqrt{2} + 1}}}} = \cdots .$$

It can be extended infinitely to become a marvelous infinite periodic continued fraction.

Exercises

(I) Multiple-choice questions:

1. If $n + 1 = 2010^2 + 2011^2$, then $\sqrt{2n+1}$ equals ( ).

   (A) 2010   (B) 2011   (C) 4021   (D) 4022

2. If the principal square root of a natural number is $a(a > 1)$, then the principal square roots of the two adjacent natural numbers of this natural number are ( ).

   (A) $a - 1, a + 1$ (B) $\sqrt{a-1}, \sqrt{a+1}$

   (C) $\sqrt{a^2 - 1}, \sqrt{a^2 + 1}$ (D) $a^2 - 1, a^2 + 1$

3. A real number a solves the equation $|2006 - a| + \sqrt{a - 2007} = a$, then the value of $a - 2006^2$ is ( ).

   (A) 2005   (B) 2006   (C) 2007   (D) 2008

4. If $\sqrt{x^2 - 4} + \sqrt{2x + y} = 0$, then $x - y$ equals ( ).

   (A) 2   (B) 6   (C) 2 or -2 (D) 6 or -6

5. If a is irrational, and $ab - a - b + 1 = 0$, then b is ( ).

   (A) a negative rational number
   (B) a positive rational number
   (C) a negative irrational number
   (D) a positive irrational number

6. If x is an irrational number such that $(x-2)(x+6)$ is rational, then ( ).

   (A) x^2 is rational   (B) $(x+6)^2$ is rational

   (C) $(x+2)(x-6)$ is irrational   (D) $(x+2)^2$ is irrational

(II) Fill in the blanks:

7. If a is a negative real number, then the square root(s) of a^2 is (are) _____ .

8. If m is a perfect square, then the smallest perfect square that is greater than m is _____ .

9. Let $a \vee b$ denote the maximum of a, b, and $a \wedge b$ denote the minimum of a, b. Then, $(\sqrt{203} \vee \sqrt[3]{2007}) \wedge (\sqrt{2006} \vee \sqrt{2008})$ = _____ .

10. Fill in with "<" and ">":
$-\sqrt{275}$_____$- 4\sqrt{11}$, $4 - \sqrt{2}$_____$1 + \sqrt{2}$,
$\sqrt{61} - 1$_____$\sqrt{6} + 3$.

11. Let m and n be the integer and fractional parts of $\sqrt{5}$, respectively. Then, $mn + 4 =$ _____ .

12. Suppose a, b, x, y are real numbers such that $y + |\sqrt{x} - \sqrt{3}| = 1 - a^2$, $|x - 3| = y - 1 - b^2$, then the value of $2^{x+y} + 2^{a+b}$ is _____ .

13. If rational numbers x, y satisfy $2x + \sqrt{3}x + (4 - \sqrt{3})y = 3 - 2\sqrt{3}$, then $x =$ _____ , $y =$ _____ .

14. If a, b, c are rational numbers and $(b + \sqrt{2})^2 = (a + \sqrt{2})(c + \sqrt{2})$, then $(a - c)^2 =$ _____ .

(III) Compute the following:

15. $(\sqrt{1.21} - \sqrt{0.0196}) \div \left[\sqrt{\frac{9625}{+}} \sqrt[3]{\left(-\frac{1}{12.5}\right)^3} \right]$.

16. $\sqrt{81 \times 82 \times 83 \times 84 + 1}$.

(IV) Answer the following questions:

17. Find a rational number x such that $x^2 + 5$ and $x^2 - 5$ are both squares of rational numbers.

18. Suppose x, y are rational numbers, and $\left(\frac{1}{2} + \frac{\pi}{3}\right) x + \left(\frac{1}{3} + \frac{\pi}{2}\right) y = 4 + \pi$, find the value of $x + y$.

19. If a, b satisfy $3\sqrt{a} + 5|b| = 7$, find the value range of $S = 2\sqrt{a} - 3|b|$.

20. Suppose a, b are nonzero real numbers and $|2a - 4| + |b + 2| + \sqrt{(a - 3)b^2} + 4 = 2a$. Find the value of $\sqrt{a + b}$.

21. Suppose a, b are rational numbers such that $(\sqrt{3}a + \sqrt{2})a + (\sqrt{3}b - \sqrt{2})b - \sqrt{2} - 25\sqrt{3} = 0$. Find the value of ab.

Chapter 9

Quadratic Radicals

In general, $\sqrt{a}(a \geq 0)$ is called a quadratic radical expression. In such expressions, the radicand is nonnegative, and the result $\sqrt{a}$ is also nonnegative.

Expressions such as $\sqrt{A + \sqrt{B}}$ have iterations of radical signs, so they are called composite quadratic radical expressions.

The following properties apply:

$$(\sqrt{a})^2 = a(a \geq 0);$$

$$\sqrt{a^2} = |a| = \begin{cases} a, & a > 0, \\ 0, & a = 0, \\ -a, & a < 0. \end{cases}$$

$$\sqrt{ab} = \sqrt{a}\sqrt{b}(a \geq 0, b \geq 0);$$

$$\sqrt{\frac{a}{b}} = \frac{\sqrt{a}}{\sqrt{b}}(a \geq 0, b > 0).$$

There are also rules of calculation, as follows:

$$a\sqrt{c} + b\sqrt{c} = (a + b)\sqrt{c}(c \geq 0);$$

$$\sqrt{a}\sqrt{b} = \sqrt{ab}(a \geq 0, b \geq 0);$$

$$\frac{\sqrt{a}}{\sqrt{b}} = \sqrt{\frac{a}{b}}(a \geq 0, b > 0);$$

$$(\sqrt{a})^n = \sqrt{a^n}(a \geq 0).$$

When using these rules and properties, it is important to check whether the premises are satisfied; otherwise, we may make mistakes, such as $\sqrt{(-25) \times (-36)} = \sqrt{-25} \times \sqrt{-36}$.

Example 1. Let a, b, c be pairwise different rational numbers. Prove that $\sqrt{\frac{1}{(a-b)^2} + \frac{1}{(b-c)^2} + \frac{1}{(c-a)^2}}$ is a rational number.

Analysis   The result follows if the radicand is a perfect square. With the identity $x^2 + y^2 + z^2 = (x+y+z)^2 - 2xy - 2yz - 2xz$, we may try to rewrite the radicand.

Proof: We have

$$\frac{1}{(a-b)^2} + \frac{1}{(b-c)^2} + \frac{1}{(c-a)^2}$$

$$= \left(\frac{1}{a-b} + \frac{1}{b-c} + \frac{1}{c-a}\right)^2$$

$$-2\left[\frac{1}{(a-b)(b-c)} + \frac{1}{(b-c)(c-a)} + \frac{1}{(c-a)(a-b)}\right]$$

$$= \left(\frac{1}{a-b} + \frac{1}{b-c} + \frac{1}{c-a}\right)^2 - 2\frac{(c-a)+(a-b)+(b-c)}{(a-b)(b-c)(c-a)}$$

$$= \left(\frac{1}{a-b} + \frac{1}{b-c} + \frac{1}{c-a}\right)^2,$$

so the given expression equals $\left|\frac{1}{a-b} + \frac{1}{b-c} + \frac{1}{c-a}\right|$, which is a rational number.

Example 2. Simplify

$$S = \sqrt{x^2 - 2x + 1} - \sqrt{x^2 - 4x + 4} + \sqrt{x^2 + 6x + 9}.$$

Solution

$$S = \sqrt{x^2 - 2x + 1} - \sqrt{x^2 - 4x + 4} + \sqrt{x^2 + 6x + 9}$$

$$= \sqrt{(x-1)^2} - \sqrt{(x-2)^2} + \sqrt{(x-3)^2}$$

$$= |x - 1| - |x - 2| + |x + 3|.$$

The zeros of $x - 1$, $x - 2$, $x + 3$ are $1, 2, -3$, respectively, so we should divide into the following four cases.

If $x \neq -3$, then $S = -(x-1) + (x-2) - (x+3) = -x - 4$.
If $-3 < x \leq 1$, then $S = -(x-1) + (x-2) + (x+3) = x + 2$.
If $1 < x \leq 2$, then $S = (x-1) + (x-2) + (x+3) = 3x$.
If $x > 2$, then $S = (x-1) - (x-2) + (x+3) = x + 4$.

Remark: Simplifying with $\sqrt{a^2} = |a|$ may create absolute values, and we should divide into several intervals based on the zeros of the absolute values.

Example 3. Suppose m, n are two consecutive positive integers with $m < n$ and $a = mn$. Prove that $\sqrt{a+n} + \sqrt{a-m}$ and $\sqrt{a+n} - \sqrt{a-m}$ are both odd numbers.

Solution It follows that $n = m + 1$, $m = n - 1$, and

$$a = m(m+1) = n(n-1).$$

Since m, n are positive integers, we have

$$\sqrt{a+n} + \sqrt{a-m}$$
$$= \sqrt{(n-1)n + n} + \sqrt{m(m+1) - m}$$
$$= \sqrt{n^2} + \sqrt{m^2} = m + n = 2m + 1.$$

Hence, $\sqrt{a+n} + \sqrt{a-m}$ is an odd number, and similarly, $\sqrt{a+n} - \sqrt{a-m} = n - m = 1$, which is also an odd number.

Example 4. Suppose x, y are real numbers such that

$$(x - \sqrt{x^2 - 2018})(y - \sqrt{y^2 - 2018}) = 2018.$$

Compare the two numbers x and y.

Solution It follows that

$$x - \sqrt{x^2 - 2018} = \frac{2018}{y - \sqrt{y^2 - 2018}}$$

$$= \frac{2018(y + \sqrt{y^2 - 2018})}{(y - \sqrt{y^2 - 2018})(y + \sqrt{y^2 - 2018})}$$

$$= y + \sqrt{y^2 - 2018},$$

so

$$x - \sqrt{x^2 - 2018} = y + \sqrt{y^2 - 2018}. \qquad \text{①}$$

For similar reasons, we also have

$$x + \sqrt{x^2 - 2018} = y - \sqrt{y^2 - 2018}. \qquad \text{②}$$

Adding ① and ②, we conclude that $x = y$.

Example 5. Suppose $c > 1$, and $x = \sqrt{c} - \sqrt{c-1}$, $y = \sqrt{c+1} - \sqrt{c}$, $z = \sqrt{c+2} - \sqrt{c+1}$. Compare the three numbers x, y, z.

Solution Since

$$x = \sqrt{c} - \sqrt{c-1} = \frac{1}{\sqrt{c} + \sqrt{c-1}},$$

$$y = \sqrt{c+1} - \sqrt{c} = \frac{1}{\sqrt{c+1} + \sqrt{c}},$$

$$z = \sqrt{c+2} - \sqrt{c+1} = \frac{1}{\sqrt{c+2} + \sqrt{c+1}},$$

and apparently $\sqrt{c} + \sqrt{c-1} < \sqrt{c+1} + \sqrt{c} < \sqrt{c+2} + \sqrt{c+1}$, so we conclude that $x > y > z$.

Example 6. Compute the following:

(1) $\frac{8+2\sqrt{15}-\sqrt{10}-\sqrt{6}}{\sqrt{5}+\sqrt{3}-\sqrt{2}}$;

(2) $\frac{\sqrt{15}+\sqrt{35}+\sqrt{21}+5}{\sqrt{3}+2\sqrt{5}+\sqrt{7}}$;

(3) $\frac{1}{2\sqrt{1}+\sqrt{2}} + \frac{1}{3\sqrt{2}+2\sqrt{3}} + \cdots + \frac{1}{100\sqrt{99}+99\sqrt{100}}$;

(4) $\frac{3\sqrt{15}-\sqrt{10}-2\sqrt{6}+3\sqrt{3}-\sqrt{2}+18}{\sqrt{5}+2\sqrt{3}+1}$.

Solution (1) It equals

$$\frac{(\sqrt{5})^2 + 2\sqrt{5} \times \sqrt{3} + (\sqrt{3})^2 - \sqrt{2} \times \sqrt{5} - \sqrt{2} \times \sqrt{3}}{\sqrt{5} + \sqrt{3} - \sqrt{2}}$$

$$= \frac{(\sqrt{5} + \sqrt{3})^2 - \sqrt{2}(\sqrt{5} + \sqrt{3})}{\sqrt{5} + \sqrt{3} - \sqrt{2}}$$

$$= \frac{(\sqrt{5} + \sqrt{3})(\sqrt{5} + \sqrt{3} - \sqrt{2})}{\sqrt{5} + \sqrt{3} - \sqrt{2}}$$

$$= \sqrt{5} + \sqrt{3}.$$

(2) Let N be the given expression, then $N = \frac{(\sqrt{3}+\sqrt{5})(\sqrt{5}+\sqrt{7})}{(\sqrt{3}+\sqrt{5})+(\sqrt{5}+\sqrt{7})}$, and

$$\frac{1}{N} = \frac{1}{\sqrt{5} + \sqrt{7}} + \frac{1}{\sqrt{3} + \sqrt{5}}$$

$$= \frac{\sqrt{7} - \sqrt{5}}{2} + \frac{\sqrt{5} - \sqrt{3}}{2}$$

$$= \frac{\sqrt{7} - \sqrt{3}}{2}.$$

Hence, $N = \frac{2}{\sqrt{7}-\sqrt{3}} = \frac{\sqrt{7}+\sqrt{3}}{2}$.

(3) The general term in the sum is

$$\frac{1}{(n+1)\sqrt{n}+n\sqrt{n+1}} = \frac{(n+1)\sqrt{n}-n\sqrt{n+1}}{(n+1)n}$$

$$= \frac{\sqrt{n}}{n} - \frac{\sqrt{n+1}}{n+1}$$

$$= \frac{1}{\sqrt{n}} - \frac{1}{\sqrt{n+1}},$$

so the given sum equals

$$\frac{1}{1} - \frac{1}{\sqrt{2}} + \frac{1}{\sqrt{2}} - \frac{1}{\sqrt{3}} + \cdots + \frac{1}{\sqrt{99}} - \frac{1}{\sqrt{100}}$$

$$= 1 - \frac{1}{10} = \frac{9}{10}.$$

(4) It equals

$$\frac{3\sqrt{3}\times\sqrt{5}-\sqrt{2}\times\sqrt{5}-2\sqrt{2}\times\sqrt{3}+3\sqrt{3}-\sqrt{2}+3\sqrt{3}\times2\sqrt{3}}{\sqrt{5}+2\sqrt{3}+1}$$

$$= \frac{3\sqrt{3}(\sqrt{5}+2\sqrt{3}+1)-\sqrt{2}(\sqrt{5}+2\sqrt{3}+1)}{\sqrt{5}+2\sqrt{3}+1}$$

$$= 3\sqrt{3}-\sqrt{2}.$$

Example 7. Simplify

$$\frac{(\sqrt{x}-\sqrt{y})^3+2x\sqrt{x}+y\sqrt{y}}{x\sqrt{x}+y\sqrt{y}} + \frac{3\sqrt{xy}-3y}{x-y}.$$

Solution   Let $\sqrt{x}=a+b$, $\sqrt{y}=a-b$. Then,

$$\sqrt{xy}=a^2-b^2, \quad \sqrt{x}+\sqrt{y}=2a, \quad \sqrt{x}-\sqrt{y}=2b.$$

The given expression equals

$$\frac{(2b)^3+2(a+b)^3+(a-b)^3}{(a+b)^3+(a-b)^3} + \frac{3(a^2-b^2)-3(a-b)^2}{4ab}$$

$$= \frac{3a^3+3a^2b+9b^3+9ab^2}{2a^3+6ab^2} + \frac{6ab-6b^2}{4ab}$$

$$= \frac{3(a+b)(a^2+3b^2)}{2a(a^2+3b^2)} + \frac{3a-3b}{2a}$$

$$= \frac{3a+3b+3a-3b}{2a} = 3.$$

Remark: The substitutions $\sqrt{x} = a + b$, $\sqrt{y} = a - b$ turn the radical expression into a rational expression, which is easier to handle. Such insight comes from experience.

Example 8. Given that $1 \leq x \leq 2$, simplify

$$\sqrt{x + 2\sqrt{x - 1}} + \sqrt{x - 2\sqrt{x - 1}}.$$

Solution   Since $1 \leq x \leq 2$, $x = (\sqrt{x-1})^2 + 1$. The given expression equals

$$\sqrt{(\sqrt{x-1})^2 + 2\sqrt{x-1} + 1} + \sqrt{(\sqrt{x-1})^2 - 2\sqrt{x-1} + 1}$$

$$= \sqrt{(\sqrt{x-1} + 1)^2} + \sqrt{(\sqrt{x-1} - 1)^2}$$

$$= |\sqrt{x-1} + 1| + |\sqrt{x-1} - 1|.$$

Since $1 \leq x \leq 2$, we have $\sqrt{x-1} - 1 \leq 0$, so the above expression equals

$$\sqrt{x-1} + 1 - (\sqrt{x-1} - 1) = 2.$$

Remark: Sometimes, we can simplify composite quadratic radicals by completing the square. It depends on whether we can find an expression like $A \pm 2\sqrt{AB} + B$. For example,

$$4 \pm 2\sqrt{3} = \sqrt{3}^2 \pm 2\sqrt{3} + 1^2,$$

$$11 + 2\sqrt{18} = 9 + 2 \times \sqrt{9 \times 2} + 2,$$

$$2 - \sqrt{3} = \frac{1}{2}(4 - 2\sqrt{3}) = \frac{1}{2}(\sqrt{3} - 1)^2.$$

Example 9. Compute $\sqrt{14 + 6\sqrt{5}} - \sqrt{14 - 6\sqrt{5}}$

Solution   Let

$$\sqrt{14 + 6\sqrt{5}} = a + b, \qquad\qquad ①$$

$$\sqrt{14 - 6\sqrt{5}} = a - b. \qquad\qquad ②$$

Then, by ① × ②, we have $a^2 - b^2 = 4$.

By ①^2 + ②^2, we have $a^2 + b^2 = 14$.

Taking the difference, we get $b^2 = 5$, and since $a + b > a - b$, we have $b > 0$, so $b = \sqrt{5}$.

Therefore, the given expression equals $(a + b) - (a - b) = 2b = 2\sqrt{5}$.

Remark: We can also complete the squares from the beginning as

$$14 \pm 6\sqrt{5} = 9 \pm 6\sqrt{5} + 5 = 3^2 \pm 2 \times 3 \times \sqrt{5} + (\sqrt{5})^2 = (3 \pm \sqrt{5})^2.$$

Example 10. Simplify $\dfrac{\sqrt{3+\sqrt{5-\sqrt{13+\sqrt{48}}}}}{\sqrt{6}+\sqrt{2}}$.

Solution Since $13+\sqrt{48}=13+4\sqrt{3}=(2\sqrt{3})^2+2\cdot 2\sqrt{3}+1=(2\sqrt{3}+1)^2$,
we have

$$5-\sqrt{13+\sqrt{48}}=5-2\sqrt{3}-1=4-2\sqrt{3}.$$

Further, $4-2\sqrt{3}=(\sqrt{3}-1)^2$, so

$$3+\sqrt{5-\sqrt{13+\sqrt{48}}}=3+(\sqrt{3}-1)=2+\sqrt{3}.$$

Hence, the given expression equals

$$\frac{\sqrt{2+\sqrt{3}}}{\sqrt{6}+\sqrt{2}}=\frac{\sqrt{\frac12(4+2\sqrt{3})}}{\sqrt{6}+\sqrt{2}}=\frac{\sqrt{\frac12}(\sqrt{3}+1)}{\sqrt{2}(\sqrt{3}+1)}=\frac12.$$

Remark: When repeatedly simplifying by completing the square, pay close attention to the signs.

Example 11. Let x,y be rational numbers. Find all pairs (x,y) such that $\sqrt{3\sqrt{3}+\frac{21}{4}}=x+\sqrt{y}$.

Solution We have

$$\sqrt{3\sqrt{3}+\frac{21}{4}}=\frac12\sqrt{21+12\sqrt{3}}$$
$$=\frac12\sqrt{(\sqrt{12})^2+2\cdot 3\cdot\sqrt{12}+3^2}$$
$$=\frac12\sqrt{(\sqrt{12}+3)^2}=\frac32+\sqrt{3}.$$

Thus, $\frac32+\sqrt{3}=x+\sqrt{y}$, and $(\frac32-x)+(\sqrt{3}-\sqrt{y})=0$.
Since x,y are rational numbers and $\sqrt{3}$ is irrational, the only pair is $(x,y)=(\frac32,3)$.

Reading

In ancient China (before the 2nd century BC), not only did people know very early to use the current "rounding" method to represent an approximate value of the actual number, but they also had approximation formulae for a long time.

The Nine Chapters on the Mathematical Art introduced a formula to approximate square roots: if $A = a^2 + r(r > 0)$, then $\sqrt{A} \approx a + \frac{r}{a}$. However, this formula has a big error.

An improved formula was introduced in *The Mathematical Classic of Sun Zi*: if $A = a^2 + r^2$, then $\sqrt{A} \approx a + \frac{r}{2a}$.

In *Arithmetic Methods in the Five Classics* and *Zhang Qiujian's Mathematical Manual*, another approximation formula was proposed: if $A = a^2 + r^2$, then $\sqrt{A} \approx a + \frac{r}{2a+1}$.

Can you try these formulas by taking specific values?

Exercises

(I) Multiple-choice questions:

1. Suppose $y = \sqrt{\frac{x^2-2}{5x-4}} - \sqrt{\frac{x^2-2}{4-5x}} + 2$, then $\sqrt{x^2+y^2}$ equals ( ).

   (A) 2   (B) 4   (C) 6   (D) $\sqrt{6}$

2. Suppose that a, b, c are represented by the following points on the number axis: Then, the expression $\sqrt{a^2} + \sqrt{(c-a)^2} - \sqrt{(a+b)^2} + \sqrt{(b+c)^2}$ equals ( ).

   (A) $2c - a$ (B) $-a$ (C) $2a - 2b$ (D) a

3. Suppose x is a real number such that $|2 - x| = 2 + |x|$, then $\sqrt{(x-3)^2}$ equals ( ).

   (A) $x - 3$ (B) $3 - x$ (C) $\pm(x - 3)$ (D) 3

4. By simplification, $(x-1)\sqrt{-\frac{1}{x-1}}$ equals ( ).

   (A) $\sqrt{1-x}$ (B) $\sqrt{x-1}$ (C) $-\sqrt{x-1}$ (D) $-\sqrt{1-x}$

5. Suppose a, m, n are positive integers such that $\sqrt{a^2 - 4\sqrt{2}} = \sqrt{m} - \sqrt{n}$, then there are(is) ( ) such triplets (a, m, n).

   (A) exactly one   (B) exactly two   (C) more than two   (D) no

6. The value of $4\sqrt{3 + 2\sqrt{2}} - \sqrt{41 + 24\sqrt{2}}$ equals ( ).

   (A) $\sqrt{2} - 1$ (B) 1   (C) $\sqrt{2}$ (D) 2

7. Let $[x]$ denote the integer part of x and $\{x\}$ denote its fractional part, so that $x = [x] + \{x\}$ (where $0 \le \{x\} < 1$). If $x = \sqrt{3 - \sqrt{5}} - \sqrt{3 + \sqrt{5}}$, then $[x]$ equals ( ).

   (A) -2 (B) -1 (C) 0   (D) 1

(II) Fill in the blanks:

8. If $\sqrt{x+2018}-\sqrt{x-2019}=2020$, then $\sqrt{x+2018}+\sqrt{x-2019}=$ _____ .

9. If a is a real number such that $\sqrt{a-1999}+|1998-a|=a$, then $a-1998^2=$ _____ .

10. Compute: $(\sqrt{30}+\sqrt{21}-3)(\sqrt{3}+\sqrt{10}-\sqrt{7})=$ _____ .

11. If a,b,c are nonzero real numbers such that $|a|+a=0, |ab|=ab, |c|-c=0$, then the simplified form of the expression $\sqrt{b}^2 - |a+b|+|a-c|-\sqrt{c^2-2bc+b^2}$ is _____ .

12. When the expression $4-x^2-\sqrt{1-x^2}$ reaches its minimum, $x=$ _____ x. When $4-x^2-\sqrt{1-x^2}$ reaches its maximum, $x=$ _____ .

13. There are _____ different integers x such that $\dfrac{4}{\sqrt{3}+\sqrt{2}} < x < \dfrac{4}{\sqrt{5}-\sqrt{3}}$.

14. Let $a>b>c>d>0$, and let $X=\sqrt{ab}+\sqrt{cd}$, $Y=\sqrt{ac}+\sqrt{bd}$, $Z=\sqrt{ad}+\sqrt{bc}$. Then, X,Y,Z can be arranged in the increasing order as _____ .

(III) Compute the following:

15. $\dfrac{2\sqrt{6}}{\sqrt{3}+\sqrt{2}-\sqrt{5}}$.

16. $\dfrac{\sqrt{3}+\sqrt{5}}{3-\sqrt{6}-\sqrt{10}+\sqrt{15}}$.

17. $\dfrac{1+\sqrt{2}+\sqrt{3}}{1-\sqrt{2}+\sqrt{3}}$.

18. $\dfrac{\sqrt{11}+4\sqrt{6}+5\sqrt{7}}{7+\sqrt{77}+\sqrt{66}+\sqrt{42}}$.

19. $\dfrac{1}{\sqrt{8}+\sqrt{11}}+\dfrac{1}{\sqrt{11}+\sqrt{14}}+\dfrac{1}{\sqrt{14}+\sqrt{17}}+\cdots+\dfrac{1}{\sqrt{6038}+\sqrt{6041}}$.

20. $\sqrt{10+8\sqrt{3+2\sqrt{2}}}-(\sqrt{2}+1)$.

21. $\sqrt{3-2\sqrt{2}}+\sqrt{5-2\sqrt{6}}+\sqrt{7-2\sqrt{12}}+\sqrt{9-2\sqrt{20}}+\sqrt{11-2\sqrt{30}}+\sqrt{13-2\sqrt{42}}+\sqrt{15-2\sqrt{56}}+\sqrt{17-2\sqrt{72}}$.

(IV) Answer the following questions:

22. Find the greatest integer that does not exceed $(\sqrt{7}+\sqrt{6})^6$.

23. Suppose that $\sqrt{11+2(1+\sqrt{5})(1+\sqrt{7})}=\sqrt{x}+\sqrt{y}+1$, where x,y are integers, find x,y.

Chapter 10

Evaluating Algebraic Expressions

An expression that connects numbers or letters representing numbers with addition, subtraction, multiplication, division, powers, or radicals is called an algebraic expression. Algebraic expressions include rational expressions and irrational expressions. Rational expressions include polynomials and rational fractions.

Evaluating an algebraic expression refers to finding the value of the algebraic expression for given values of variables or given constraints. In order to evaluate, we sometimes need to simplify the given expressions, rewrite the given constraints, or sometimes both.

Example 1. Given $x = \frac{1}{\sqrt{3}-\sqrt{2}}$, evaluate

$$x^6 - 2\sqrt{2}x^5 - x^4 + x^3 - 2\sqrt{3}x^2 + 2x - \sqrt{2}.$$

Analysis The expression to be evaluated is quite complicated, so direct computation is cumbersome. Here, we may obtain a partial evaluation first based on the value of x and then rewrite the given expression for evaluation.

Solution It follows that $x = \sqrt{3} + \sqrt{2}$, so $(x - \sqrt{3})^2 = 2$, or $x^2 - 2\sqrt{3}x + 1 = 0$.

Also, since $(x - \sqrt{2})^2 = 3$, we have $x^2 - 2\sqrt{2}x - 1 = 0$.

Hence, the given expression equals

$$x^4(x^2 - 2\sqrt{2}x - 1) + x(x^2 - 2\sqrt{3}x + 1) + x - \sqrt{2}$$
$$= x - \sqrt{2} = \sqrt{3} + \sqrt{2} - \sqrt{2} = \sqrt{3}.$$

Example 2. Given $a = \frac{\sqrt{5}-1}{2}$, evaluate $\frac{a^5+a^4-2a^3-a^2-a+2}{a^3-a}$.

Solution We have $a^2 = \left(\frac{\sqrt{5}-1}{2}\right)^2 = \frac{3-\sqrt{5}}{2} = 1 + \frac{1-\sqrt{5}}{2} = 1 - a$, so $a^2 + a = 1$. Therefore,

$$\frac{a^5 + a^4 - 2a^3 - a^2 - a + 2}{a^3 - a}$$

$$= \frac{a^3(a^2 + a) - 2a^3 - (a^2 + a) + 2}{a \cdot a^2 - a}$$

$$= \frac{a^3 - 2a^3 + 1}{a(1-a) - a} = \frac{-a^3 + 1}{-a^2}$$

$$= \frac{-(a^3 - 1)}{-(1-a)} = \frac{(a-1)(a^2 + a + 1)}{1 - a}$$

$$= -(a^2 + a + 1) = -2.$$

Remark: Examples 1 and 2 are common problems that often appear in junior high school mathematics competitions. When solving such problems, it is not appropriate to directly replace a with its value in the expression to be evaluated. Instead, we should transform the given numbers to find hidden conditions, and use the conditions repeatedly to simplify our expression, so as to find its value.

Example 3. If x, y are real numbers such that $x^3 + y^3 + 3xy = 1$, find the minimum of $x^2 + y^2$.

Solution We have $x^3 + y^3 + 3xy - 1 = 0$. Equivalently,

$$(x + y)^3 - 1^3 - 3x^2y - 3xy^2 + 3xy = 0,$$

Factorizing the left-hand side, we get

$$(x + y - 1)[(x + y)^2 + (x + y) + 1] - 3xy(x + y - 1) = 0,$$

so that $(x + y - 1)(x^2 + y^2 - xy + x + y + 1) = 0$, and

$$\frac{1}{2}(x + y - 1)[(x - y)^2 + (x + 1)^2 + (y + 1)^2] = 0.$$

Thus, $x + y = 1$ or $x = y = -1$.

If $x + y = 1$, then $x^2 + y^2 = \frac{1}{2}[(x + y)^2 + (x - y)^2]$

$$= \frac{1}{2}[1 + (x - y)^2] \geq \frac{1}{2}.$$

The equality is attained when $x = y = \frac{1}{2}$.

If $x = y = -1$, then $x^2 + y^2 = 2$. In summary, the minimum of $x^2 + y^2$ is $\frac{1}{2}$.

Example 4. If $x = \frac{\sqrt{3}-\sqrt{2}}{\sqrt{3}+\sqrt{2}}$, $y = \frac{\sqrt{3}+\sqrt{2}}{\sqrt{3}-\sqrt{2}}$, evaluate $\frac{2\sqrt{10(x+y)}-3\sqrt{xy}}{3x^2-5xy+3y^2}$.

Solution We have $x + y = \frac{(\sqrt{3}-\sqrt{2})^2+(\sqrt{3}+\sqrt{2})^2}{(\sqrt{3}+\sqrt{2})(\sqrt{3}-\sqrt{2})} = 10$ and $xy = \frac{\sqrt{3}-\sqrt{2}}{\sqrt{3}+\sqrt{2}} \cdot \frac{\sqrt{3}+\sqrt{2}}{\sqrt{3}-\sqrt{2}} = 1$. Thus,

$$\frac{2\sqrt{10(x+y)}-3\sqrt{xy}}{3x^2-5xy+3y^2} = \frac{2\sqrt{100}-3}{3(x+y)^2-11xy} = \frac{17}{300-11} = \frac{1}{17}.$$

Remark: Observing the characteristic of the given expression, it is better to use the values of $x + y$ and xy rather than plugging in x, y immediately.

Example 5. Suppose $x + \frac{1}{x} = 3$, evaluate $x^{10} + x^5 + \frac{1}{x^5} + \frac{1}{x^{10}}$.

Solution It follows that $x^2 + \frac{1}{x^2} = \left(x + \frac{1}{x}\right)^2 - 2 = 7$.
Also, $\left(x + \frac{1}{x}\right)\left(x^2 + \frac{1}{x^2}\right) = 21$, so

$$x^3 + \frac{1}{x^3} = 21 - \left(x + \frac{1}{x}\right) = 18.$$

Further, $\left(x^2 + \frac{1}{x^2}\right)\left(x^3 + \frac{1}{x^3}\right) = 126$, so

$$x^5 + \frac{1}{x^5} = 126 - \left(x + \frac{1}{x}\right) = 123,$$

$$x^{10} + \frac{1}{x^{10}} = \left(x^5 + \frac{1}{x^5}\right)^2 - 2 = 15127.$$

Therefore,

$$x^{10} + x^5 + \frac{1}{x^5} + \frac{1}{x^{10}} = 123 + 15127 = 15250.$$

Example 6. Given $a^2 - 3a + 1 = 0$, evaluate $2a^2 - 5a - 2 + \frac{3}{1+a^2}$.

Solution It follows that $a^2 - 3a = -1$, and $a^2 + 1 = 3a$. Dividing both sides by $a(a \neq 0)$, we have $a + \frac{1}{a} = 3$. Hence,

$$2a^2 - 5a - 2 + \frac{3}{1+a^2} = 2(a^2 - 3a) + a - 2 + \frac{1}{a}$$

$$= -4 + \left(a + \frac{1}{a}\right) = -4 + 3 = -1.$$

Remark: According to the characteristics of the expression to be evaluated, we deform the known conditions and then substitute as a whole. This method can greatly simplify the calculation.

Example 7. Suppose $abc \neq 0, a + b + c = 0$, evaluate $\frac{a^2}{bc} + \frac{b^2}{ac} + \frac{c^2}{ab}$.

Solution   The given expression equals

$$\frac{-(b+c) \cdot a}{bc} + \frac{-(a+c) \cdot b}{ac} + \frac{-(a+b) \cdot c}{ab}$$

$$= -\left(\frac{a}{b} + \frac{a}{c}\right) - \left(\frac{b}{a} + \frac{b}{c}\right) - \left(\frac{c}{a} + \frac{c}{b}\right)$$

$$= -\frac{b+c}{a} - \frac{a+c}{b} - \frac{a+b}{c}$$

$$= \frac{a}{a} + \frac{b}{b} + \frac{c}{c} = 3.$$

Example 8. Suppose a, b, c are nonzero real numbers such that $a+b+c \neq 0$ and $\frac{a+b-c}{c} = \frac{a-b+c}{b} = \frac{-a+b+c}{a}$. Find the value of $\frac{(a+b)(b+c)(c+a)}{abc}$.

Solution   Assume $\frac{a+b-c}{c} = \frac{a-b+c}{b} = \frac{-a+b+c}{a} = k$. Then,

$$a + b - c = kc, \quad a - b + c = kb, \quad -a + b + c = ka,$$

so $a + b = (k + 1)c, \; a + c = (k + 1)b, \; b + c = (k + 1)a$.

Taking the sum, we have $2(a + b + c) = (k + 1)(a + b + c)$. Since $a + b + c \neq 0$, necessarily $k + 1 = 2$, so $k = 1$.

Therefore, $\frac{(a+b)(b+c)(c+a)}{abc} = \frac{2c \cdot 2a \cdot 2b}{abc} = 8$.

Example 9. Suppose x, y, z are real numbers such that

$$\frac{y}{x - y} = \frac{x}{y + z}, \qquad \qquad \text{①}$$

$$z^2 = x(y + z) - y(x - y). \qquad \qquad \text{②}$$

Find the value of $\frac{y^2 + z^2 - x^2}{2yz}$.

Solution   From ①, we get $y^2 - x^2 = -xy - yz$.

From ②, we get $z^2 = xz + y^2$, and $z^2 - y^2 = xz$. Hence,

$$\frac{y^2 + z^2 - x^2}{2yz} = \frac{xz + y^2 - xy - yz}{2yz}$$

$$= \frac{(x - y)(z - y)}{2yz} = \frac{1}{2} \cdot \frac{x - y}{y} \cdot \frac{z - y}{z}$$

$$= \frac{1}{2} \cdot \frac{y + z}{x} \cdot \frac{z - y}{z} = \frac{1}{2} \cdot \frac{z^2 - y^2}{xz} = \frac{1}{2}.$$

Remark: The characteristics of the expression to be evaluated are the starting point for rewriting the conditions.

Example 10. Suppose $2004x^3 = 2005y^3 = 2006z^3, xyz > 0$, and

$$\sqrt[3]{2004x^2 + 2005y^2 + 2006z^2} = \sqrt[3]{2004} + \sqrt[3]{2005} + \sqrt[3]{2006}.$$

Find the value of $\frac{1}{x} + \frac{1}{y} + \frac{1}{z}$.

Solution Assume $2004x^3 = 2005y^3 = 2006z^3 = k$. Then, apparently, $k \neq 0$. Thus,

$$2004 = \frac{k}{x^3}, \quad 2005 = \frac{k}{y^3}, \quad 2006 = \frac{k}{z^3}.$$

Plugging into the given expression, we have

$$\sqrt[3]{\frac{k}{x} + \frac{k}{y} + \frac{k}{z}} = \sqrt[3]{\frac{k}{x^3}} + \sqrt[3]{\frac{k}{y^3}} + \sqrt[3]{\frac{k}{z^3}}.$$

Equivalently, $\sqrt[3]{k} \cdot \sqrt[3]{\frac{1}{x} + \frac{1}{y} + \frac{1}{z}} = \sqrt[3]{k} \left(\frac{1}{x} + \frac{1}{y} + \frac{1}{z} \right)$.

Since $k \neq 0$ and $xyz > 0$, it follows that $x > 0, y > 0, z > 0$. Hence,

$$\sqrt[3]{\frac{1}{x} + \frac{1}{y} + \frac{1}{z}} = \frac{1}{x} + \frac{1}{y} + \frac{1}{z}.$$

Since $\frac{1}{x} + \frac{1}{y} + \frac{1}{z} > 0$, we conclude that $\frac{1}{x} + \frac{1}{y} + \frac{1}{z} = 1$.

Example 11. Let x be a real number such that $\sqrt{x^3 + 2020} - \sqrt{2030 - x^3} = 54$. Find the value of $28\sqrt{x^3 + 2020} + 27\sqrt{2030 - x^3}$.

Solution Let $x^3 + 2020 = a, 2030 - x^3 = b$, then $a + b = 4050$.
It then follows that

$$\sqrt{a} - \sqrt{b} = 54. \qquad \qquad ①$$

Squaring both sides yields $a + b - 2\sqrt{ab} = 2916$, so we have $2\sqrt{ab} = a + b - 2916 = 1134$.

Thus, $(\sqrt{a} + \sqrt{b})^2 = a + b + 2\sqrt{ab} = 5184$, so that

$$\sqrt{a} + \sqrt{b} = \sqrt{5184} = 72. \qquad \qquad ②$$

From ① + ②, we get $\sqrt{a} = 63$, and from ② − ①, we get $\sqrt{b} = 9$.
Therefore, $28\sqrt{x^3 + 2020} + 27\sqrt{2030 - x^3} = 28 \times 63 + 27 \times 9 = 2007$.

Example 12. Suppose $x > 0, y > 0$, and

$$\sqrt{x}(\sqrt{x} + 2\sqrt{y}) = \sqrt{y}(6\sqrt{x} + 5\sqrt{y}).$$

Find the value of $\frac{x + \sqrt{xy} - y}{2x + \sqrt{xy} + 3y}$.

Solution   It follows that $(\sqrt{x})^2 - 4\sqrt{x}\sqrt{y} - 5(\sqrt{y})^2 = 0$. Equivalently,

$$(\sqrt{x} - 5\sqrt{y})(\sqrt{x} + \sqrt{y}) = 0.$$

Since $x > 0, y > 0$, we have $\sqrt{x} + \sqrt{y} > 0$, so necessarily $\sqrt{x} - 5\sqrt{y} = 0$, and $x = 25y$.

Therefore, $\frac{x + \sqrt{xy} - y}{2x + \sqrt{xy} + 3y} = \frac{25y + \sqrt{25y^2} - y}{50y + \sqrt{25y^2} + 3y} = \frac{29y}{58y} = \frac{1}{2}$.

Reading

George Polya, a Hungarian–American mathematician, had studied how to solve problems for decades and summarized a famous "table of how to solve problems." He thought problem-solving training should emphasize four steps:

First, clarify the problem. What are the unknowns? What are the known data? What are the conditions? and so on. Draw a picture and introduce appropriate symbols. Separate the various parts of the conditions, and write them down.

Second, draw up a plan, and find connections between the known and the unknown. If you cannot find a direct connection, you may have to consider auxiliary problems. You should finally come up with a plan for a solution.

Third, realize the plan, and check each step.

Fourth, review and test the obtained solution. Consider whether the result can be derived from other methods, and whether the method or result can be applied to other problems.

In his book, *How to Solve It*, Polya explained these four steps in detail, and in particular, many suggestions were given on how to find problem-solving ideas.

Exercises

(I) Multiple-choice questions:

1. If $a = \sqrt{7} - 1$, then the value of $3a^3 + 12a^2 - 6a - 12$ is (   ).

   (A) 24   (B) $4\sqrt{7} + 10$ (C) 25   (D) $4\sqrt{7} + 12$

2. Suppose a, b are positive real numbers such that $ab = a + b$, then $\frac{a}{b} + \frac{b}{a} - ab$ equals (   ).

   (A) -2 (B) $-\frac{1}{2}$ (C) $\frac{1}{2}$ (D) 2

3. If $\sqrt{24 - a} - \sqrt{8 - a} = 2$, then the value of $\sqrt{24 - a} + \sqrt{8 - a}$ is (   ).

   (A) 7   (B) 8   (C) 9   (D) 10

4. If $a + b + c = 0, \frac{1}{a} + \frac{1}{b} + \frac{1}{c} = -4$, then $\sqrt{\frac{1}{a^2} + \frac{1}{b^2} + \frac{1}{c^2}}$ equals (   ).

   (A) 3   (B) 4   (C) 8   (D) 16

5. If natural numbers a, b, c, d satisfy that $\frac{1}{a^2} + \frac{1}{b^2} + \frac{1}{c^2} + \frac{1}{d^2} = 1$, then $\sqrt{\frac{1}{a^3} + \frac{1}{b^4} + \frac{1}{c^5} + \frac{1}{d^6}}$ equals (   ).

   (A) $\frac{1}{8}$ (B) $\frac{\sqrt{7}}{8}$ (C) $\frac{\sqrt{15}}{8}$ (D) $\frac{\sqrt{15}}{16}$

(II) Fill in the blanks:

6. If $a = \sqrt{3} - 1$, then $a^{2014} + 2a^{2013} - 2a^{2012} = $ _____ .

7. If $a = -2 + \sqrt{2}$, then $1 + \frac{1}{2 + \frac{1}{3 + a}} = $ _____ .

8. Suppose a, b, c are real numbers such that $\frac{ab}{a + b} = \frac{1}{3}, \frac{bc}{b + c} = \frac{1}{4}, \frac{ca}{c + a} = \frac{1}{5}$, then $ab + bc + ca = $ _____ .

9. If $\sqrt{x} = \frac{1}{\sqrt{a}} - \sqrt{a}$, then $\sqrt{4x + x^2} = $ _____ .

10. Let $M = \frac{8}{\sqrt{2008} - 44}$, and a, b are the integer and fractional parts of M, respectively. Then, $a^2 + 3(\sqrt{2008} + 37)ab + 10 = $ _____ .

11. If $\sqrt[3]{x} - \frac{1}{\sqrt[3]{x}} = 3$, then $x^3 - \frac{1}{x^3} = $ _____ .

(III) Answer the following questions:

12. Given $x = \frac{2 + \sqrt{3}}{2 - \sqrt{3}}, y = \frac{2 - \sqrt{3}}{2 + \sqrt{3}}$, evaluate $2x^2 - 3xy + 2y^2$.

13. If $\frac{x}{3y} = \frac{y}{2x - 5y} = \frac{6x - 15y}{x}$, evaluate $\frac{4x^2 - 5xy + 6y^2}{x^2 - 2xy + 3y^2}$.

14. If $\frac{1}{a} - \frac{1}{b} = 4$, evaluate $\frac{a - 2ab - b}{2a + 7ab - 2b}$.

15. If x, y, z are real numbers such that $x + y + z \neq 0$, and $a = \frac{x}{y + z}, b = \frac{y}{z + x}, c = \frac{z}{x + y}$, find the value of $\frac{a}{1 + a} + \frac{b}{1 + b} + \frac{c}{1 + c}$.

16. Given $a + 2b + 3c = 12, a^2 + b^2 + c^2 = ab + bc + ca$, find the value of $a + b^2 + c^3$.

17. Suppose $\sqrt{a} + \sqrt{b} = 1$, $\sqrt{a} = m + \frac{a-b}{2}$, $\sqrt{b} = n - \frac{a-b}{2}$, find the value of $m^2 + n^2$.

18. If x, y are real numbers such that $(x + \sqrt{x^2 + 1})(y + \sqrt{y^2 + 1}) = 1$, find the value of $x + y$.

19. Suppose x, y, z, w are real numbers such that $2013x^2 = 2014y^2 = 2015z^2 = 2016w^2$ and $\frac{1}{x} + \frac{1}{y} + \frac{1}{z} + \frac{1}{w} = 1$. Find the value of $\frac{\sqrt{2013x + 2014y + 2015z + 2016w}}{\sqrt{2013} + \sqrt{2014} + \sqrt{2015} + \sqrt{2016}}$.

Chapter 11

Symmetric Polynomials

A polynomial is called a k-variate polynomial if it has k variables. For example, $8x^6 - 7x^4 + 5$ is a univariate polynomial, and $x^3y^3 - y^4z + 8yz^5$ is a trivariate polynomial.

Some polynomials have special properties that make them convenient when solving problems. Symmetric polynomials and cyclic polynomials are among these special polynomials.

1. Symmetric Polynomials

   If a polynomial is unchanged if we interchange any two variables in its expression, then it is called a symmetric polynomial (in these variables). For example, in $x+y+z$, if we interchange x and y, or y and z, or z and x, the resulting polynomial is still $x + y + z$, so $x + y + z$ is a symmetric polynomial.

   Another example is $(x+y)^6 + x^3y^3$, where the polynomial is unchanged if we interchange x and y. This is also a symmetric polynomial.

   Polynomials $x+y+z$, $xy+yz+zx$, xyz are called trivariate symmetric polynomials.

   It is trivial to check that the sum, difference, and product of two symmetric polynomials are still symmetric. For example, $x^3 + y^3 + z^3$ and $x^2(y + z) + y^2(x + z) + z^2(x + y)$ are both symmetric polynomials, and it is easy to prove that their sum, difference, and product are also symmetric polynomials.

2. Cyclic Polynomials

   In a multivariate polynomial, if we can arrange the variables in some order (usually in alphabetical order) and substitute the first variable for the second, the second variable for the third, and so on, and finally substitute the last variable for the first, so that the resulting polynomial is still the original one, then we call it a cyclic polynomial in these variables.

For example, in the polynomial $ab^2 + bc^2 + ca^2$, we arrange the variables in the order a, b, c and apply a cyclic substitution. Then, we get $bc^2 + ca^2 + ab^2$, which is exactly the original polynomial. Therefore, $ab^2 + bc^2 + ca^2$ is a cyclic polynomial in a, b, c.

We see easily that symmetric polynomials are always cyclic polynomials, but not the converse. For example, $(a - b)^3 + (b - c)^3 + (c - a)^3$ is a cyclic polynomial, but not a symmetric polynomial. However, bivariate cyclic polynomials are always symmetric.

Example 1. Let $\alpha = x_1 + x_2 + x_3, \beta = x_1 x_2 + x_2 x_3 + x_3 x_1, \gamma = x_1 x_2 x_3$. Use an expression of α, β, γ to represent the symmetric polynomial $x_1^2 x_2 + x_1 x_2^2 + x_1^2 x_3 + x_1 x_3^2 + x_2^2 x_3 + x_2 x_3^2$.

Solution

$$\begin{aligned}
&x_1^2 x_2 + x_1 x_2^2 + x_1^2 x_3 + x_1 x_3^2 + x_2^2 x_3 + x_2 x_3^2 \\
&= x_1(x_1 x_2 + x_1 x_3) + x_2(x_1 x_2 + x_2 x_3) + x_3(x_1 x_3 + x_2 x_3) \\
&= x_1(\beta - x_2 x_3) + x_2(\beta - x_1 x_3) + x_3(\beta - x_1 x_2) \\
&= x_1 \beta + x_2 \beta + x_3 \beta - 3 x_1 x_2 x_3 \\
&= (x_1 + x_2 + x_3)\beta - 3\gamma = \alpha\beta - 3\gamma.
\end{aligned}$$

Remark: Any trivariate symmetric polynomial (in $x_1 x_2 x_3$) can be represented by an expression of α, β, γ, which needs some trying.

Example 2. Factorize $(ax + by)^3 + (ay + bx)^3 - (a^3 + b^3)(x^3 + y^3)$.

Analysis   Note that the polynomial is unchanged if we interchange a and b, or if we interchange x and y. Also, if $a = 0$, then the expression equals 0; if $x = 0$, then it equals 0; if $a = -b$, then it equals 0; and if $x = -y$, then it equals 0. These, combined with symmetry, allow us to factorize the expression.

Solution   Note that $a, x, b, y, a + b, x + y$ are all factors of the given polynomial. Combining with its degree, we claim that the polynomial equals $kabxy(a + b)(x + y)$, where k is an undetermined coefficient.

Let $a = b = x = y = 1$. We may calculate the two expressions and obtain $k = 3$.

Therefore, the original polynomial equals $3abxy(a + b)(x + y)$.

Example 3. Factorize $(a + b + c)^5 - a^5 - b^5 - c^5$.

Analysis   This is a fifth-degree symmetric polynomial. If we find a factor that is not symmetric, then we can immediately find two other factors of the same type. Note that if we let $a = -b$, the polynomial equals 0, so $a+b$ is its factor according to the factor theorem. Similarly, $b+c, c+a$ are also its factors.

Solution   When $a = -b$, we have $(a + b + c)^5 - a^5 - b^5 - c^5 = 0$, so the given polynomial has the factor $(a + b)(b + c)(c + a)$.

Since it is a fifth-degree symmetric polynomial, we may assume that

$$(a + b + c)^5 - a^5 - b^5 - c^5$$
$$= (a + b)(b + c)(c + a)[k(a^2 + b^2 + c^2) + m(ab + bc + ca)]. \qquad ①$$

Here, k, m are undetermined coefficients.

Let $a = 1, b = 1, c = 0$, and by ①, we have $30 = 2(2k + m)$, so

$$2k + m = 15. \qquad ②$$

Let $a = 0, b = 1, c = 2$, and by ②, we have $210 = 6(5k + 2m)$, so

$$5k + 2m = 35. \qquad ③$$

By ② and ③, we obtain $k = 5$, $m = 5$. Therefore,

$$(a + b + c)^5 - a^5 - b^5 - c^5$$
$$= 5(a + b)(b + c)(c + a)(a^2 + b^2 + c^2 + ab + bc + ca).$$

Example 4. Factorize $a^2(b - c) + b^2(c - a) + c^2(a - b)$.

Analysis   It is easy to check that this is a cyclic polynomial. If $a = b$, then the polynomial reduces to 0, so $a - b$ is one of its factors. Since it is cyclic, $b - c$ and $c - a$ are also its factors, and there are already three factors. By using the method of undetermined coefficients, we may obtain the factorization.

Solution   If $a = b$, we have $a^2(b - c) + b^2(c - a) + c^2(a - b) = 0$, so, by the factor theorem, $a - b$ is a factor. According to cyclicity, $b - c, c - a$ are also its factors.

Since the given polynomial has a degree of 3, we may assume that

$$a^2(b - c) + b^2(c - a) + c^2(a - b) = k(a - b)(b - c)(c - a), \qquad ①$$

where k is an undetermined coefficient.

Let $a = 1$, $b = 0$, $c = -1$, and by ①, we have $k = -1$.
Therefore, $a^2(b - c) + b^2(c - a) + c^2(a - b) = -(a - b)(b - c)(c - a)$.

Example 5. Simplify

$$(x+y+z)^3 - (y+z-x)^2 - (z+x-y)^2 - (x+y-z)^3.$$

Analysis This is a symmetric polynomial, so after simplification, it is still symmetric.

Solution When $x = 0$, we have

$$(x+y+z)^3 - (y+z-x)^2 - (z+x-y)^2 - (x+y-z)^3$$
$$= (y+z)^3 - (y+z)^3 - (z-y)^3 - (y-z)^3 = 0.$$

By the factor theorem, the original polynomial has a factor $x - 0$, which is simply x.

Similarly, it has factors y, z.

Since the degree is 3, it cannot have other nontrivial factors, so we assume that

$$(x+y+z)^3 - (y+z-x)^2 - (z+x-y)^2 - (x+y-z)^3 = kxyz, \qquad ①$$

for the undetermined k.

Let $x = 1, y = 1, z = -1$, and by ①, we have $k = 24$.

Therefore, the original polynomial equals $24xyz$.

Example 6. Let a, b, c be nonzero real numbers and

$$M = \frac{a}{|a|} + \frac{b}{|b|} + \frac{c}{|c|} + \frac{ab}{|ab|} + \frac{bc}{|bc|} + \frac{ac}{|ac|} + \frac{abc}{|abc|}.$$

Find the possible values of M.

Analysis Since M involves three variables, there are eight cases, as shown in the following:

$$a>0 \begin{cases} b>0 \begin{cases} c>0 \\ c<0 \end{cases} \\ b<0 \begin{cases} c>0 \\ c<0 \end{cases} \end{cases} \qquad a<0 \begin{cases} b>0 \begin{cases} c>0 \\ c<0 \end{cases} \\ b<0 \begin{cases} c>0 \\ c<0 \end{cases} \end{cases}$$

Note that a, b, c are symmetric in M, and $\frac{ab}{|ab|} + \frac{ac}{|ac|} + \frac{bc}{|bc|}$ depends on the sign of the product of two numbers, while $\frac{abc}{|abc|}$ depends on the sign of the product of three numbers. Therefore, we may divide the cases based on the number of positive numbers in a, b, c.

This is the general method. However, we can also handle this problem from the view of symmetric polynomials.

Let $x_1 = \frac{a}{|a|}, x_2 = \frac{b}{|b|}, x_3 = \frac{c}{|c|}$, then

$$M = x_1 + x_2 + x_3 + x_1 x_2 + x_2 x_3 + x_3 x_1 + x_1 x_2 x_3.$$

Apparently, M is a symmetric polynomial in x_1, x_2, x_3, and the value of the three variables can only be ± 1. This simplifies our discussion.

Solution As in the analysis, we can write

$$M = (1 + x_1)(1 + x_2)(1 + x_3) - 1$$

$$= \left(1 + \frac{a}{|a|}\right)\left(1 + \frac{b}{|b|}\right)\left(1 + \frac{c}{|c|}\right) - 1.$$

If at least one of a, b, c is negative, then at least one of $1 + \frac{a}{|a|}, 1 + \frac{b}{|b|}, 1 + \frac{c}{|c|}$ is 0, so that $M = -1$. If a, b, c are all positive, then $M = -7$.

Therefore, all possible values of M are -1 and 7.

Example 7. Given $a = 2017, b = 2018, c = 2019$, evaluate

$$\frac{a^2 \left(\frac{1}{b} - \frac{1}{c}\right) + b^2 \left(\frac{1}{c} - \frac{1}{a}\right) + c^2 \left(\frac{1}{a} - \frac{1}{b}\right)}{a \left(\frac{1}{b} - \frac{1}{c}\right) + b \left(\frac{1}{c} - \frac{1}{a}\right) + c \left(\frac{1}{a} - \frac{1}{b}\right)}.$$

Solution The given expression equals

$$\frac{a^3(c - b) + b^3(a - c) + c^3(b - a)}{a^2(c - b) + b^2(a - c) + c^2(b - a)}.$$

We can check that the numerator and denominator are both unchanged under the cyclic substitution $a \to b \to c \to a$. Therefore, both the numerator and the denominator are cyclic polynomials.

Let $a = b$. Then, the numerator equals 0, so it has a factor $a - b$, and by cyclicity, it also has $b - c$, $c - a$ as factors. Thus, we can factorize the numerator as $(a - b)(b - c)(c - a)(a + b + c)$.

Similarly, the denominator can be factorized as $(a - b)(b - c)(c - a)$. Therefore, the original expression equals $a + b + c$. For $a = 2017, b = 2018, c = 2019$, its value is 6054.

Example 8. Let x_1, x_2, x_3, x_4, x_5, x_6 be a permutation of $1, 2, 3, 4, 5, 6$, and let

$$S = |x_1 - x_2| + |x_2 - x_3| + |x_3 - x_4| + |x_4 - x_5| + |x_5 - x_6| + |x_6 - x_1|,$$

Find the minimum of S.

Solution   We note that S is cyclic in variables $x_1, x_2, x_3, x_4, x_5, x_6$, so without loss of generality assume that $x_1 = 6$, and $x_n = 1, n \neq 1$. Thus,

$$S = |6 - x_2| + \cdots + |x_{n-1} - 1| + |1 - x_{n+1}| + \cdots + |x_6 - 6|$$

$$\geq |(6 - x_2) + \cdots + (x_{n-1} - 1)| + |(1 - x_{n-1}) + \cdots + (x_6 - 6)|$$

$$= |5| + |-5| = 10.$$

This shows $S \geq 10$. On the other hand, let $x_1 = 6, x_2 = 5, x_3 = 4$, $x_4 = 3, x_5 = 2, x_6 = 1$, then

$$S = 1 + 1 + 1 + 1 + 1 + 5 = 10.$$

Therefore, the minimum of S is 10.

Remark: Here, we have used the triangle inequality:

$$|x_1| + |x_2| \geq |x_1 + x_2|,$$

$$|x_1| + |x_2| + |x_3| \geq |x_1 + x_2 + x_3|.$$

Think about it   For n numbers $a_1, a_2, \ldots, a_n$, we define:

$\frac{a_1 + a_2 + \cdots + a_n}{n}$ as the arithmetic mean of these numbers;

$\sqrt[n]{a_1 a_2 \cdots a_n}$ as the geometric mean of these numbers;

$\dfrac{n}{\frac{1}{a_1} + \frac{1}{a_2} + \cdots + \frac{1}{a_n}}$ as the harmonic mean of these numbers.

Try some specific numbers, and calculate the three means of these numbers. Then, guess whether there is a fixed relationship among their sizes.

Exercises

(I) Multiple-choice questions:

1. There are (   ) cyclic polynomials in the following:

   (1) $3x + 2y + z$.   (2) $x^2 + y^3 + z^4 + x^4 y^3 z^2$.

   (3) $xy^2 + y^2 z^3 + z^3 x$.   (4) $x^3 + y^3 + z^3 - x^2 - y^2 - z^2$.

   (A) 0   (B) 1   (C) 2   (D) 3

2. If $x^2 y + x y^2 + y^2 z + y z^2 + z^2 x + z x^2 + 3xyz = k(x + y + z)(xy + yz + zx)$, then $k = ($ $)$.

   (A) $\frac{1}{2}$ (B) 1   (C) 3   (D) -1

3. Let $\alpha = x_1 + x_2 + x_3$, $\beta = x_1 x_2 + x_2 x_3 + x_3 x_1$, $\gamma = x_1 x_2 x_3$, then $x_1^3 + x_2^3 + x_3^3$ equals (　).

(A) $\alpha^3 - 3\alpha\beta + 3\gamma$ (B) $\beta^3 - 3\alpha\gamma + 3\gamma$

(C) $\alpha^3 + 3\alpha\beta - 3\gamma$ (D) $\beta^2 - 3\alpha\beta + 3\gamma$

(II) Factorize the following polynomials:

4. $(a + b + c)^3 - a^3 - b^3 - c^3$.
5. $(x + y)^4 + x^4 + y^4$.
6. $(a - b)^3 + (b - c)^3 + (c - a)^3$.
7. $(ab + bc + ca)(a + b + c) - abc$.
8. $xy(x^2 - y^2) + yz(y^2 - z^2) + zx(z^2 - x^2)$.
9. $x^2(y + z) + y^2(x + z) + z^2(x + y) - (x^3 + y^3 + z^3) - 2xyz$.

(III) Answer the following questions:

10. Let $a = x + y$, $b = x^3 + y^3$, and $a \neq 0$. Express $x^2 + y^2$ in terms of a, b.

11. Simplify $a(b + c - a)^2 + b(c + a - b)^2 + c(a + b - c)^2 + (b + c - a)$ $(c + a - b)(a + b - c)$.

12. Given $a + b + c + d = 0$, $a^3 + b^3 + c^3 + d^3 = 3$, prove that:

(1) $(a + b)^3 + (c + d)^3 = 0$.
(2) $ab(c + d) + cd(a + b) = 1$.

13. If a, b, c, d are nonzero, and

$$A = \frac{|a|}{a} + \frac{|b|}{b} + \frac{|c|}{c} + \frac{|d|}{d} - \frac{|ab|}{ab} - \frac{|ac|}{ac} - \frac{|ad|}{ad} - \frac{|bc|}{bc} - \frac{|bd|}{bd}$$
$$- \frac{|cd|}{cd} + \frac{|abc|}{abc} + \frac{|abd|}{abd} + \frac{|acd|}{acd} + \frac{|bcd|}{bcd} - \frac{|abcd|}{abcd},$$

find the value of A.

14. Given $xyz = 1$, $x + y + z = 2$, $x^2 + y^2 + z^2 = 16$, find the values of

(1) $xy + yz + zx$.
(2) $\frac{1}{xy + 2z} + \frac{1}{yz + 2x} + \frac{1}{zx + 2y}$.

Chapter 12

Proof of Identities

If two algebraic expressions take the same value within the allowed range of values of their variables, then we say that they are identical, denoted as $A \equiv B$, or simply $A = B$. Such equalities are called identities. Converting an expression into its identical expression is called an identical transformation.

Proving identities refers to proving two expressions are identical through identical transformations. Common methods include:

(1) converting from left to right or converting from right to left, where, usually, we convert from the complicated side to the simple side;

(2) simultaneously converting two sides and proving they are equal to the same expression;

(3) proving that $\mathrm{LHS} - \mathrm{RHS} = 0$ or $\frac{\mathrm{LHS}}{\mathrm{RHS}} = 1$, where $\mathrm{RHS} \neq 0$.

To prove identities for given constraints, we need to use the constraints flexibly.

Example 1. Given $\frac{1}{a} + \frac{1}{b} = \frac{1}{c}$, prove that $(a + b - c)^2 = a^2 + b^2 + c^2$.

Proof: It follows from the given condition that $bc + ac = ab$. We have

$$(a + b - c)^2 = a^2 + b^2 + c^2 + 2ab - 2ac - 2bc$$

$$= a^2 + b^2 + c^2 + 2ab - 2(bc + ac)$$

$$= a^2 + b^2 + c^2 + 2ab - 2ab$$

$$= a^2 + b^2 + c^2.$$

Example 2. Given $abc = 1$, prove that
$$\frac{a}{ab+a+1} + \frac{b}{bc+b+1} + \frac{c}{ac+c+1} = 1.$$

Proof: Since $abc = 1, a, b, c$ are all nonzero. Hence, the left-hand side equals

$$\frac{a}{ab+a+1} + \frac{b \cdot a}{abc+ab+a} + \frac{c \cdot ab}{ab \cdot ac + c \cdot ab + ab}$$

$$= \frac{a}{ab+a+1} + \frac{ab}{ab+a+1} + \frac{abc}{ab+a+1}$$

$$= \frac{a+ab+1}{ab+a+1} = 1.$$

Remark: A generalization of this problem is if $abcd = 1$, then

$$\frac{a}{abc+ab+a+1} + \frac{b}{bcd+bc+b+1} + \frac{c}{cda+cd+c+1}$$

$$+ \frac{d}{dab+da+d+1} = 1.$$

Example 3. Suppose $ax = by = cz$ and $\frac{1}{x} + \frac{1}{y} + \frac{1}{z} = 1$, prove that
$$a^3 x^2 + b^3 y^2 + c^3 z^2 = (a+b+c)^3.$$

Proof: Let $ax = by = cz = k$, then $a = \frac{k}{x}, b = \frac{k}{y}, c = \frac{k}{z}$. Thus,

$$a^3 x^2 + b^3 y^2 + c^3 z^2 = \frac{k^3}{x^3} \cdot x^2 + \frac{k^3}{y^3} \cdot y + \frac{k^3}{z^3} \cdot z$$

$$= \left(\frac{1}{x} + \frac{1}{y} + \frac{1}{z} \right) k^3 = k^3.$$

On the other hand, $(a+b+c)^3 = \left(\frac{k}{x} + \frac{k}{y} + \frac{k}{z} \right)^3 = \left(\frac{1}{x} + \frac{1}{y} + \frac{1}{z} \right)^3 k^3 = k^3$, so we have

$$a^3 x^2 + b^3 y^2 + c^3 z^2 = (a+b+c)^3.$$

Example 4. Suppose $x + y = a + b, x^2 + y^2 = a^2 + b^2$, prove that
$$x^{2019} + y^{2019} = a^{2019} + b^{2019}.$$

Proof: From $x+y = a+b$ we have $y = a+b-x$. Plugging into $x^2 + y^2 = a^2 + b^2$, we have
$$x^2 + (a+b)^2 - 2(a+b)x + x^2 = a^2 + b^2,$$

which reduces to $x^2 - (a+b)x + ab = 0$.

This implies $(x-a)(x-b) = 0$, so $x = a$ or $x = b$.

If $x = a$, then $y = b$, and if $x = b$, then $y = a$.

Either case, we have $x^{2019} + y^{2019} = a^{2019} + b^{2019}$.

Example 5. Suppose a, b, c are nonzero and $a^2 - b^2 = bc, b^2 - c^2 = ac$.
Prove that $a^2 - c^2 = ab$.

Proof: Adding the two given equalities, we have

$$a^2 - c^2 = c(a + b). \qquad\qquad ①$$

Since $a^2 - b^2 = bc$, we have $a^2 = b(b + c)$, so that

$$b + c = \tfrac{a^2}{b}. \qquad\qquad ②$$

Since $b^2 - c^2 = ac$, we also have $(b + c)(b - c) = ac$.

Combining with ②, we get $\frac{a^2}{b}(b - c) = ac$, which implies $ab - ac = bc$,
or, equivalently, $ab = c(a + b)$.

Comparing with ①, we conclude that $a^2 - c^2 = ab$, which is exactly
what we want to prove.

Remark: Adding the conditions, we have $a^2 - c^2 = bc + ac$, and comparing
with the desired result, we see that it suffices to prove that $bc + ac = ab$,
which is equivalent to $\frac{1}{a} + \frac{1}{b} = \frac{1}{c}$.

Example 6. Let k be an integer and a, b, c be pairwise distinct real numbers. Prove that

$$\frac{(a + b - kc)^2}{(a - c)(b - c)} + \frac{(b + c - ka)^2}{(b - a)(c - a)} + \frac{(c + a - kb)^2}{(c - b)(a - b)} = (k + 1)^2.$$

Proof: Reducing to a common denominator, the left-hand side equals

$$\frac{(a + b - kc)^2(b - a) + (b + c - ka)^2(c - b) + (c + a - kb)^2(a - c)}{(a - b)(b - c)(c - a)}$$

$$= \frac{\begin{array}{l}(b - c)[(a + b - kc)^2 - (b + c - ka)^2] \\ + (c - a)[(a + b - kc)^2 - (c + a - kb)^2]\end{array}}{(a - b)(b - c)(c - a)}$$

$$= \frac{(b - c)(a + 2b + c - kc - ka)(a - c + ka - kc)}{(a - b)(b - c)(c - a)}$$

$$\quad + \frac{(c - a)(2a + b + c - kb - kc)(b - c + kb - kc)}{(a - b)(b - c)(c - a)}$$

$$= \frac{-(a + 2b + c - kc - ka)(1 + k)}{a - b} + \frac{(2a + b + c - kb - kc)(1 + k)}{a - b}$$

$$= \frac{1+k}{a-b}[(2a+b+c-kb-kc)-(a+2b+c-kc-ka)]$$

$$= \frac{1+k}{a-b}(a-b-kb+ka)$$

$$= \frac{1+k}{a-b}(a-b)(1+k)$$

$$= (1+k)^2.$$

Thus, the desired identity is proven.

Remark: The key idea in this proof is the identity $b - a = (b - c) + (c - a)$. Taking various values of k, we may obtain a series of identities. For example, if $k = 0$, then

$$\frac{(a+b)^2}{(a-c)(b-c)} + \frac{(b+c)^2}{(b-a)(c-a)} + \frac{(c+a)^2}{(c-b)(a-b)} = 1.$$

Example 7. If a, b, c are real numbers such that $\frac{1}{a} + \frac{1}{b} + \frac{1}{c} = \frac{1}{a+b+c}$, prove that

$$\frac{1}{a^{2n-1}} + \frac{1}{b^{2n-1}} + \frac{1}{c^{2n-1}} = \frac{1}{a^{2n-1} + b^{2n-1} + c^{2n-1}},$$

where n is a positive integer.

Proof: From $\frac{1}{a} + \frac{1}{b} + \frac{1}{c} = \frac{1}{a+b+c}$, we have

$$(bc + ac + ab)(a + b + c) = abc,$$

which reduces to $[c(a+b) + ab][(a+b) + c] = abc$.

Further, it is equivalent to $c(a+b)^2 + ab(a+b) + c^2(a+b) = 0$,

$$(a+b)(ca + cb + ab + c^2) = 0,$$

$$(a+b)(c+a)(b+c) = 0.$$

Hence, $a + b = 0$, or $c + a = 0$, or $b + c = 0$. In other words, $a = -b$ or $c = -a$ or $c = -b$.

Without loss of generality, we assume $a = -b$, then $a^{2n-1} = -b^{2n-1}$. Thus,

$$\frac{1}{a^{2n-1}} + \frac{1}{b^{2n-1}} + \frac{1}{c^{2n-1}} = \frac{1}{c^{2n-1}},$$

$$\frac{1}{a^{2n-1} + b^{2n-1} + c^{2n-1}} = \frac{1}{c^{2n-1}},$$

and the desired result follows.

Example 8. Suppose $\frac{x}{a} + \frac{y}{b} + \frac{z}{c} = 1$ and $\frac{a}{x} + \frac{b}{y} + \frac{c}{z} = 0$. Prove that

$$\frac{x^2}{a^2} + \frac{z^2}{y^2} + \frac{z^2}{c^2} = 1.$$

Analysis Both the conditions and the result involve $\frac{x}{a}\frac{y}{b}\frac{z}{c}$ and their reciprocals, so we may change the variables as $\frac{x}{a} = u$, $\frac{y}{b} = v$, $\frac{z}{c} = w$.

Proof: Let $\frac{x}{a} = u$, $\frac{y}{b} = v$, $\frac{z}{c} = w$, then the conditions reduce to

$$u + v + w = 1,$$

$$\frac{vw + uw + uv}{uvw} = 0.$$

Thus,

$$(u + v + w)^2 = 1,$$

$$uv + vw + uw = 0.$$

Consequently, we have

$$u^2 + v^2 + w^2 = (u + v + w)^2 - 2(uv + vw + wu) = 1,$$

which is exactly

$$\frac{x^2}{a^2} + \frac{z^2}{y^2} + \frac{z^2}{c^2} = 1.$$

Example 9. Suppose $ax^3 = by^3 = cz^3$, and $\frac{1}{x} + \frac{1}{y} + \frac{1}{z} = 1$. Prove that

$$\sqrt[3]{ax^2 + by^2 + cz^2} = \sqrt[3]{a} + \sqrt[3]{b} + \sqrt[3]{c}.$$

Proof: Let $ax^3 = by^3 = cz^3 = t^3$, then

$$a = \frac{t^3}{x^3}, \quad b = \frac{t^3}{y^3}, \quad c = \frac{t^3}{z^3}.$$

Thus,

$$\sqrt[3]{a} + \sqrt[3]{b} + \sqrt[3]{c} = t\left(\frac{1}{x} + \frac{1}{y} + \frac{1}{z}\right) = t.$$

Also, we have

$$\sqrt[3]{ax^2 + by^2 + cz^2} = \sqrt[3]{ax^3 \cdot \frac{1}{x} + by^3 \cdot \frac{1}{y} + cz^3 \cdot \frac{1}{z}}$$

$$= \sqrt[3]{t^3\left(\frac{1}{x} + \frac{1}{y} + \frac{1}{z}\right)} = t.$$

Therefore, $\sqrt[3]{ax^2 + by^2 + cz^2} = \sqrt[3]{a} + \sqrt[3]{b} + \sqrt[3]{c}$.

Remark: If we let $\sqrt[3]{a} = p$, $\sqrt[3]{b} = q$, $\sqrt[3]{c} = r$, then the condition becomes $px = qy = rz$, and we can still prove that both sides are equal to px.

Think about it The problem of writing a natural number as the sum of the squares of n natural numbers has been attracting attention since it was first posed more than 2000 years ago in ancient Greece. Around the 3rd century AD, Greek mathematician Diophantus conjectured that each natural number can be written as the sum of four perfect squares. For example,

$$5 = 1^2 + 2^2 + 0^2 + 0^2, \quad 6 = 1^2 + 1^2 + 2^2 + 0^2, \quad 7 = 1^2 + 1^2 + 1^2 + 2^2.$$

This conjecture was studied by many mathematicians. Among them, Euler found that

$$(a^2 + b^2 + c^2 + d^2)(x^2 + y^2 + z^2 + t^2)$$
$$= (ax - by - cz - dt)^2 + (ay + bx + ct - dz)^2 + (az - bt + cx + dy)^2$$
$$+ (at + bz - cy + dx)^2.$$

This means that the product of two integers that are the sum of four squares is also the sum of four squares. Can you derive this identity?

Exercises

(II) Fill in the blanks:

1. If $S_1 = 1 + \frac{1}{1^2} + \frac{1}{2^2}$, $S_2 = 1 + \frac{1}{2^2} + \frac{1}{3^2}, \ldots, S_k = 1 + \frac{1}{k^2} + \frac{1}{(k+1)^2}$, then in terms of n, $\sqrt{S_1} + \sqrt{S_2} + \cdots + \sqrt{S_n} = $ _____ .

2. If $xyz \neq 0, x + y + z = 0$, then $\sqrt{\frac{1}{x^2} + \frac{1}{y^2} + \frac{1}{z^2}}$ can be simplified to _____ . (The result should not include a radical sign.)

3. If a, b, c are nonzero real numbers such that $a + b + c = 0$, then $\frac{abc}{a^3 + b^3 + c^3} = $ _____ .

4. If a, b, c, d are positive numbers and $a^4 + b^4 + c^4 + d^4 = 4abcd$, then the four numbers can be arranged in the increasing order as _____ .

(II) Answer the following questions:

5. Prove that $1 + 4\left(\frac{a^2 b^2}{a^4 - b^4}\right) = \left(\frac{a^4 + b^4}{a^4 - b^4}\right)^2$.

6. Prove that

$$\left(a + \frac{1}{a}\right)^2 + \left(b + \frac{1}{b}\right)^2 + \left(ab + \frac{1}{ab}\right)^2$$

$$= \left(a + \frac{1}{a}\right)\left(b + \frac{1}{b}\right)\left(ab + \frac{1}{ab}\right) + 4.$$

7. Prove that $\frac{a^3(c-b)+b^3(a-c)+c^3(b-a)}{a^2(c-b)+b^2(a-c)+c^2(b-a)} = a + b + c.$

8. Prove that $\frac{b}{a(a+b)} + \frac{c}{(a+b)(a+b+c)} + \frac{d}{(a+b+c)(a+b+c+d)} = \frac{b+c+d}{a(a+b+c+d)}.$

9. Prove that $\frac{x}{ax-a^2} + \frac{y}{ay-a^2} + \frac{z}{az-a^2} = \frac{1}{x-a} + \frac{1}{y-a} + \frac{1}{z-a} + \frac{3}{a}.$

10. Prove that $\frac{a\sqrt{a}+b\sqrt{b}}{\sqrt{a}+\sqrt{b}} - \sqrt{ab} = \left(\frac{a-b}{\sqrt{a}+\sqrt{b}}\right)^2.$

11. Suppose that A, B, C, D, x, y, z, t are positive numbers and $\frac{A}{x} = \frac{B}{y} = \frac{C}{z} = \frac{D}{t}$. Prove that

$$\sqrt{Ax} + \sqrt{By} + \sqrt{Cz} + \sqrt{Dt} = \sqrt{(A+B+C+D)(x+y+z+t)}.$$

12. If $\frac{a}{b} = \frac{c}{d} = 3$, prove that

$$\frac{a^2 + c^2}{a + c} + \frac{b^2 + d^2}{b + d} = \frac{(a+b)^2 + (c+d)^2}{a+b+c+d}.$$

13. If $a + b = 1$, prove that

$$\frac{b}{a^3 - 1} - \frac{a}{b^3 - 1} = \frac{2(a-b)}{a^2 b^2 + 3}.$$

14. If $c > b > a > 0$ and

$$\frac{(b+c)^2 - a^2}{bc} + \frac{(c+a)^2 - b^2}{ca} + \frac{(a+b)^2 - c^2}{ab} = 8,$$

prove that $a + b = c$.

15. Suppose a, b, c, x, y, z are pairwise distinct nonzero real numbers, and

$$\frac{yz}{bz + cy} = \frac{xz}{cx + az} = \frac{xy}{ay + bx} = \frac{x^2 + y^2 + z^2}{a^2 + b^2 + c^2}.$$

Prove that

$$a + b + c = 2(x + y + z).$$

Chapter 13

Linear Functions

Functions like $y = kx + b$ (where k, b are constants and $k \neq 0$) are called linear functions. If $b = 0$, then the function is $y = kx(k \neq 0)$, which is called a proportional function, and we say that y is proportional to x.

The graph of a linear function is a line. When $k > 0$, the graph looks like Figure 13.1, where y increases as x increases. When $k < 0$, the graph looks like Figure 13.2, where y decreases as x increases. The graph of a proportional function is centrally symmetric with respect to the origin.

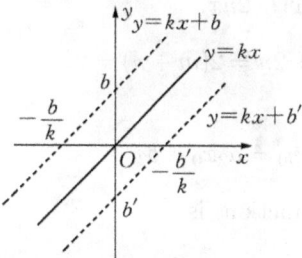
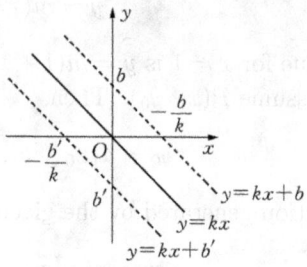

Fig. 13.1   Fig. 13.2

When solving problems with linear functions, we tend to combine numbers and shapes and do some casework. Undetermined coefficient method is a common method to find the formula for a linear function. In certain problems, the constraints on the variable are another concern.

Example 1. Let A, B be points on the graphs of $y = x, y = 8x$, respectively, and their x-coordinates are $a, b(a > 0, b > 0)$, respectively. The line AB is the graph of function $y = kx + m$. If $\frac{b}{a}$ and k are both integers, find all possible values of k.

Solution   We have $A(a, a), B(b, 8b)$. Since A, B both lie on the graph of $y = kx + m$, it follows that

$$a = ka + m, \quad 8b = kb + m.$$

Taking the difference between both sides, we have $k = \frac{8b-a}{b-a} = \frac{8(b-a)+7a}{b-a} = 8 + \frac{7}{\frac{b}{a}-1}$.

Since $\frac{b}{a}$ and k are both integers, $\frac{b}{a}$ must be equal to 2 or 8. Therefore, k equals 15 or 9.

Example 2. Let $y = a_1 x + b$ and $y = a_2 x + b$ be two linear functions. We say that $y = m(a_1 x + b_1) + n(a_2 x + b_2)$ is the function generated by these two functions, where m, n are given constants with $m + n = 1$.

(1) If $x = 1$, find the corresponding value of y in the function generated by $y = x + 1$ and $y = 2x$.
(2) If the functions $y = a_1 x + b_1$ and $y = a_2 x + b_2$ intersect at a point P, decide whether P lies on the function generated by these two, and explain why.

Solution   (1) The function generated by $y = x + 1$ and $y = 2x$ is

$$y = m(x + 1) + 2nx.$$

so its value for $x = 1$ is $y = m(1 + 1) + 2n = 2(m + n) = 2$.
   (2) Assume $P(x_0, y_0)$. Then,

$$y_0 = a_1 x_0 + b_1, \quad y_0 = a_2 x_0 + b_2.$$

The function generated by the given functions is

$$y = m(a_1 x + b_1) + n(a_2 x + b_2).$$

And when $x = x_0$, we have

$$m(a_1 x + b_1) + n(a_2 x + b_2) = m(a_1 x_0 + b_1) + n(a_2 x_0 + b_2)$$
$$= my_0 + ny_0 = (m + n)y_0 = y_0.$$

Therefore, P always lies on the graph of the generated function.

Example 3. Let S_k be the area of a triangle enclosed by the lines $y = kx + k - 1, y = (k + 1)x + k$ and the x-axis (where k is a positive integer). Find the value of $S_1 + S_2 + \cdots + S_{2018}$.

Solution   In $y = kx + k - 1$, if $y = 0$, then $x = \frac{1}{k} - 1$.

In $y = (k+1)x + k$, if $y = 0$, then $x = \frac{1}{k+1} - 1$.

Thus, the intersections of these two lines with the x-axis are $\left(\frac{1}{k} - 1, 0\right)$ and $\left(\frac{1}{1+k} - 1, 0\right)$.

Let (x, y) be the intersection of the two lines, then

$$\begin{cases} y = kx + k - 1, \\ y = (k+1)x + k. \end{cases}$$

which yields $\begin{cases} x = -1, \\ y = -1. \end{cases}$ Thus, the triangle enclosed by the three lines has one side equal to $\left(\frac{1}{k} - 1\right) - \left(\frac{1}{k+1} - 1\right) = \frac{1}{k} - \frac{1}{k+1}$, and the corresponding altitude is $|y| = |-1| = 1$, so its area is

$$S_k = \frac{1}{2}\left(\frac{1}{k} - \frac{1}{k+1}\right).$$

Therefore,

$$S_1 + S_2 + \cdots + S_{2018} = \frac{1}{2}\left[\left(1 - \frac{1}{2}\right) + \left(\frac{1}{2} - \frac{1}{3}\right) + \cdots + \left(\frac{1}{2018} - \frac{1}{2019}\right)\right]$$

$$= \frac{1}{2}\left(1 - \frac{1}{2019}\right) = \frac{1009}{2019}.$$

Remark: To find the intersection of the graphs of two functions, we simply put their formulas together and solve the system of equations.

Example 4. Suppose the convex quadrilateral enclosed by the lines $x = 1, y = -1, y = 3$, and $y = kx - 3$ has an area of 12. Find k.

Analysis   The region enclosed is as shown in Figure 13.3, which can be either a right triangle or a right trapezoid. However, the condition specifies

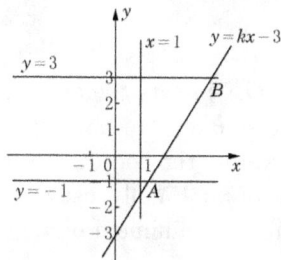

Fig. 13.3

that it is a quadrilateral, so we may calculate the side lengths of the trapezoid by finding the coordinates of its vertices.

Solution   The intersection of $y = -1$ and $y = kx - 3$ is $A\left(\frac{2}{k}, -1\right)$.

The intersection of $y = 3$ and $y = kx - 3$ is $B\left(\frac{6}{k}, 3\right)$. According to the condition, A, B lie on the same side of the line $x = 1$, so we have

$$\frac{1}{2}\left(\left|\frac{2}{k} - 1\right| + \left|\frac{6}{k} - 1\right|\right) \times 4 = 12.$$

Equivalently, $\left|\frac{2}{k} - 1\right| + \left|\frac{6}{k} - 1\right| = 6$.

Here, $\frac{2}{k} - 1$ and $\frac{6}{k} - 1$ are either both positive or both negative. This implies $\frac{2}{k} - 1 + \frac{6}{k} - 1 = 6$ or $1 - \frac{2}{k} + 1 - \frac{6}{k} = 6$. The former gives $k = 1$, and the latter gives $k = -2$, where both are valid solutions.

Remark: The area of a trapezoid equals the product of the height and the midsegment, so we may also derive that the midsegment equals 3. Since the midsegment lies on the line $y = 1$, we see that the intersection of $y = 1$ and $y = kx - 3$ must be $(4, 1)$ or $(-2, 1)$. This also allows us to obtain $k = 1$ or $k = -2$.

Example 5. As in Figure 13.4, the square $ABCD$ is divided into two parts with the same area by the line OE. Suppose the lengths of OD, AD are both positive integers, and $\frac{CE}{BE} = 20$. Find the minimal area of the square $ABCD$.

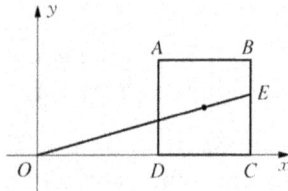

Fig. 13.4

Analysis   Since the line OE bisects the area of the square, it must pass through the center of the square, which we denote as O'. Let $BE = a$, $OD = m$, then we may express the coordinates of O' and plug into OE: $y = kx$, and find the value of m. Finally, use the assumption that OD, AD are positive integers to find the minimum of a.

Solution   It follows that OE passes through the center of $ABCD$, which is O'. Assume $BE = a, OD = m$. Thus, $CE = 20a$, and the side length of the square is $21a$.

Further, we have $O'(m+10.5a, 10.5a)$, $E(m+21a, 20a)$. If the equation of the line OE is $y = kx$, then $k(m+10.5a) = 10.5a$ and $k(m+21a) = 20a$. Hence,

$$\frac{m+10.5a}{m+21a} = \frac{10.5a}{20a}.$$

This implies $m = \frac{21}{19}a$.

Since the lengths of OD, AD are positive integers, so that $m, 21a$ are positive integers, we see that the minimum of $21a$ is 19, in which case $m = 1$.

Therefore, the minimal area of the square $ABCD$ is $(21a)^2 = 19^2 = 361$.

Example 6. Given function $y = |x+1| - 2|x-1| + |x+2|$,

(1) Draw the graph of the function in the coordinate plane.
(2) Suppose the equation (for x) $kx + 3 = |x+1| - 2|x-1| + |x+2|$ has three solutions, where $k \neq 0$. Find the value range of k.

Solution (1) Remove the absolute value symbol by dividing into pieces, and we have

$$y = \begin{cases} -5, & x \leq -2, \\ 2x - 1, & -2 < x \leq -1, \\ 4x + 1, & -1 < x \leq 1, \\ 5, & x > 1. \end{cases}$$

The graph is shown in Figure 13.5.

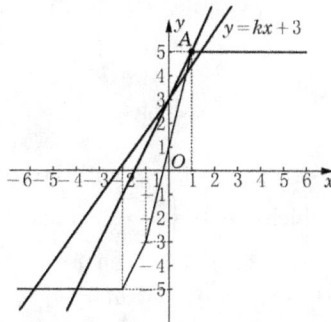

Fig. 13.5

(2) Set $A(1, 5)$, and consider the function $y = kx + 3$. The solutions of the equation are the x-coordinates of the intersections of $y = kx + 3$ and $y = |x+1| - 2|x-1| + |x+2|$.

Note that the graph of $y = kx + 3$ passes through $(0, 3)$. If it passes through A, then the two functions have only two intersections, in which case $k = 2$.

We can see that the two functions have three intersections exactly when $0 < k < 2$, so the value range of k is $0 < k < 2$.

Remark: Functions and equations are related by nature. When solving such problems, we should try to gain insight from the graphs and combine numbers with shapes.

Example 7. Given $A(-6, 1), B(-1, 5)$, a point $C(m, 0)$ lies on the x-axis, and a point $D(0, n)$ lies on the y-axis, such that $AB + BD + CD + CA$ is minimal. Find the value of $\frac{m}{n}$.

Solution   Let A' be the reflection of A with respect to the x-axis and B' be the reflection of B with respect to the y-axis. The line $A'B'$ intersects the x-axis at C and intersects the y-axis at D, as in Figure 13.6. It is well known that such C, D give the minimum of $AB + BD + CD + CA$.

Let the equation of $A'B'$ be $y = kx + b (k \neq 0)$.

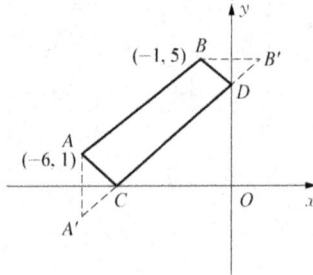

Fig. 13.6

Since the coordinate of A' is $(-6, -1)$ and the coordinate of B' is $(1, 5)$, we have $\begin{cases} 5 = k + b \\ -1 = -6k + b \end{cases}$, which yields $\begin{cases} k = \frac{6}{7} \\ b = \frac{29}{7} \end{cases}$. Thus, $y = \frac{6}{7}x + \frac{29}{7}$.

If $x = 0$, then $y = \frac{29}{7}$, and if $y = 0$, then $x = -\frac{29}{6}$.

Therefore, $C\left(-\frac{29}{6}, 0\right), D\left(0, \frac{29}{7}\right)$, and $m = -\frac{29}{6}, n = \frac{29}{7}$ $\frac{m}{n} = -\frac{7}{6}$.

Example 8. As in Figure 13.7, a parallelogram $ABCD$ has $\angle DAB = 60°, AB = 5, BC = 3$. A point P starts at D and travels at a constant speed along sides DC, CB. The distance P has traveled is x, and the area of the region enclosed by AD, AP and the locus of P is y. During this process, y is clearly a function of x. Which one of the following graphs most precisely describes this relation? (   ).

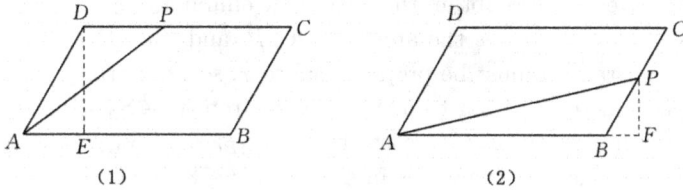

(1)   (2)

Fig. 13.7

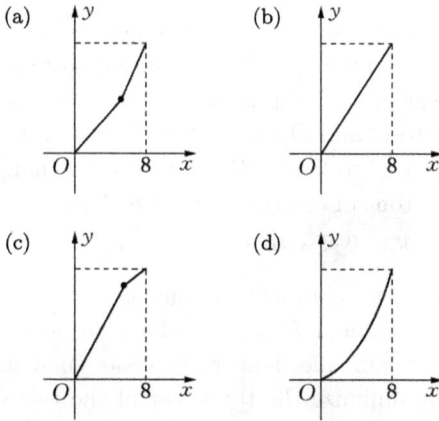

(a)   (b)   (c)   (d)

Solution We first establish the function.

When $0 < x \leq 5$, the region enclosed is $\triangle ADP$, as in Figure 13.7(1). Let E be the projection of D on AB. In the right triangle ADE, we have $AD = BC = 3, \angle DAE = 60°$, so

$$AE = \frac{3}{2}, DE = \sqrt{AD^2 - AE^2} = \frac{3}{2}\sqrt{3}$$

Hence, $y = \frac{1}{2} \cdot x \cdot \frac{3}{2}\sqrt{3} = \frac{3}{4}\sqrt{3}x$.

When $5 < x \leq 8$, the region enclosed is $ADCP$, whose area is the difference between S_{ABCD} and S_{ABP}. As in Figure 13.7(2), let F be the projection of P on the line AB.

In the right triangle BFP, $BP = 8 - x, \angle PBF = 60°$, so we have

$$BF = \frac{8 - x}{2}$$

$$PF = \sqrt{BP^2 - BF^2} = \frac{\sqrt{3}}{2}(8 - x)$$

Hence, $y = 5 \times \frac{3}{2}\sqrt{3} - \frac{1}{2} \times 5 \times \frac{\sqrt{3}}{2}(8 - x) = \frac{5}{4}\sqrt{3}x - \frac{5}{2}\sqrt{3}$.

Note that the line corresponding to $5 < x \leq 8$ should intersect the negative half of the y-axis, so the answer is (a).

Remark: We can also obtain the answer by elimination.

When $0 < x \leq 5$, y is the area of $\triangle ADP$, and since the altitude over DP is a constant, y must be proportional to x, so D is wrong.

When x increases from to 5, y increases from to $\frac{1}{2}S_{ABCD}$, and when x increases from 5 to 8, y increases from $\frac{1}{2}S_{ABCD}$ to S_{ABCD}, so y is not proportional to x over the whole interval $0 < x \leq 8$. Thus, B is wrong.

Finally, the value of y for $x = 5$ is half of the value of y for $x = 8$, so C is wrong. Therefore, the answer is (a).

Example 9. Two villages A, B produce oranges. A has 200 tons of oranges, and B has 300 tons. These oranges are carried to two warehouses C, D. We know that C can hold 240 tons and D can hold 260 tons. The costs of carrying from A to C and D are 20 and 25 yuan/ton, respectively. The costs of carrying from B to C and D are 15 and 18 yuan/ton, respectively. Suppose we carry x tons of oranges from A to C, and the cost of carrying from A (resp., B) is y_A (resp., y_B):

(1) Find the formula of y_A, y_B as functions of x.
(2) Try to decide which of A, B pays less for transportation.
(3) Suppose that B can afford at most 4830 yuan for transportation. How should we minimize the total cost of the two villages? Find the minimum.

Solution   Based on the given conditions, we construct Table 13.1.

Table 13.1

Warehouse Village	C	D	Total
A	x tons	$(200 - x)$ tons	200 tons
B	$(240 - x)$ tons	$(60 + x)$ tons	300 tons
Total	240 tons	260 tons	500 tons

Hence, $y_A = 20x + 25(200 - x) = -5x + 5000 (0 \leq x \leq 200)$,

$$y_B = 15(240 - x) + 18(60 + x) = 3x + 4680 (0 \leq x \leq 200).$$

(2) If $y_A = y_B$, then $-5x + 5000 = 3x + 4680$, and $x = 40$.
    If $y_A > y_B$, then $-5x + 5000 > 3x + 4680$, and $x < 40$.
    If $y_A < y_B$, then $-5x + 5000 < 3x + 4680$, and $x > 40$.

Therefore, when $x = 40, y_A = y_B$, so the two villages pay the same amount; when $\le x < 40, y_A > y_B$, so B pays less; when $40 < x \le 200, y_A < y_B$, so A pays less.

(3) We have $y_B \le 4830$, so $3x + 4680 \le 4830$, and $\le x \le 50$.

Let $y = y_A + y_B$ be the total cost, then $y = -2x + 9680$. Thus, y decreases as x increases, and the minimum of y is 9580 when $x = 50$.

Therefore, the minimal total cost is attained when village A carries 50 tons to warehouse C and 150 tons to warehouse D, while village B carries 190 tons to C and 110 tons to D. In this case, the minimal cost is 9580 yuan.

Remark: (1) When constructing the function, constructing a table may help us sort out the conditions.

(2) If we use linear functions in application problems, we need to pay attention to the domain of x.

Example 10. A cuboid sink has a base area of 100 cm^2 and a height of 20 cm. We put a cylinder-shaped beaker in the sink (where the weight and volume of the beaker are neglected), as shown in Figure 13.8. Now, we pour water into the beaker at some speed and continue pouring after the beaker is full until the whole sink is full. Here, we assume that the beaker cannot move in the sink. The level of water in the sink is h, and its relation with time t is shown in Figure 13.9.

(1) Find the base area of the beaker.
(2) If the height of the beaker is 9 cm, find the speed of pouring and the time needed to fill the whole sink.
(3) Write h as a function of t, and specify the corresponding domain.

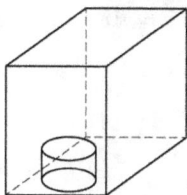

Fig. 13.8

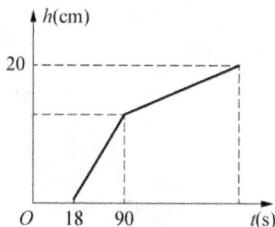

Fig. 13.9

Analysis (1) The graph of the function shows that when the water level equals the height of the beaker, the volume of water added is $\frac{90V}{18}$.

(2) After pouring for 18 s, the beaker is full. After pouring for 90 s, the water level in the sink is h cm, and the volume of water added is $100h$. From $100h = \frac{90Sh}{18}$, we may find S.

(3) According to the result of (2), we can write h as a function of t in each corresponding interval.

Solution   Assume the base area of the beaker is S cm^2 and its height is h_1 cm. Also, assume the speed of pouring is v cm^3/s, and the time needed to fill the whole sink is t_0s.

(1) It follows from the graph that after 18 s, the beaker is full. After 90 s, the water level in the sink is exactly h_1 cm, so we have $Sh_1 = 18v, 100h_1 = 90v$. Hence,

$$100h_1 = 90 \times \frac{1}{18}Sh_1,$$

which yields $S = 20$. Therefore, the base area of the beaker is 20 cm^2.

(2) If $h_1 = 9$, then $v = \frac{Sh_1}{18} = \frac{1}{18} \times 20 \times 9 = 10$, so the speed of pouring is 10 cm^3/s.

Since $vt_0 = 100 \times 20$, we solve that $t_0 = 200$. Therefore, the time needed to fill the whole sink is 200 s.

(3) ① $h = 0(0 \le t < 18)$.

② From the result of (2), we see that the graph passes through $(90, 9)$ and $(18, 0)$. Plugging into $y = kx + b$, we get

$$\begin{cases} 18k + b = 0, \\ 90k + b = 9, \end{cases}$$

and solving the equations, we have $\begin{cases} k = \frac{1}{8}, \\ b = -\frac{9}{4}. \end{cases}$ Hence, $h = \frac{t}{8} - \frac{9}{4}(18 \le t < 90)$.

③ From (2). we see that when $90 \le t \le 200$, the graph passes through $(200, 20)$ and $(90, 9)$. Plugging into $y = kx + b$, we have

$$\begin{cases} 200k + b = 20, \\ 90k + b = 9, \end{cases}$$

and solving the equations, we get $\begin{cases} k = \frac{1}{10}, \\ b = 0. \end{cases}$ Hence, $h = \frac{t}{10}(90 \le t \le 200)$.

In summary, $h = \begin{cases} 0, & 0 \le t < 18, \\ \frac{t}{8} - \frac{9}{4}, & 18 \le t < 90, \\ \frac{t}{10}, & 90 \le t \le 200. \end{cases}$

Exercises

(I) Multiple-choice questions:

1. Suppose that points $(-7, m)$ and $(-8, n)$ both lie on the line $y = -2x - 6$, then ( ).

   (A) $m > n$ (B) $m < n$ (C) $m = n$
   (D) unable to decide

2. If the lines $x + 2y = 2m$ and $2x + y = 2m + 3$ (m is a constant) intersect in the fourth quadrant, then the value of m can be ( ).

   (A) $-3, -2, -1, 0$ (B) $0, 1, 2, 3$
   (B) $-1, 0, 1, 2$ (D) $-2, -1, 0, 1$

3. Suppose that y is proportional to $x + 2$ and $z - 4$ is proportional to y. When $x = -3$, we have $y = -6$, and when $y = 6$ we have $z = -2$. Then, ( ).

   (A) $z = 6x - 8$ (B) $z = -6x + 8$
   (B) $z = -6 - 8x$ (D) $z = -6x - 8$

4. If the graph of $y = ax + b$ passes through quadrants I, II, III, and intersects the x-axis at $(-2, 0)$, then the solution set to $ax > b$ is ( ).

   (A) $x > -2$ (B) $x < -2$ (C) $x > 2$ (D) $x < 2$

5. In cryptography, raw information is called plaintext, and after some cryptographic algorithms, it becomes ciphertext. For example, an encryption method first pairs the 26 alphabets with the integers $1, 2, \ldots, 26$ (as shown in the following).

Alphabet	a	b	c	d	e	f	g	h	i	j	k	l	m
Number	1	2	3	4	5	6	7	8	9	10	11	12	13
Alphabet	n	o	p	q	r	s	t	u	v	w	x	y	z
Number	14	15	16	17	18	19	20	21	22	23	24	25	26

If the number corresponding to the plaintext is odd (which is x), then let the ciphertext be the alphabet corresponding to $y = \frac{x+1}{2}$; if the number corresponding to the plaintext is even (which is x), then the ciphertext corresponds to $y = \frac{x}{2} + 13$.

According to the rules described above, if the plaintext is "love," then the ciphertext is (   ).

(A) gawq   (B) shxc   (C) sdri   (D) love

6. If $\frac{a}{b+c} = \frac{b}{a+c} = \frac{c}{a+b} = t$, then the graph of $y = tx + t^2$ must pass through quadrants (   ).

(A) I, II   (B) I, II, III   (C) II, III, IV   (D) III, IV

7. Suppose $b > a$. If we draw the graphs of $y = bx + a$ and $y = ax + b$ in the same coordinate plane, they may look like (   ).

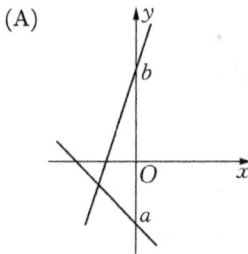

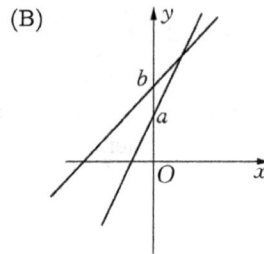

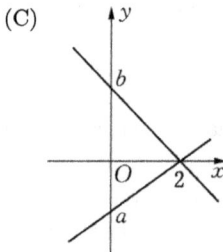

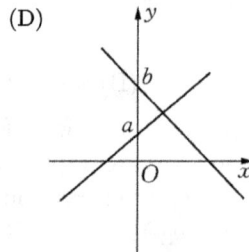

8. The story of a crow drinking water goes as follows. Once there is a crow who wants to drink water, and there is some water in a narrow-mouth bottle, which is too narrow, so the crow cannot reach the water. Then, the crow puts some pebbles in the bottle, and the level of the water rises. Thus, the crow manages to reach the water, but before it drinks enough, the level goes down again, and the crow needs to bring more pebbles. Finally, the crow drinks enough water and flies away. If the total volume of the pebbles added is x and the water level in the bottle is y, the relationship between y and x is most accurately described in (   ).

(A)

(B)

(C)

(D)

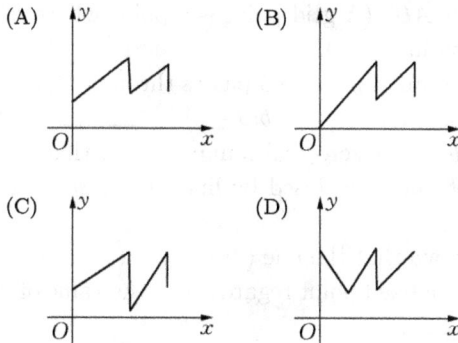

9. As shown in the figure, a line $l : y = x + 1$ intersects the y-axis at A_1, and point B_1 lies on the positive half of x-axis, with $OB_1 = OA_1$. Further, point A_2 lies on l and $A_2B_1 \perp x$-axis, point B_2 lies on x-axis with $B_1B_2 = B_1A_2$, point A_3 lies on l and $A_3B_2 \perp x$-axis, point B_3 lies on x-axis with $B_2B_3 = B_2A_3$, and so on. The area of $\triangle OA_1B_1$ is denoted as S_1, the area of $\triangle B_1A_2B_2$ is denoted as S_2, the area of $\triangle B_2A_3B_3$ is denoted as S_3, and so on. Then, $S_{2017} = ($ $)$.

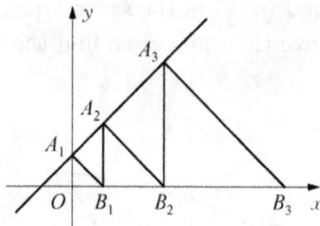

(A) 2^{4030} (B) 2^{4031} (C) 2^{4032} (D) 2^{4033}

10. Suppose the point $B(1, n)$ lies on the line $y = mx - 1$, and the distance between B and the origin is $\sqrt{10}$. Then, the area of the triangle enclosed by this line and the two coordinate axes is (   ).

   (A) $\frac{1}{2}$ (B) $\frac{1}{4}$ or $\frac{1}{2}$ (C) $\frac{1}{4}$ or $\frac{1}{8}$ (D) $\frac{1}{8}$ or $\frac{1}{2}$

(II) Fill in the blanks:

11. The line $y = \frac{5}{4}x - \frac{95}{4}$ intersects the x-axis at A and intersects the y-axis at B. Then, there are _____ grid points on the line

segment AB. (A grid point is a point whose x and y coordinates are both integers.)

12. If the graph of $y = x + 5$ passes through $P(a, b)$ and $Q(c, d)$, then the value of $a(c - d) + b(d - c)$ is _____ .

13. Let a, b be nonzero real numbers such that $a \neq b$, then the area of the triangle enclosed by lines $y = a, y = ax + b, y = bx + a$ is _____ .

14. It is known that the line $(2k - 1)x - (k + 3)y - (k - 11) = 0$ passes through a fixed point regardless of the value of k. This fixed point is _____ .

15. We have two linear functions $y = ax + b$ and $y = cx + 5$. Student A calculates their intersection and gives the correct answer, which is $(3, -2)$. Student B misheard the value of c, and his answer is $\left(\frac{3}{4}, \frac{1}{4}\right)$. Then, the formula of $y = ax + b$ is _____ .

(III) Answer the following questions:

16. In the coordinate plane, $M(x, 0)$ is a moving point on the x-axis, and $P(5, 5), Q(2, 1)$ are fixed points. When is $MP + MQ$ minimal?

17. As shown in the following figure, the line $y = -\frac{\sqrt{3}}{3}x + 1$ intersects the x-axis and y-axis at A and B, respectively. $\triangle ABC$ is an isosceles right triangle where AB is a leg and $\angle BAC = 90°$. If there is a point $P(a, \frac{1}{2})$ in the second quadrant such that $\triangle ABP$ and $\triangle ABC$ have the same area, find the value of a.

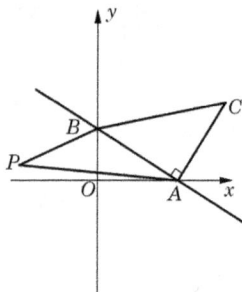

18. We choose two numbers $p, q (p \neq q)$ among $1, 2, 3, 4, 5$, and write the functions $y = px - 2$ and $y = x + q$. How many pairs (p, q) are there such that the intersection of the two functions lies to the left of the line $x = 2$?

19. Two cars start from the same point and travel in the same direction at the same speed. Each car carries at most 24 barrels of gasoline, and a barrel of gasoline allows the car to travel 60 km.

Suppose that both cars need to go back to the starting point, but not necessarily at the same time. Each car can borrow gasoline from the other.

Now, we want to let one car travel as far from the starting point as possible. Then, where should the other car turn around and start going back? In this case, what is the total distance traveled by the car that travels farther?

20. In a dormitory, some students are collecting water from the boiler house, and every student collects 2 L. They first turn on two taps, and later, one tap stops working and is turned off. Suppose the time gap between two collectors is neglected, and water is not wasted in any way. The remaining water in the boiler house is y (L) and the time elapsed is x (min). The relationship between y and x is shown in the following figure. Answer the following questions based on the graph.

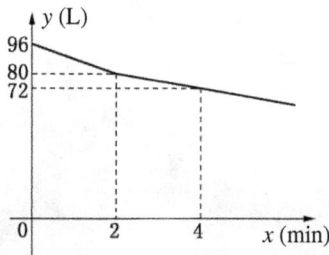

(1) Write down one result that you can obtain from the graph.
(2) How much time does it take so that 15 students can finish collecting?
(3) Mary says that eight students in her dormitory used exactly 3 min to collect water. Is it possible? Explain.

21. An electrical appliance business owner is planning to purchase some air conditioners and some electric fans (each of the same type). If he purchases 8 air conditioners and 20 electric fans, the total cost is 17400 yuan, and if he purchases 10 air conditioners and 30 electric fans, the total cost is 22500 yuan.

(1) How much is one air conditioner and one electric fan, respectively?
(2) The business owner wants to purchase 70 electrical appliances of these two types, and the total cost he can afford

is 30000 yuan. According to market research, the profit of selling an air conditioner is 200 yuan, and the profit of selling an electric fan is 30 yuan. The owner hopes that after selling all the appliances, the total profit will be at least 3500 yuan. Based on these assumptions, how many different plans does the owner have? Which brings the maximal profit, and what is the maximal profit?

Chapter 14

Inversely Proportional Functions

Functions such as $y = \frac{k}{x}$ (where k is a nonzero constant) are called inversely proportional functions, in which we say y is inversely proportional to x. Apparently, $x \neq 0$.

The graph of an inversely proportional function is a hyperbola, and both of its branches approximate the coordinate axes infinitely (but never touch the axes). If $k > 0$, then the hyperbola lies in the first and third quadrants, and y decreases as x increases in each branch (as shown in Figure 14.1). If $k < 0$, the hyperbola lies in the second and fourth quadrants, and y increases as x increases in each branch (as shown in Figure 14.2).

The graph of an inversely proportional function is centrally symmetric with respect to the origin O and is also symmetric with respect to the lines $y = x$ and $y = -x$.

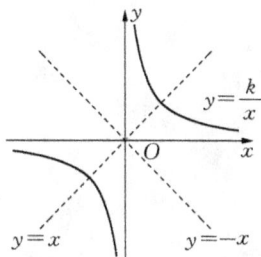
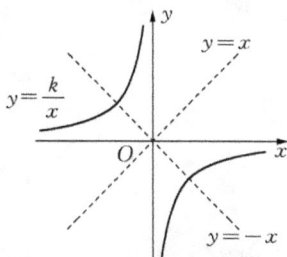

Fig. 14.1   Fig. 14.2

Similar to linear functions, when we deal with inversely proportional functions, it is common to combine numbers and shapes and do casework.

Again, the method of undetermined coefficients is useful when finding the formula, and in specific problems, we need to consider the constraints on the variables.

Example 1. As in Figure 14.3, we successively choose points $A_1, A_2, \ldots, A_6$ on the positive half of the x-axis, so that $OA_1 = A_1 A_2 = A_2 A_3 = A_3 A_4 = A_4 A_5 = A_5 A_6$, and we mark $P_1, P_2, \ldots, P_6$ on the graph of $y = \frac{2}{x}$, so that $P_i A_i$ is perpendicular to the x-axis. Thus, we obtain right triangles $OA_1 P_1, A_1 P_2 A_2, A_2 P_3 A_3, A_3 P_4 A_4, A_4 P_5 A_5, A_5 P_6 A_6$. Find the sum of the areas of the six triangles.

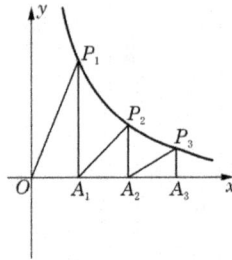

Fig. 14.3

Solution Let $OA_1 = A_1 A_2 = A_2 A_3 = A_3 A_4 = A_4 A_5 = A_5 A_6 = a$, then

$$OA_2 = 2a, \quad OA_3 = 3a, \ldots, \quad OA_6 = 6a,$$

$$A_1 P_1 = \frac{2}{a}, \quad A_2 P_2 = \frac{2}{2a} = \frac{1}{a},$$

$$A_3 P_3 = \frac{2}{3a}, \ldots, \quad A_6 P_6 = \frac{2}{6a} = \frac{1}{3a}.$$

Thus,

$$S_{\triangle OP_1 A_1} = \frac{1}{2} OA_1 \cdot A_1 P_1 = \frac{1}{2} \cdot a \cdot \frac{2}{a} = 1,$$

$$S_{\triangle A_1 P_2 A_2} = \frac{1}{2} A_1 A_2 \cdot A_2 P_2 = \frac{1}{2} \cdot a \cdot \frac{1}{a} = \frac{1}{2},$$

$$S_{\triangle A_2 P_3 A_3} = \frac{1}{2} A_2 A_3 \cdot A_3 P_3 = \frac{1}{2} \cdot a \cdot \frac{2}{3a} = \frac{1}{3},$$

$$S_{\triangle A_3 P_4 A_4} = \frac{1}{4}, \quad S_{\triangle A_4 P_5 A_5} = \frac{1}{5}, \quad S_{\triangle A_5 P_6 A_6} = \frac{1}{6}.$$

Therefore, the sum of these areas is $1 + \frac{1}{2} + \frac{1}{3} + \frac{1}{4} + \frac{1}{5} + \frac{1}{6} = 2\frac{27}{60}$.

Remark: For a point $A(x_0, y_0)$ on the graph of $y = \frac{k}{x} (k > 0)$, if E is its projection on the x-axis, then $S_{\triangle OAE} = \frac{1}{2} \cdot OE \cdot AE = \frac{1}{2} x_0 y_0 = \frac{1}{2} k$. When $k < 0$, the corresponding area is $\frac{1}{2}|k|$.

Example 2. As in Figure 14.4, the graphs of inversely proportional functions $y_1 = \frac{k_1}{x}$ and $y_2 = \frac{k_2}{x} (k_1 > k_2 > 0)$ in the first quadrant are the curves C_1 and C_2, respectively. Let P be a point on C_1, E be its projection on the x-axis, and PE intersect C_2 at A. Let D be the projection of P on the y-axis, and PD intersects C_2 at B. Find the area of the quadrilateral $PAOB$.

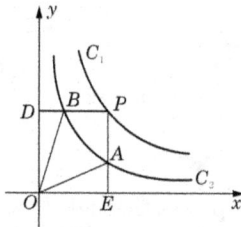

Fig. 14.4

Solution Assume $P(m, n)$, then $mn = k_1$, and

$$A\left(m, \frac{k_2}{m}\right), \quad B\left(\frac{k_2}{n}, n\right),$$

$$S_{PAOB} = S_{PEOD} - S_{\triangle OEA} - S_{\triangle OBD}$$

$$= mn - \frac{1}{2} \cdot m \cdot \frac{k_2}{m} - \frac{1}{2} \cdot \frac{k_2}{n} \cdot n$$

$$= k_1 - k_2.$$

Example 3. In Figure 14.5, the curve ABC is part of a rollercoaster rail in an amusement park. Assume that AB is a line segment and BC is part of the graph of $y = \frac{k}{x}$. It is known that $A(10, 1)$, $B(8, 2)$, $C(2, y_C)$. In order to secure the rail, we need to choose a point on the rail and draw perpendicular line segments to the coordinate axes. The point where the total lengths of the two segments (which is denoted as S) is minimal is called the optimal point of support.

(1) Find the equation of line AB, and the value of k.
(2) Find the coordinates of the optimal point of support.

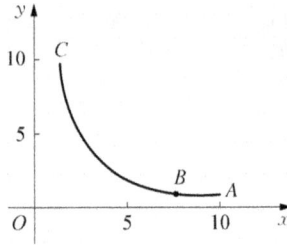

Fig. 14.5

Solution (1) Let $y = kx + b$ be the equation of AB. Plug in $A(10, 1)$, $B(8, 2)$, and we get

$$\begin{cases} 10k + b = 1, \\ 8k + b = 2, \end{cases} \text{ which yields } \begin{cases} k = -\dfrac{1}{2}, \\ b = 6. \end{cases}$$

Hence, $y = -\frac{1}{2}x - 6$.

Since the graph of $y = \frac{k}{x}$ passes through $B(8, 2)$, we have $k = 8 \times 2 = 16$.

(2) Consider two cases:

① If $P(x, -\frac{1}{2}x + 6)$ lies on AB, then $8 \leq x \leq 10$, and the distances from P to the x and y axes are $-\frac{1}{2}x + 6$ and x, respectively. Thus,

$$S = -\frac{1}{2}x + 6 + x = \frac{1}{2}x + 6.$$

Since $\frac{1}{2} > 0$, S increases as x increases, so S is minimal when $x = 8$, in which case $S = \frac{1}{2} \times 8 + 6 = 10$.

② Let $P(x, \frac{16}{x})$ be a point on the curve BC, then $2 \leq x \leq 8$, and the distances from P to the coordinate axes are $\frac{16}{x}$ and x. Thus, $S = \frac{16}{x} + x = (\sqrt{x} - \frac{4}{\sqrt{x}})^2 + 8 \geq 8$.

The equality is attained when $\sqrt{x} - \frac{4}{\sqrt{x}} = 0$, in which case $x = 4$.

Since $10 > 8$, we conclude that the optimal point of support is $(4, 4)$.

Example 4. As shown in Figure 14.6, $P_1(x_1, y_1)$, $P_2(x_2, y_2)$ are two points on the graph of $y = \frac{k}{x}(k > 0, x > 0)$, and $x_1 < x_2$. Let B, D be the projections of P_1, P_2 on the x-axis, respectively, and let A, C be their projections on the y-axis, respectively.

(1) Let S_1, S_2 be the areas of quadrilaterals AP_1BO and CP_2DO, respectively, and C_1, C_2 be their perimeters. Try to compare S_1 with S_2 and compare C_1 with C_2.

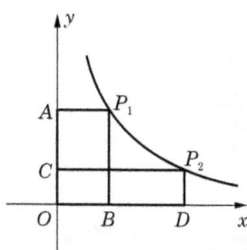

Fig. 14.6

(2) If P is a point on the graph of $y = \frac{k}{x}(k > 0, x > 0)$ and M, N are its projections on the x and y axes, respectively, when is the perimeter of $PMON$ minimal and what is the minimum?

Solution (1) It is easy to see that $S_1 = x_1 y_1 = k$, $S_2 = x_2 y_2 = k$, so $S_1 = S_2$.

Also, $C_1 = 2(x_1 + y_1) = 2(x_1 + \frac{k}{x_1})$, and $C_2 = 2(x_2 + \frac{k}{x_2})$, so

$$C_2 - C_1 = 2\left(\frac{x_2^2 + k}{x_2} - \frac{x_1^2 + k}{x_1}\right) = 2 \cdot \frac{x_1 x_2^2 - x_2 x_1^2 + kx_1 - kx_2}{x_1 x_2}.$$

Since $x_1 > 0$, $x_2 > 0$, $x_1 x_2 > 0$ and $x_2 - x_1 > 0$, we have

$$x_1 x_2^2 - x_2 x_1^2 + kx_1 - kx_2$$
$$= x_1 x_2 (x_2 - x_1) - k(x_2 - x_1)$$
$$= (x_1 x_2 - k)(x_2 - x_1).$$

Therefore, we have $C_2 = C_1$ when $x_1 x_2 = k$, $C_2 > C_1$ when $x_1 x_2 > k$, and $C_2 < C_1$ when $x_1 x_2 < k$.

(2) Let C be the perimeter of $PMON$, then $C = 2(x + y)$. Since $xy = k$, $x > 0$, $k > 0$, we have

$$C = 2\left(x + \frac{k}{x}\right) = 2\left[\left(\sqrt{x} - \frac{\sqrt{k}}{\sqrt{x}}\right)^2 + 2\sqrt{k}\right].$$

When $\sqrt{x} - \frac{\sqrt{k}}{\sqrt{x}} = 0$, or, equivalently, $x = \sqrt{k}$, C attains its minimum which is $4\sqrt{k}$, and in this case, $P(\sqrt{k}, \sqrt{k})$.

Example 5. As shown in Figure 14.7, the line $y_1 = x + m$ intersects the x- and y-axes at A, and B, respectively, and intersects the hyperbola $y_2 = \frac{k}{x}(x < 0)$ at C, D. Suppose the coordinate of C is $(-1, 2)$.

(1) Find the formulas of the line and the hyperbola.
(2) Find the coordinate of D.
(3) Answer the following question without calculation: for what values of x does $y_1 > y_2$ hold?

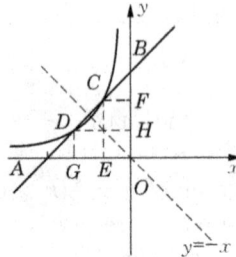

Fig. 14.7

Solution   (1) Since $C(-1, 2)$ lies on the hyperbola $y_2 = \frac{k}{x} (x < 0)$. We have $k = (-1) \times 2 = -2$. Thus, the formula of the hyperbola is $y_2 = -\frac{2}{x} (x < 0)$.

Since $C(-1, 2)$ lies on the line $y = x + m$, we have $m = 2 = 3$, and the formula of the line is $y_1 = x + 3$.

(2) Construct the line $y = -x$. Let E, F be the projections of C on the x- and y-axes, respectively, and let G, H be the projections of D on the x- and y-axes, respectively. Since the line and the hyperbola are both symmetric with respect to $y = -x$, it follows that F and G as well as H and E are both symmetric pairs with respect to $y = -x$. Hence, $OG = OF$, $OH = OE$, and since $C(-1, 2)$, we may obtain $D(-2, 1)$.

(3) It follows from the graph that $y_1 > y_2$ exactly when $-2 < x < -1$.

Remark: (1) The symmetric point of (x, y) with respect to the line $y = x$ is (y, x), and its symmetric point with respect to $y = -x$ is $(-y, -x)$.

(2) When comparing the values of two functions, we may draw a line parallel to the y-axis that intersects the two graphs, and the function whose intersection lies above has greater function value.

Example 6. Suppose the line $y = kx + b$ intersects the hyperbola $y = \frac{k^2}{x}$ at $M(m, -1)$ and $N(n, 2)$. Find the solution set to the inequality $\frac{x-b}{k} > \frac{k^2}{x}$.

Analysis   If (x_0, y_0) lies on the graph of $y = kx + b$, then $y_0 = kx_0 + b$, and the point (y_0, x_0) lies on the graph of $x = yk + b$, or, equivalently, $y = \frac{x-b}{k}$.

If (x_0, y_0) lies on the graph of $y = \frac{k}{x}$, then (y_0, x_0) still lies on this graph.

Solution Since M, N lie on the graph of $y = kx + b$, the points $(-1, m)$, $(2, n)$ lie on the graph of $y = \frac{x-b}{k}$. Also, since M, N lie on the graph of $y = \frac{k^2}{x}$, the points $(-1, m)$, $(2, n)$ also lie on the graph of $y = \frac{k^2}{x}$.

Hence, the intersections of $y = \frac{x-b}{k}$ and $y = \frac{k^2}{x}$ are $P(-1, m)$ and $N(2, n)$, respectively.

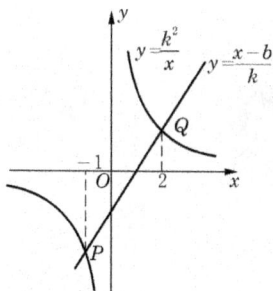

Fig. 14.8

As shown in Figure 14.8, $\frac{x-b}{k} > \frac{k^2}{x}$ exactly when $-1 < x < 0$ or $x > 2$. Therefore, the solution set to the inequality is $-1 < x < 0$ or $x > 2$.

Example 7. (1) As shown in Figure 14.9, M, N are points on the graph of $y = \frac{k}{x}(k > 0)$, and E is the projection of M on the y-axis, while F is the projection of N on the x-axis. Prove that $EF \parallel MN$.

(2) If we change the positions of M, N as in Figure 14.10, while other conditions are unchanged, determine whether $MN \parallel EF$ still holds.

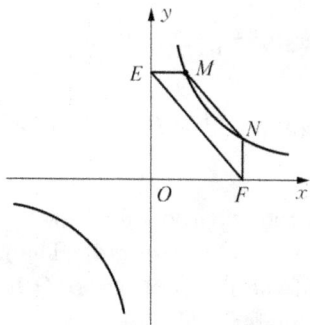

Fig. 14.9

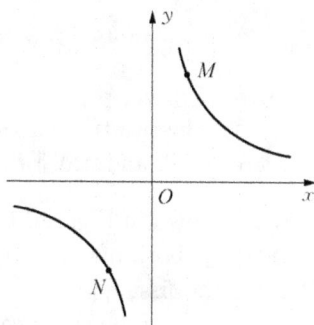

Fig. 14.10

Proof: (1) As shown in Figure 14.11, we connect MF, NE. Suppose the coordinates of M and N are (x_1, y_1) and (x_2, y_2), respectively.

Since M, N lie on the graph of $y = \frac{k}{x}(k > 0)$, we have $x_1 y_1 = x_2 y_2 = k$.

Since $ME \perp y$-axis, $NF \perp y$-axis, we have $OE = y_1$, $OF = x_2$.

Hence, $S_{\triangle EFM} = \frac{1}{2}x_1 y_1 = \frac{1}{2}k$ and $S_{\triangle EFN} = \frac{1}{2}x_2 y_2 = \frac{1}{2}k$, so $S_{\triangle EFM} = S_{\triangle EFN}$.

Note that the two triangles share the side EF, so their corresponding altitudes are equal, which implies $MN \parallel EF$.

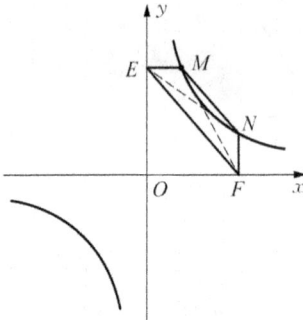 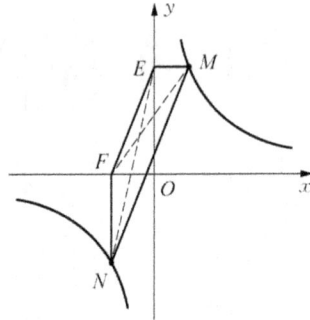

Fig. 14.11   Fig. 14.12

(2) $MN \parallel EF$ still holds, and we shall prove it.

As shown in Figure 14.12, we connect MF and NE. Let the coordinates of M, N be (x_1, y_1), (x_2, y_2), respectively.

Since M, N lie on the graph of $y = \frac{k}{x}(k > 0)$, we have $x_1 y_1 = x_2 y_2 = k$.

Since $ME \perp y$-axis and $MF \perp x$-axis, we have $OE = y_1$ and $OF = -x_2$. Thus,

$$S_{\triangle EFM} = \frac{1}{2}x_1 y_1 = \frac{1}{2}k,$$

$$S_{\triangle EFN} = \frac{1}{2}x_2 y_2 = \frac{1}{2}k,$$

so $S_{\triangle EFM} = S_{\triangle EFN}$.

Again, since the two triangles share the side EF, their corresponding altitudes are equal, and $MN \parallel EF$.

Example 8. As shown in Figure 14.13, the hyperbola $y = \frac{k}{x}(k > 0)$ intersects the lines l_1, l_2 (both of which pass through the origin). The intersections A, P lie in the first quadrant, and the intersections B, Q lie in the third quadrant. Let m, n be the x-coordinates of A, P, respectively.

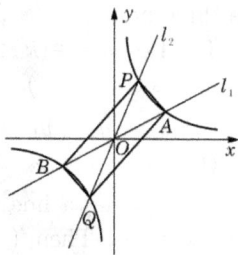

Fig. 14.13

(1) Find the coordinate of B;

(2) Can $APBQ$ be a rectangle? Can it be a square? If possible, specify the conditions that m, n need to satisfy, and if impossible, explain the reasons.

Solution (1) It follows that $A(m, \frac{k}{m})$, and A, B are centrally symmetric with respect to the origin, which means $B(-m, -\frac{k}{m})$.

(2) By symmetry, we have $OA = OB$, $OP = OQ$, so $APBQ$ is a parallelogram. When $OA = OP$, it is also a rectangle. Since $P(n, \frac{k}{n})$, Pythagorean theorem shows that

$$m^2 + \frac{k^2}{m^2} = n^2 + \frac{k^2}{n^2},$$

so $m^2 - n^2 = \frac{k^2}{m^2 n^2}(m^2 - n^2)$. Equivalently, $(m^2 - n^2)(1 - \frac{k^2}{m^2 n^2}) = 0$.

Since $m \neq n$, we have $k^2 = m^2 n^2$.

Since A, P both lie in the first quadrant and $k > 0$, we conclude that $APBQ$ is a rectangle when $mn = k$.

For the second part, $APBQ$ cannot be a square since $\angle AOP < 90°$.

Exercises

(I) Multiple-choice questions:

    1. Which of the following is an axis of symmetry of the graph of $y = \frac{k}{x}(k \neq \pm 1)$? ( ).

    (A) $y = kx$ (B) $y = -kx$ (C) $y = |k|x$ (D) $y = \frac{k}{|k|}x$

2. Suppose there are three points $A_1(x_1, y_1)$, $A_2(x_2, y_2)$, $A_3(x_3, y_3)$ on the graph of the function $y = \frac{k}{x}(k < 0)$, and $x_1 < x_2 < 0 < x_3$. Which of the following is true? ( ).

(A) $y_1 < y_2 < y_3$ (B) $y_3 < y_2 < y_1$
(C) $y_3 < y_1 < y_2$ (D) $y_2 < y_1 < y_3$

3. In the same coordinate plane, a line $y = k_1 x$ and a hyperbola $y = \frac{k_2}{x}$ have no intersections. Then, ( ).

(A) k_1, k_2 necessarily have the same sign
(B) k_1, k_2 necessarily have different signs
(C) $k_1 k_2$ are necessarily opposite numbers
(D) k_1, k_2 are necessarily reciprocals of each other

4. If y is proportional to x, and z is inversely proportional to y, then z is ( ).

(A) proportional to x (B) inversely proportional to x
(C) neither proportional nor   (D) proportional to x^2
                          inversely proportional to x

5. If the function $y = \frac{k}{x}(k \neq 0)$ is increasing for $x < 0$, then the graph of $y = kx - k$ passes through quadrants ( ).

(A) I, II, III   (B) I, II, IV   (C) I, III, IV   (D) II, III, IV

6. If the graphs of $y = ax(a \neq 0)$ and $y = \frac{b}{x}(b \neq 0)$ intersect at two points, one of which is $(-3, -2)$, then the other intersection is ( ).

(A) $(2, 3)$ (B) $(3, -2)$ (C) $(-2, 3)$ (D) $(3, 2)$

7. Which of the following shows the graphs of $y = kx$ and $y = \frac{k}{x}$ in the same coordinate plane? ( ).

(A)   (B)   (C)   (D)

8. Which of the following can be the graphs of $y = \frac{m}{x}$ and $y = mx - m$ in the same coordinate plane? (   ).

(A)

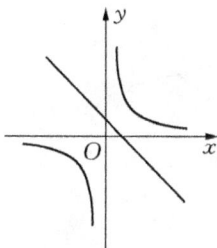

(B)

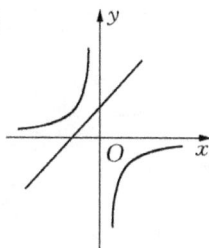

(C)

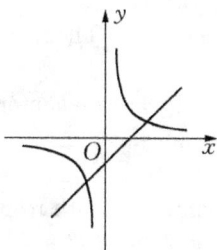

(D)

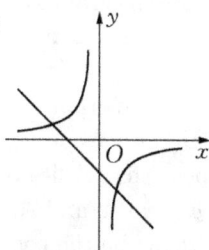

9. As shown in the figure, P is a moving point on the graph of $y = \frac{1}{2x}(x > 0)$, and M, N are its projections on the x- and y-axes, respectively. The line segments PM, PN intersect the line $AB : y = -x + 1$ at E, F, respectively. Here, A lies on the x-axis and B lies on the y-axis. Then, $AF \cdot BE$ equals (   ).

(A) 4   (B) 2   (C) 1   (D) $\frac{1}{2}$

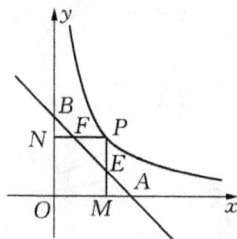

10. As shown in the figure, the graphs of $y = -\frac{4}{x}$ and $y = -\frac{1}{3}x$ intersect at points A, B, respectively. The line through A and parallel to y-axis intersects the line through B and parallel to x-axis at C. Then, the area of $\triangle ABC$ is (   ).

(A) 8   (B) 6   (C) 4   (D) 2

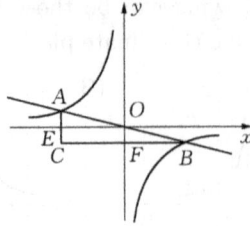

(II) Fill in the blanks:

11. Suppose the graphs of two functions $y = 2x$ and $y = \frac{4}{x}$ intersect at $A(x_1, y_1)$ and $B(x_2, y_2)$ with $x_1 > x_2$. Then, the area of $\triangle ABC$ is _____ .

12. Suppose the graphs of $y = \frac{k}{x}$ and $y = ax + b$ intersect at $A(-2, m)$ and $B(5, n)$, respectively, then $3a + b =$ _____ .

13. Suppose that A lies on the hyperbola $y = -\frac{2}{x}$ and B lies on the line $y = x - 4$, and A, B are symmetric with respect to the y-axis. Let (m, n) be the coordinate of A, then $\frac{m}{n} + \frac{n}{m} =$ _____ .

14. The graphs of $y = x$ and $y = \frac{1}{x}$ intersect at A, C, respectively, where A lies in the first quadrant. Let C, D be the projections of A, B on the x-axis, respectively. Then, the area of the quadrilateral $ABCD$ is _____ .

15. The graphs of $y = \frac{3}{x}$ and $y = \frac{6}{x}$ in the first quadrant are shown in the figure. Let

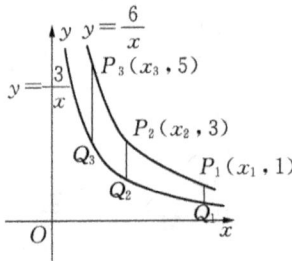

$P_1, P_2, \ldots, P_{2005}$ be points on the graph of $y = \frac{6}{x}$. Their x-coordinates are $x_1, x_2, \ldots, x_{2005}$, and their y-coordinates are consecutive odd numbers $1, 3, 5, \ldots$. Let $Q_1, Q_2, Q_3, \ldots, Q_{2005}$ be points on the graph of $y = \frac{3}{x}$, such that P_iQ_i is parallel to the y-axis ($i = 1, 2, \ldots, 2005$). If the coordinate of Q_i is (x_i, y_i), then $y_{2005} =$ _____ .

16. As shown in the figure, A is a point on the line $y = x$, and B lies on the hyperbola $y = \frac{k}{x}$ such that $BA \perp OA$. If $OA^2 - AB^2 = 8$, then $k =$ _____ .

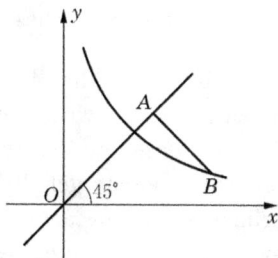

(III) Answer the following questions:

17. As shown in the figure, the graphs of $y_1 = kx + b$ and $y_2 = \frac{m}{x}$ intersect at $A(-2, 1)$ and $B(1, n)$.

    (1) Find the formulas of both functions.
    (2) Use the graph to determine the value range of x when $y_1 > y_2$.

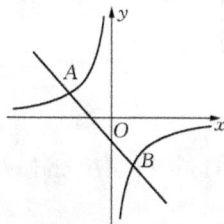

18. Let $AOBC$ be a rectangle with $OB = 4$, $OA = 3$. We construct a coordinate plane where OB and OA are the x- and y-axes, respectively, as shown in the figure. Let F be a moving point on the side BC (which does not coincide with the endpoints). The graph of $y = -\frac{k}{x}(k > 0)$ passes through F and intersects the side AC at E.

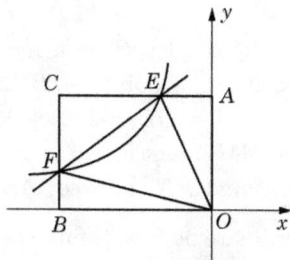

(1) Find the relationship between the areas of $\triangle AOE$ and $\triangle BOF$ and explain why.

(2) Let $S = S_{\triangle OEF} - S_{\triangle ECF}$, write S as a function of k, and find the value of k when S reaches its maximum.

19. Psychologists have discovered that in a 40 min class, the students' level of attention varies with time. In the beginning, the students' attention increases and stays at an optimal level for some time. Later, the students get distracted and start to lose attention. Experiments have revealed the relationship between y, the attention level, and x, the time elapsed, which is shown in the figure. Here AB, BC are line segments, and CD is part of an inversely proportional function.

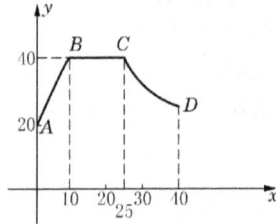

(1) Find the equations of AB and CD, and specify their corresponding domains.

(2) When is the students' attention level higher: 5 min or 30 min since the beginning of the class?

(3) Suppose there is a hard problem that takes the teacher 19 minutes to explain. It is necessary that the students' attention level is never below 36 during this period. Is it possible for the teacher to finish this task within one class? Explain.

20. As shown in the figure, let P be a moving point on the graph of $y = -\frac{2}{x} (x < 0)$. The points A, B lie on the x- and y-axes, respectively, with $OA = OB = 2$. Let M be the projection of P on the x-axis and PM intersect AB at E. Let N be the projection of P on the y-axis and PN intersect AB at F.

(1) If the y-coordinate of P is $\frac{5}{3}$, find the coordinates of E, F and the area of $\triangle OEF$.

(2) Assume $P(a, b)$ with $-2 < a < 0, 0 < b < 2$ and $|a| \neq |b|$, with other conditions unchanged. If there is a triangle whose side lengths are equal to AE, EF, BF, what kind of triangle is it? Prove your result.

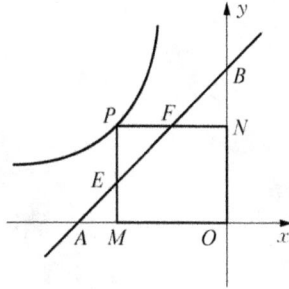

Chapter 15

Statistics

In the real world, many problems involve collecting data through surveys, experiments, measurements, and consulting references. Then, we reorganize the data, use statistical charts to present them intuitively, try to gain insight from the data, and modify our decision-making process. For example, we use the median and the mode to describe the average properties of the data and use the standard deviation to measure the dispersion of the data. The process of collecting data, acquiring information, and analyzing problems based on the data is called statistics.

Statistics generally includes five procedures:

(1) design, which means to decide what data to collect and how to organize these data based on the general objective;
(2) survey, which means to figure out the source of the data and find appropriate methods to collect them;
(3) reorganization, which means to review the data collected and reorganize them in a scientific way;
(4) analysis, which means to analyze the reorganized data and calculate the statistical quantities involved;
(5) statistical inference, which means to obtain results based on the analysis and make a statement.

Example 1. Peter is calculating the arithmetic mean of three numbers, $a, b, c\,(a < b < c)$. However, he first writes the average of a and b (which is x) and then writes the average of x and c (which is y). Then, y is (   ) the actual average of a, b and c.

(A) greater than   (B) equal to
(C) less than   (D) none of the above

Solution   The result according to Peter is

$$y = \frac{\left(\frac{a+b}{2} + c\right)}{2} = \frac{a+b+2c}{4},$$

while the actual average of a, b, and c is $\frac{a+b+c}{3}$, and $y - \frac{a+b+c}{3} = \frac{2c-a-b}{12}$.

Since $a < b < c$, we have $2c - a - b > 0$, and $y > \frac{a+b+c}{3}$. Therefore, the answer is (C).

Example 2. Suppose that 25 marbles labeled as $1, 2, \ldots, 25$ are divided into two baskets, A, B, and marble number 15 is in A. If we move number 15 into basket B, the average of the labels in A is increased by $\frac{1}{4}$ and the average of the labels in B is also increased by $\frac{1}{4}$. How many marbles are there in A in the beginning?

Solution   Suppose there are x marbles in A in the beginning, then there are $(25 - x)$ marbles in B. Also, let a (resp., b) be the original average of the labels in A (resp., B). Thus,

$$\begin{cases} ax + b(25 - x) = 1 + 2 + 3 + \cdots + 25, \\[2mm] \dfrac{ax - 15}{x - 1} = a + \dfrac{1}{4}, \\[2mm] \dfrac{b(25 - x) + 15}{(25 - x) + 1} = b + \dfrac{1}{4}, \end{cases}$$

which is simplified to

$$\begin{cases} (a - b)x + 25b = 325, & \qquad ① \\ x - 4a = -59, & \qquad ② \\ x - 4b = -34. & \qquad ③ \end{cases}$$

By ③$-$②, we have $a - b = \frac{25}{4}$, and plugging into ①, we obtain $x + 4b = 52$.

Combining with ③, we have $x = 9$. Therefore, in the beginning, there are 9 marbles in A.

Example 3. In a math competition, there are 15 problems. Table 15.1 shows the number of students who solve n problems ($n = 0, 1, 2, \ldots, 15$). Suppose that for students who have solved at least 4 problems, the average number of problems they solve is 6, and for students who have solved at most 10 problems, the average number of problems they solve is 4. How many students does the table include at least?

Table 15.1

n	0	1	2	3	...	12	13	14	15
Number of students who solve n problems	7	8	10	21	...	15	6	3	1

Solution From the table, we see that the number of students who solve at most 3 problems is $7 + 8 + 10 + 21 = 46$. The total number of problems they solve is $0 \times 7 + 1 \times 8 + 2 \times 10 + 3 \times 21 = 91$.

The number of students who solve at least 12 problems is $15 + 6 + 3 + 1 = 25$, and the total number of problems they solve is $12 \times 15 + 13 \times 6 + 14 \times 3 + 15 \times 1 = 315$.

Let x_i be the number of students who solve exactly i problems, $i = 0, 1, \ldots, 15$. Then,

$$\begin{cases} \dfrac{4x_4 + 5x_5 + \cdots + 15x_{15}}{x_4 + x_5 + \cdots + x_{15}} = 6, & \text{①} \\[2mm] \dfrac{0x_0 + 1x_1 + \cdots + 10x_{10}}{x_0 + x_1 + \cdots + x_{10}} = 4. & \text{②} \end{cases}$$

Hence, $\begin{cases} 4x_4 + 5x_5 + \cdots + 15x_{15} = 6(x_4 + x_5 + \cdots + x_{15}), \\ x_1 + 2x_2 + \cdots + 10x_{10} = 4(x_0 + x_1 + \cdots + x_{10}). \end{cases}$

Taking the difference between both sides, we have

$$11x_{11} + 12x_{12} + \cdots + 15x_{15} - (x_1 + 2x_2 + 3x_3)$$

$$= 2(x_4 + x_5 + \cdots + x_{10}) + 6(x_{11} + x_{12} + \cdots + x_{15})$$

$$- 4(x_0 + x_1 + x_2 + x_3)$$

Since

$$12x_{12} + 13x_{13} + 14x_{14} + 15x_{15} = 315,$$

$$x_0 + 1x_1 + 2x_2 + 3x_3 = 91,$$

$$x_0 + x_1 + x_2 + x_3 = 46,$$

$$x_{12} + x_{13} + x_{14} + x_{15} = 25,$$

we have

$$11x_{11} + 315 - 91 = 2(x_0 + x_1 + \cdots + x_{15}) + 4x_1 + 4 \times 25 - 6 \times 46.$$

Hence, $2(x_0 + x_1 + \cdots + x_{15}) = 7x_{11} + 400$, and $x_0 + x_1 + \cdots + x_{15} = 3.5x_{11} + 200$.

Since $x_{11} \geq 0$, we see that the number of students included is at least 200.

Example 4. There are seven positive integers, namely $10, 2, 5, 2, 4, 2, x$. If we arrange the average, the median, and the mode of these numbers in increasing order, then the number in the middle is the average of the other two. Find all possible values of x.

Solution   Regardless of the value of x, the mode of these numbers is always 2, and the average is $\frac{25+x}{7}$.

The six numbers other than x are $2, 2, 2, 4, 5, 10$ in the increasing order, and since x is a positive integer, the median of these numbers can be $2, 3$, or 4:

(1) If the median is 2, then there is no such x.
(2) If the median is 3, then it follows that $2 + \frac{25+x}{7} = 6$, so $x = 3$.
(3) If the median is 4, then we have $x \geq 4$ and necessarily $2 + \frac{25+x}{7} = 4 \times 2$, so $x = 17$.

Therefore, the possible values of x are 3 and 17.

Example 5. A store sold $10, 8, x, 10$ microwave ovens of types A, B, C, D, respectively. Suppose the median and average of these numbers are the same. Find the median.

Analysis   In order to get the median, we need to sort the numbers, and since x is unknown, we need to consider different cases.

Solution   The average of $10, 8, x, 10$ is $\frac{28+x}{4}$.

(1) If $x \leq 8$, then $10, 8, x, 10$ can be sorted as $x, 8, 10, 10$, and the median is 9. In this case, $\frac{28+x}{4} = 9$, and $x = 8$, which is valid.
(2) If $8 < x < 10$, then the numbers are sorted as $8, x, 10, 10$, so that $\frac{28+x}{4} = \frac{x+10}{2}$. This yields $x = 8$, which is invalid.
(3) If $x \geq 10$, then the numbers are sorted as $8, 10, 10, x$, so that $\frac{28+x}{4} = 10$. This yields $x = 12$, which is valid.

In summary, we have $x = 8$ or $x = 12$, and the median is 9 or 10.

Example 6. A biologist wants to estimate the number of fish in a lake. On May 1, he randomly caught 60 fish from the lake, which he put back into the lake after leaving marks on them. On September 1, he again catches 70 fish randomly from the lake and finds that 3 of them have marks. By estimation, 25% of the May 1st fish no longer stay in the lake on September 1st due to natural death and human activities, and 40% of the fish on September 1st entered the lake after May 1st. If the September 1st sample

Table 15.2

x	x_1	x_2	x_3	x_4	x_5	x_6	x_7
y	51	107	185	285	407	549	717

Table 15.3

k	1	2	3	4	5	6	7
y_k	51	107	185	285	407	549	717
T_k	56	78	100	122	142	168	
$T_{k+1} - T_k$	22	22	22	20	26		

perfectly represents the whole population, then how many fish were there in the lake on May 1, approximately?

Solution Suppose there were x fish in the lake on May 1. Then,

$$\frac{3}{70} = \frac{60(1 - 25\%)}{x(1 - 25\%) \div (1 - 40\%)}.$$

Solving the equation, we have $x = 840$. Therefore, there are approximately 840 fish in the lake on May 1.

Example 7. A student is trying to draw the graph of $y = ax^2 + bx + c\,(a \neq 0)$ by plotting points. He chooses seven values of x, such that $x_1 < x_2 < \cdots < x_7$ and $x_2 - x_1 = x_3 - x_2 = \cdots = x_7 - x_6$. The corresponding values of y are shown in Table 15.2.

However, one of the values of y is miscalculated. Find the wrong value and give the correct answer.

Solution Let $x_2 - x_1 = \cdots = x_7 - x_6 = d$, and the function value corresponding to x_i be y_i, $i = 1, 2, \ldots, 7$. Then,

$$T_k = y_{k+1} - y_k = (ax_{k+1}^2 + bx_{k+1} + c) - (ax_k^2 + bx_k + c)$$
$$= [a(x_k + d)^2 - ax_k^2] + b(x_k + d - x_k)$$
$$= 2adx_k + ad^2 + bd.$$

Hence, we have $T_{k+1} - T_k = 2ad(x_{k+1} - x_k) = 2ad^2$.

Since a, d are fixed, it follows that $T_{k+1} - T_k$ is a constant. Based on the given data, we may construct Table 15.3:

Thus, we can see that 20 and 26 are problematic, which should be 22 instead. Therefore, T_5 should be 144, T_6 should be 166, and y_6 should be 551.

In summary, y_6 is the wrong value, and the correct value is 551.

Reading

Do mathematicians live a long life?

Someone analyzed the life expectancy of 153 mathematicians in the book *Biography of World-Famous Mathematicians*, authored by Wu Wenjun and published by Science Press, Beijing, in 1995. Except for 9 mathematicians with unknown years of birth and death, 21 Chinese mathematicians, and 3 female mathematicians, the remaining 120 foreign male mathematicians have an average life expectancy of 70.3 years, as shown in Table 15.4.

Table 15.4

Lifespan	20–29	30–39	40–49	50–59	60–69	70–79	80–89	90–100
Frequency	2	3	9	12	23	34	30	7

Table 15.5

Lifespan	30–39	40–49	50–59	60–69	70–79	80–89	90–100	≥ 100
Frequency	1	2	11	21	29	38	12	1

Another study analyzed 115 deceased mathematicians in modern China. Their average life expectancy is 74.85 years, the standard deviation is 1.22 years, the median is 76 years, the mode is 86 years, and the one with the longest life span is 101-year old Su Buqing. The segmented statistics are shown in Table 15.5.

Exercises

(I) Multiple-choice questions:

1. Several students are making cartoon cards for the Olympic Games. The number of cards everyone makes is shown in the following

chart. Let a, b, and c be the average, median, and mode of the numbers. Then, ( ).

(A) $a > b > c$ (B) $b > a > c$ (C) $c > a > b$ (D) $a > c > b$

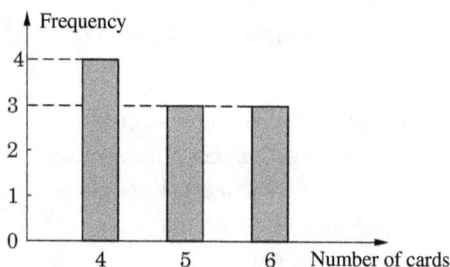

2. Suppose the variance of positive numbers x_1, x_2, x_3, x_4, x_5 is $S^2 = \frac{1}{5}(x_1^2 + x_2^2 + x_3^2 + x_4^2 + x_5^2 - 20)$. We have the following statements regarding the numbers $x_1 + 2, x_2 + 2, x_3 + 2, x_4 + 2, x_5 + 2$: (1) the variance is S^2; (2) the average is 2; (3) the average is 4; (4) the variance is $4S^2$. The true statements are ( ).

(A) (1) and (2)   (B) (1) and (3)
(C) (2) and (4)   (D) (3) and (4)

3. Suppose the average of x_1, x_2, x_3 is a and the average of y_1, y_2, y_3 is b. Then, the average of $2x_1 + 3y_1, 2x_2 + 3y_2, 2x_3 + 3y_3$ is ( ).

(A) $2a + 3b$ (B) $\frac{2}{3}a + b$ (C) $6a + 9b$ (D) $2a + b$

4. A collection of data consists of pairwise different numbers, whose median is 80, and the average of the numbers below the median is 70, while the average of the numbers above the median is 96. (Here, above and below mean strictly greater than and less than, respectively.) Let $\bar{x}$ be the average of all the numbers, then necessarily ( ).

(A) $\bar{x} = 82$ (B) $80 \le \bar{x} \le 82$ (C) $\bar{x} = 83$ (D) $82 \le \bar{x} \le 83$

5. A team of 20 students is planning to buy some shoes for the sports meeting. The team leader collects everyone's size, as shown in the following table:

Size	38	39	40	41	42
Frequency	5			3	2

However, the paper gets dirty, and two numbers are unrecognizable. Which of the following statements is true? ( ).

(A) The median of these sizes is 40 and the mode is 39.

(B) The median and mode of these sizes must be equal.

(C) The average of these sizes P satisfies $39 < P < 40$.

(II) Fill in the blanks:

6. A test consists of five problems, where each is worth 3 points. Suppose 26 people take the test, the average score is at least 4.8, the lowest score is 3, and at least 3 people get 4 points. Then, the number of people who get 5 points is _____.

7. In a basketball shooting exercise, everyone shoots 10 times and gets one point for every goal. The scoring table is shown as follows:

Score	0	1	2	...	8	9	10
Frequency	7	5	4	...	3	4	1

Suppose that the average score of the shooters who score at least 3 points is 6 points, and the average score of the shooters who score at most 8 points is 3 points. Then, the total number of shooters is

_____.

8. Suppose the variance of numbers $x_1, x_2, \ldots, x_n$ is S^2. Let k, a be constants, then the variance of $kx_1 + a, kx_2 + a, \ldots, kx_n + a$ is

_____.

9. If a, b are given real numbers such that $1 < a < b$, then for the numbers $1, a + 1, 2a + b, a + b + 1$, the absolute difference between their average and their median is _____.

(III) Answer the following questions:

10. A tourist area has five scenic spots. Recently, the ticket prices of these spots were modified, and it is assumed that the daily numbers of tourists for each spot remain unchanged after the modification. The data are shown in the following table:

Scenic Spot	A	B	C	D	E
Original price	10	10	15	20	25
Current price	5	5	15	25	30
Daily number of tourists (thousand)	1	1	2	3	2

(1) The administration has claimed that the average ticket price of the five spots remains unchanged after the modification. How do they calculate the average?

(2) On the other hand, the tourists have complained that the average price has risen by 9.4% after the modification. How is this average obtained?

(3) In your opinion, which of the above claims better reflects the truth?

11. A class of students takes a health knowledge test. The scores of the test are integers, and the full score is 100. The following is a histogram of the students' scores. Answer the following questions according to the diagram:

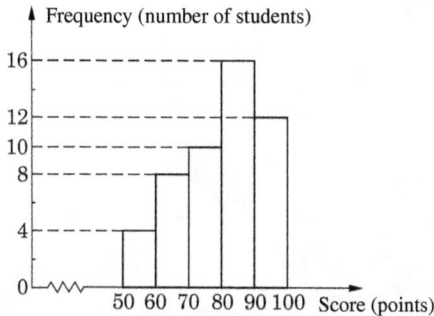

(Each group contains the upper bound but not the lower bound.)

(1) Find the number of students in the class.

(2) Suppose scores greater than or equal to 80 are excellent, and there are 5 students whose scores are exactly 80. What is the excellence rate?

(3) If there are 20 students in this class who score higher than 82, what might be the median of the students' scores? (Simply write the answer.)

12. In order to estimate the consumption of disposable chopsticks in 2016, we have chosen 10 restaurants among the 600 in the town as a sample. The numbers of boxes of disposable chopsticks these restaurants consume every day are listed as follows:

$$0.6, 3.7, 2.2, 1.5, 2.8, 1.7, 1.2, 2.1, 3.2, 1.0.$$

(1) Based on the sample, estimate the number of boxes of disposable chopsticks consumed by the whole town in 2016. (Assume a restaurant remains open for 350 days every year.)

(2) In 2018, we do the same sample survey as in 2016, and the result is that the 10 chosen restaurants consume an average of 2.42 boxes every day for each restaurant. Suppose the number of restaurants in the town and the number of opening days are unchanged in 2017 and 2018. Find the average annual growth rate of disposable chopstick consumption in 2017 and 2018.

(3) Under the same assumption as in (2), suppose that producing a set of desks and chairs costs $0.07\,\text{m}^3$ of wood. Then, with the same amount of wood used for chopsticks in 2018 in this town, how many sets of desks and chairs can be produced?

Relevant data are listed as follows:

Every box of disposable chopsticks contains 100 pairs, each pair weigh $5\,\text{g}$, and the density of wood is $0.5 \times 10^3\,\text{kg/m}^3$.

Chapter 16

The Sides and Angles of a Triangle

A triangle is a shape enclosed by three line segments that are connected to each other at the endpoints. The sum of any two sides is greater than the third side, and the difference of any two sides is less than the third side. Conversely, if we have three segments such that the sum of any two is greater than the third, then they can be the sides of a triangle.

A triangle whose side lengths are all integers is called an integer triangle.

The sum of the three angles of a triangle is $180°$, and consequently, an external angle of the triangle equals the sum of the other two internal angles.

Diophantine equations and inequalities are common tools in problems regarding integer triangles, while casework and enumeration are also frequently involved. Make sure to check the validity of the solutions.

Example 1. If two sides of an isosceles triangle are 6 and 7, find its perimeter.

Analysis The perimeter has different values, depending on whether the leg has length 6 or 7.

Solution If the leg has length 6, then the perimeter is $6 + 6 + 7 = 19$.

If the leg has length 7, then the perimeter is $6 + 7 + 7 = 20$.

Example 2. In triangle ABC, $\angle ABD = \angle DBE = \angle EBC$, $\angle ACD = \angle DCE = \angle ECB$, as shown in Figure 16.1. If $\angle BEC = 145°$, determine the size of $\angle BDC$.

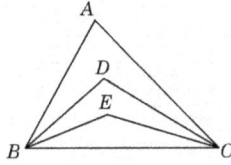

Fig. 16.1

Analysis How can a connection be made between $\angle BDC$ and $\angle BEC$? The sum of the internal angles of $\triangle BCD$ and $\triangle BCE$ and the fact that BE and CE cut the angles at a given ratio give us some hint.

Solution In triangle BCE, we have $\angle CBE + \angle BCE = 180° - \angle BEC = 180° - 145° = 35°$.

Since BE, CE bisect $\angle CBD$, $\angle BCD$, respectively, we have

$$\angle CBD + \angle BCD = 2(\angle CBE + \angle BCE) = 2 \times 35° = 70°.$$

Hence, $\angle BDC = 180° - 70° = 110°$.

Example 3. The number of different integer triangles whose perimeter is 24 is _____.

Analysis We can write down all the different cases to split 24 into the sum of three positive integers and see which of them can make a triangle. However, this method is cumbersome. A better way is to determine the greatest side first and enumerate with the help of inequalities.

Solution Assume the side lengths of the triangle are $a, b, c (a \geq b \geq c)$, then $3a \geq a + b + c = 24$, $2a < a + (b + c) = 24$. Thus, $8 \leq a < 12$, and the possible values of a are $8, 9, 10, 11$.

All valid tuples (a, b, c) are then listed as follows:

$$(8, 8, 8), (9, 9, 6), (9, 8, 7), (10, 10, 4), (10, 9, 5), (10, 8, 6),$$

$$(10, 7, 7), (11, 11, 2), (11, 10, 3), (11, 9, 4), (11, 8, 5), (11, 7, 6).$$

Therefore, the total number of such triangles is 12.

Example 4. Suppose the side lengths a, b, c of a triangle are all integers, and $a \leq b < c$. If $b = 7$, then there are ( ) such triangles.
(A) 21   (B) 28   (C) 49   (D) 54

Analysis It follows that $a \leq 7 < c$, so the maximum of a is 7 and the minimum of a is 1, while the minimum of c is 8. For $a = 1, 2, \ldots, 7$, we determine the number of different triangles in each case.

Solution Since $a \leq b < c$ and $b = 7$, we have $a \leq 7 < c$, so that $a = 1, 2, \ldots, 7$.

If $a = 1$, then $b = 7$, $c \geq 8$, which is impossible.

If $a = 2$, then $c = 8$, and there is one such triangle.

If $a = 3$, then $c = 8, 9$, and there are 2 such triangles.

If $a = 4$, then $c = 8, 9, 10$, and there are 3.

If $a = 5$, then $c = 8, 9, 10, 11$, and there are 4.

If $a = 6$, then $c = 8, 9, 10, 11, 12$, and there are 5.

If $a = 7$, then $c = 8, 9, 10, 11, 12, 13$, and there are 6.

Together, there are $1 + 2 + 3 + 4 + 5 + 6 = 21$ such triangles, and the answer is (A).

Reading

Given an equilateral triangle with side length 1, we divide each side into three segments of equal length and construct equilateral triangles projecting outward based on the middle segment of each side. The original triangle then becomes a hexagonal star. Further, we divide each side of the hexagonal star into three equal parts and construct equilateral triangles projecting outward based on the middle segment of each side. If we continue like this, we will find that the boundary of the figure becomes more and more intricate, like a snowflake (see Figure 16.2). Such a figure is called a snowflake curve, or Koch snowflake, named after Swedish mathematician Helge von Koch, who

Fig. 16.2

described the curve for the first time in 1904.

The snowflake curve has some interesting characteristics. For example, its enclosed area is always smaller than the area of the circumcircle of the original equilateral triangle, but its total length is infinite because the length is multiplied by $\frac{4}{3}$ after each operation, which tends to infinity if the operation iterates infinitely.

In 1999, the China Junior High School Mathematics Competition featured a problem with this background. In that problem, we had a triangle with perimeter 3, where in each step we do the same operation as above, and the question was to find the perimeter of the 2000th figure.

We can think this way: the first figure has perimeter 3, the second figure has perimeter $3 \times \frac{4}{3}$, the third figure has perimeter $3 \times \frac{4}{3} \times \frac{4}{3}$, and so on. Thus, the 2000th figure has perimeter $3 \times \left(\frac{4}{3}\right)^{1999}$.

Exercises

(I) Multiple-choice questions:

1. Which of the following tuples cannot be the side lengths of a triangle? (   ).

   (A) $\left(\frac{1}{2}, \frac{1}{3}, \frac{1}{4}\right)$ (B) $\left(\frac{2}{3}, \frac{3}{4}, \frac{4}{5}\right)$ (C) $(3^2, 4^2, 5^2)$ (D) $(4^2, 5^2, 6^2)$

2. If the perimeter of a triangle is an even number and two of its sides have lengths 4 and 2007, then there are (   ) such triangles.

   (A) 1   (B) 3   (C) 5   (D) 7

3. If two side lengths of a triangle are 20 and 15, then the perimeter of the triangle cannot be (   ).

   (A) 52   (B) 57   (C) 67   (D) 72

4. Observe the following shapes:

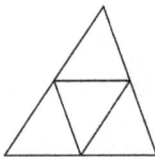   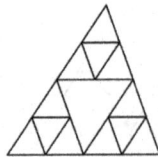   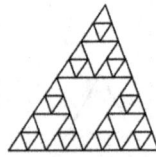

   (1)   (2)   (3)

   According to the patterns in (1), (2), and (3), the number of triangles in (4) is (   ).

   (A) 159   (B) 106   (C) 161   (D) 160

(II) Fill in the blanks:

5. If two sides of an isosceles triangle have lengths 11 and 9, then its perimeter is _____ .

6. If the side lengths of a triangle are $3, 1-2x, 4$, then the value range of x is _____ .

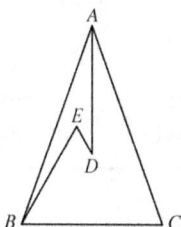

7. As shown in the figure, D, E are two points inside $\triangle ABC$, AD bisects $\angle BAC$, $\angle BAC = 40°$, $\angle ABE = 10°$, $\angle D = 30°$. Then, the size of $\angle E$ is _____ .

8. As shown in the figure, suppose $BG \parallel CE$, then $\angle A + \angle B + \angle C + \angle D + \angle E + \angle F + \angle G =$ _____ .

(III) Answer the following questions:

9. Suppose we have four triangles, all with side lengths of 3, 4, and 5 cm. How many different convex polygons can we obtain by concatenating these triangles? (Only draw the polygons; proof is not required.)

Note: The internal angles of a convex polygon are all less than 180°, and it has no holes inside. For example, the first concatenation in the following yields a convex polygon, while the second doesn't.

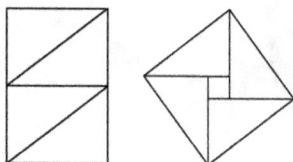

10. By connecting $3, 5, 6, \ldots$ matches at the endpoints, what kind of triangles can we build? After trying, we construct the following table:

Number of matches	3	5	6	...
Sketch Shape	Equilateral triangle	Isosceles triangle	Equilateral triangle	... ...

(1) Is it possible to build a triangle with four matches?

(2) How many different triangles can we build with 8 and 12 matches? Sketch them.

11. Suppose three line segments have lengths of $4a^2 + 5$, $2a + 3$, $4a^2 + 3a + 1$, where a is a nonnegative integer. Is it possible that they are the sides of a triangle? If possible, what are their lengths?

12. Two altitudes of $\triangle ABC$ have lengths of 5 and 20. If the length of the third altitude is also an integer, find the maximum of the third altitude.

Chapter 17

Congruent Triangles

There are four methods we may use to determine that two triangles are congruent: SAS, AAS, ASA, and SSS. For congruence of right triangles, there is another method called HL.

For two congruent triangles, not only are their corresponding sides and angles equal, but the corresponding medians, altitudes, and angle bisectors are also equal.

With congruent triangles, we can prove the equality of segments and angles, as well as other problems such as the perpendicularity of lines. The basic idea in such problems is to reduce the problem to proving the congruence of triangles and try to find equal sides and angles in the pair of triangles. When proving congruence, we can make good use of auxiliary lines, which arise either from the shape itself or from the conditions.

Example 1. Let $\triangle ABC$ and $\triangle A_1B_1C_1$ be given triangles, and D, D_1 are midpoints of BC, B_1C_1, respectively. Which of the following conditions imply $\triangle ABC \cong \triangle A_1B_1C_1$? (   )

(A) $\angle C = \angle C_1, AC = A_1C_1, AD = A_1D_1$;
(B) $\angle B = \angle B_1, AB = A_1B_1, AC = A_1C_1$;
(C) $AD = A_1D_1, BD = B_1D_1, CD = C_1D_1$;
(D) $\angle BAD = \angle B_1A_1D_1, AB = A_1B_1, AD = A_1D_1$.

Analysis Neither (A) nor (B) implies $\triangle ABC \cong \triangle A_1B_1C_1$ because both are "SSA." Condition (C) does not imply $\triangle ABD \cong \triangle A_1B_1D_1$ or $\triangle ACD \cong \triangle A_1C_1D_1$. Next, we show that (D) is correct.

Proof: In triangles ABD and $A_1B_1D_1$, we have $AB = A_1B_1, \angle BAD = \angle B_1A_1D_1, AD = A_1D_1$, so $\triangle ABD \cong \triangle A_1B_1D_1$. Hence, $\angle B = \angle B_1, BD = B_1D_1$.

Since D, D_1 are midpoints, we have $BC = 2BD = 2B_1D_1 = B_1C_1$. Thus, in triangles ABC and $A_1B_1C_1$, we have $AB = A_1B_1, \angle B = \angle B_1. BC = B_1C_1$. Therefore, $\triangle ABC \cong \triangle A_1B_1C_1$.

Remark: The "SSA" condition does not imply congruence.

Example 2. In triangle ABC, E lies on BC, D lies on AE, and $\angle ABD = \angle ACD$, as shown in Figure 17.1. Please add one more condition that makes $BE = CE$ hold.

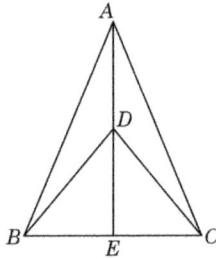

Fig. 17.1

Analysis Observe the shape, and we see that BE lies in triangles BDE and ABE, while CE lies on triangles CDE and ACE. Thus, $\triangle BDE$ and $\triangle CDE$ or $\triangle ABE$ and $\triangle ACE$ may be congruent. Next, we need to find conditions based on the criteria for congruence.

Solution Adding the condition $\angle BAD = \angle CAD$ can make $BE = CE$ hold.

Since $\angle BAD = \angle CAD, AD = AD, \angle ABD = \angle ACD$, we have $\triangle ABD \cong \triangle ACD$, which implies $AB = AC$.

Next, since $AB = AC, \angle BAD = \angle CAD, AE = AE$, we have $\triangle ABE = \triangle ACE$, so that $BE = CE$.

Remark: We can also add other conditions: $\angle BDE = \angle CDE, \angle ADB = \angle ADC$, etc. If we use the properties of isosceles triangles, conditions regarding side lengths work as well.

Example 3. As shown in Figure 17.2, it is known that $\angle ABC = \angle ACB = 40°$, BD is the angle bisector of $\angle ABC$, and E lies on the line BD with $DE = AD$. Find the size of $\angle ECA$.

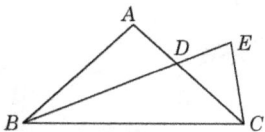

Fig. 17.2

Analysis In order to find $\angle ECA$, we need to inspect its relationship with the known angles. The conditions are $\angle ABC = \angle ACB = 40°, \angle BAC = 100°, \angle ABD = \angle CBD = 20°$. Observing that $\angle ECA$ is an internal angle of $\triangle CDE$, if we can find a triangle congruent to $\triangle CDE$, which also has a close relationship with the given conditions, then the problem can be solved. However, this does not seem possible at first glance. Thus, we need to construct such a triangle.

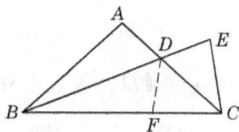

Fig. 17.3

Solution Choose a point F on side BC such that $BF = AB$, and connect DF, as in Figure 17.3.

Since BD is the angle bisector of $\angle ABC$, we have $\angle ABD = \angle FBD$. Combining with $AB = FB, BD = BD$, we have $\triangle ABD \cong \triangle FBD$. Hence, $AD = FD, \angle ADB = \angle FDB$.

On the other hand, since $\angle ABC = \angle ACB = 40°$, we have $\angle BAC = 100°, \angle ABD = \angle CBD = 20°$, and $\angle ADB = \angle FDB = 60°$.

Thus, $\angle CDE = \angle CDF = 60°$. Also $AD = DF = DE, CD = CD$, so we have $\triangle CDF \cong \triangle CDE$.

Therefore, $\angle ECA = \angle ACB = 40°$.

Remark: It is important to observe the shape, guess which triangles may be congruent, and add auxiliary lines to construct congruent triangles.

Example 4. As shown in Figure 17.4, the opposite angles of the quadrilateral $ABCD$ sum to $180°$, and $\angle BAC = \angle DAC, AB = 15, AD = 12$. Let E be the projection of C on AB. Find the value of $\frac{AE}{BE}$.

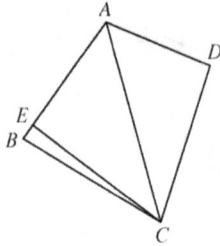 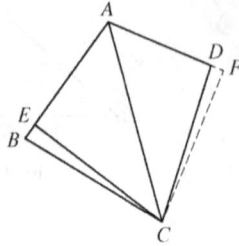

Fig. 17.4   Fig. 17.5

Analysis It suffices to find AE, and BE can be determined from AE. In order to build the relationship between AB and AD, we consider constructing congruent triangles.

Solution Let F be the projection of C on AD, as in Figure 17.5.

Since $\angle BAC = \angle DAC, CE \perp AB$, we may obtain $\triangle ACE \cong \triangle ACF$, and consequently $AE = AF, CE = CF$.

Also, since $\angle ABC + \angle ADC = 180°$, we have

$$\angle CDF = 180° - \angle ADC = \angle ABC = \angle EBC,$$

Thus, $\triangle BCE \cong \triangle DCF$, and $BE = DF$.

On the other hand, we have $AB = AE + BE, AD = AF - DF$, so

$$AB + AD = AE + BE + AF - DF = 2AE,$$

$$AE = \frac{1}{2}(AB + AD) = 13.5,$$

$$BE = AB - AE = 1.5.$$

Therefore, $\frac{AE}{BE} = \frac{13.5}{1.5} = 9$.

Example 5. As in Figure 17.6, BD is the angle bisector of $\angle ABC$ in triangle ABC. Let E be a point outside $\triangle ABC$ such that $\angle EAB = \angle ACB, AE = DC$. If the segments ED, AB intersect at K, prove that $KE = KD$.

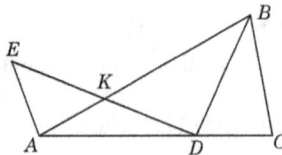

Fig. 17.6

Proof: As in Figure 17.7, let F be the projection of E on AB, let P be the projection of D on AB, and let H be the projection of D on BC.

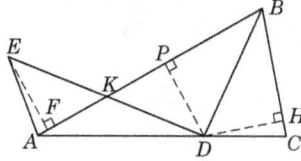

Fig. 17.7

Since BD bisects $\angle ABC$, we have $DP = DH$.

Since $\angle EAB = \angle ACB$, we have $\angle AEF = \angle CDH$. Combining with $AE = CD$, it follows that $\triangle AEF \cong \triangle CDH$. This implies $EF = DH$, and further, $EF = DP$.

Then, it follows easily that $\triangle KEF \cong \triangle KDP$, so $KE = KD$.

Remark: With the angle bisector condition, we may draw perpendicular lines from the bisector to both sides of the angle, and use the property of the angle bisector to prove congruence.

Example 6. In the quadrilateral $ABCD$, it is known that $AB = a$, $AD = b, BC = CD$, and the diagonal AC bisects $\angle BAD$. What condition of a, b needs to be satisfied so that $\angle B + \angle D = 180°$ can be derived? Please draw the shapes and prove your assertion.

Solution If $a = b$, we have $\triangle ABC \cong \triangle ADC$, and $\angle B = \angle D$, as in Figure 17.8. Thus, $\angle B + \angle D = 180°$ only if $\angle B = \angle D = 90°$, which does not always hold.

If $a \neq b$, then without loss of generality, we assume that $a > b$, as in Figure 17.9.

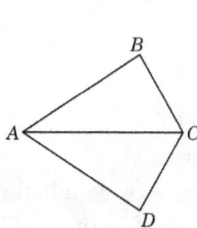 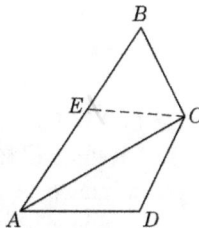

Fig. 17.8   Fig. 17.9

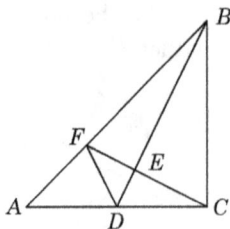

Fig. 17.10

Let E be a point on AB such that $AE = b$. Connect EC, and it follows easily that $\triangle AEC \cong \triangle ADC$. Thus, $\angle AEC = \angle D, EC = DC = BC$, and consequently $\angle B + \angle D = \angle CEB + \angle AEC = 180°$.

Therefore, if $a = b$, then $\angle B + \angle D = 180°$ only when $\angle B = \angle D = 90°$. If $a \neq b$, then $\angle B + \angle D = 180°$ always holds.

Remark: This is a geometric exploration problem in which we can first see what happens in special cases (such as $a = b$) and then consider more general cases.

Example 7. Let $\triangle ABC$ be an isosceles right triangle with $\angle ACB = 90°$, as shown in Figure 17.10. Let D be the midpoint of AC. F lies on AB such that $\angle ADF = \angle CDB$ and CF intersects BD at E. Prove that $BD \perp CF$.

Proof: Let G be a point on the line DF such that $GA \perp AC$, as shown in Figure 17.11.

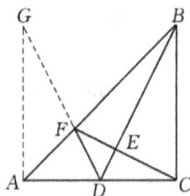

Fig. 17.11

Since $\angle GAD = \angle BCD = 90°, \angle ADG = \angle CDB, AD = CD$, we have $\triangle ADG \cong \triangle CDB$, and $AG = BC, \angle G = \angle DBC$.

Since $AC = BC$, we have $AG = AC$, combining with $\angle GAF = \angle CAF = 45°$ and AF being a common side of $\triangle AGF$ and $\triangle ACF$, it follows that $\triangle AGF \cong \triangle ACF$. This implies $\angle G = \angle ACF$, and hence $\angle ACF = \angle CBD$.

Note that $\angle ACF + \angle BCF = 90°$, so we have $\angle BCF + \angle CBD = 90°$, and $\angle BEC = 90°$. Equivalently, $BD \perp CF$.

Next, we present another solution to this problem.

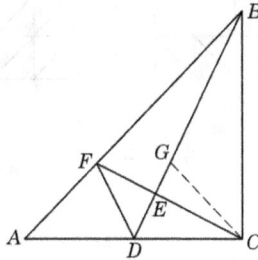

Fig. 17.12

As in Figure 17.12, let the angle bisector of $\angle ACB$ intersect BD at G.

Since CG bisects $\angle ACB$, $\angle ACB = 90°$, we have $\angle ACG = 45°$. From $\angle BAC = \angle ACG = 45°, \angle ADF = \angle CDG, AD = CD$, we obtain $\triangle ADF \cong \triangle CDG$, so that $CG = AF$.

On the other hand, since $CG = AF, \angle BAC = \angle DCG, AC = BC$, we have $\triangle ACF \cong \triangle CBG$, so $\angle CBD = \angle ACF$.

Combining with $\angle ACF + \angle BCF = 90°$, we have $\angle BCD + \angle BCF = 90°$. Therefore, $\angle BEC = 90°$, and, equivalently, $BF \perp CF$.

Remark: The idea of Jiang Han's method is that it suffices to prove $\angle CBE + \angle BCE = 90°$, which is equivalent to $\angle DCE = \angle CBE$ by observation. Thus, we proceed to construct congruent triangles by adding the angle bisector of $\angle ACB$.

Have a Try

We know that if two geometric objects have the same shape and size, then they are congruent. We already know the methods of determining whether two triangles are congruent, but for more complex objects, it is difficult to apply such simple methods. However, we can determine whether two objects are congruent by translation, rotation, and reflection.

Can you find which two objects are congruent in Figure 17.13 and Figure 17.14?

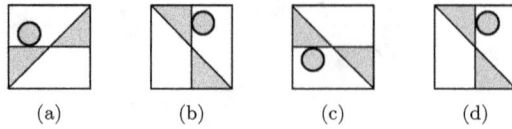

(a)	(b)	(c)	(d)

Fig. 17.13

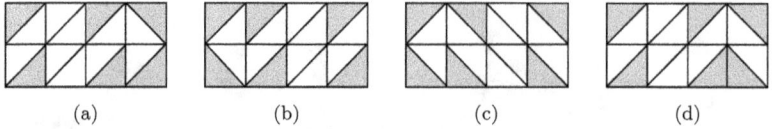

(a)	(b)	(c)	(d)

Fig. 17.14

Exercises

(I) Multiple-choice questions:

1. In quadrilateral $ABCD$, AC is the perpendicular bisector of BD, and they intersect at O, as shown in the figure. Then there are ( ) pairs of congruent triangles in this figure.

(A) 1   (B) 2   (C) 3   (D) 4

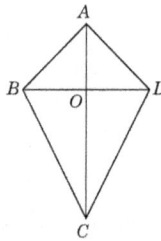

2. As shown in the figure, $AB = AC$, $AP = BQ$, $AO = BO = CO$, $\angle AQO = 16°$. Then, $\angle CPO = ($ $)$.

(A) 16°   (B) 32°   (C) 45°   (D) 46°

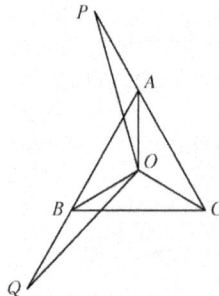

3. As shown in the figure, AD is the angle bisector of $\angle BAC$ in triangle ABC, and E, F are the projections of D on AB, AC, respectively. Then, which of the following is wrong? ( ).

   (A) $AD \perp EF$ (B) $DE = DF$ (C) $AE = AF$ (D) $DE \perp DF$

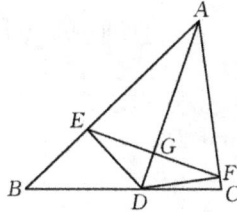

4. In an equilateral triangle ABC, D, F are points on sides BC, AB, respectively, and $BD = AF$. Let E be the intersection of AD, CF, then $\angle CED$ equals ( ).

   (A) $30°$ (B) $60°$ (C) $45°$ (D) $75°$

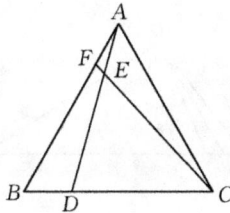

(II) Fill in the blanks:

5. As shown in the figure, $AB = AD, \angle 1 = \angle 2$. In order to derive $\triangle ABC \cong \triangle ADE$, we can add a condition _____ (one condition is enough).

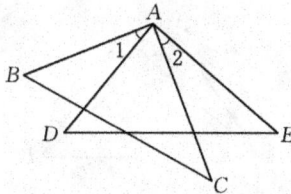

6. As shown in the figure, A lies on DE and F lies on AB, such that $AC = CE, DE = 3, \angle 1 = \angle 2 = \angle 3$. Then, $AB = $ _____ .

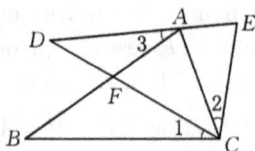

7. As shown in the figure, $AD \parallel BC$, $\angle 1 = \angle 2$, $\angle 3 = \angle 4$, $AD = 4$, $BC = 2$. Then, $AB = $ _____.

8. As shown in the figure, $\angle EDF = 15°$, DP is the angle bisector of $\angle EDF$ in triangle DEF, and $DP \perp DQ$. If $FQ = ED + DF$, then $\angle Q = $ _____, $\angle E = $ _____.

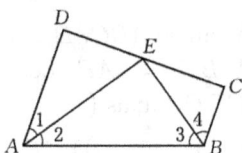

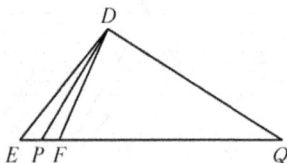

(III) Answer the following questions:

9. As shown in the figure, we are given five conditions: ① $AD = BC$, ② $AC = BD$, ③ $CE = DE$, ④ $\angle D = \angle C$, ⑤ $\angle DAB = \angle CBA$. Please choose two as assumptions and one of the remaining three as a conclusion which composes a correct statement, and prove the statement.

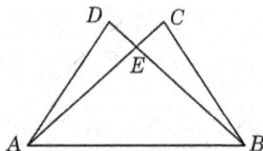

10. As shown in the figure, $\angle ACB = 90°$, D is the midpoint of AC, F lies on the extension of BC with $\angle CDF = \angle A$, and $DE \parallel BC$ with E lying on AB. If $\angle B = 51°$. Find the size of $\angle F$.

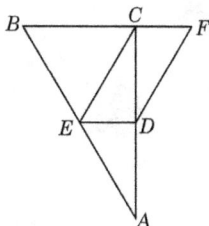

11. In a right triangle ABC, $\angle BAC = 90°$, $AC < AB$, the perpendicular bisector of BC intersects AC at K, and the perpendicular bisector of BK intersects AB at L. If CL bisects $\angle ACB$, find the size of $\angle ABC, \angle ACB$.

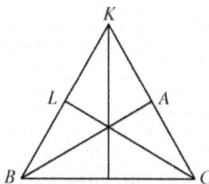

12. As shown in the figure, D is the midpoint of BC, and $DE \perp DF$. Determine the relationship between $BE + CF$ and EF, and prove your result.

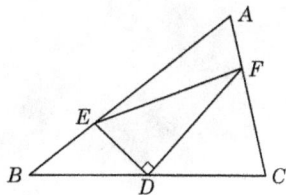

Chapter 18

Isosceles Triangles

A triangle with a pair of equal sides (or angles) is called an isosceles triangle. Isosceles triangles are axially symmetric. The two base angles are equal, and the altitude from the base, the median through the midpoint of the base, and the bisector of the vertex angle coincide with each other. The coincidence of these three segments is an important property of isosceles triangles.

The sides of an equilateral triangle are all equal, and the internal angles are all 60°.

When trying to determine the sides or angles of an isosceles triangle, it is common to discuss cases such as which side is the base and which angle is the vertex angle.

In problems with isosceles triangles, in addition to congruent triangles, we may use special properties of isosceles triangles. For example, in order to prove two sides (angles) are equal, we can put them into the same triangle and prove that this triangle is isosceles.

Example 1. Suppose that in an isosceles triangle, the angle between one leg and the altitude from the other leg is 20°. Find the internal angles of this triangle.

Analysis The conditions do not specify whether this triangle is acute, right, or obtuse. This means that the altitude from one leg of the isosceles triangle may be inside the triangle, be outside the triangle, or coincide with the other leg (which is impossible in this problem). Hence, we need to consider different cases.

Solution   According to the conditions, there are two possible cases, as in Figures 18.1 and 18.2.

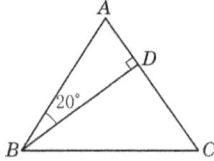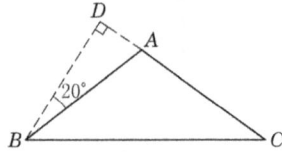

Fig. 18.1   Fig. 18.2

In Figure 18.1, we have $\angle A = 70°$, so that $\angle ABC = \angle C = 55°$.

In Figure 18.2, we have $\angle BAD = 70°$, so $\angle BAC = 110°$, and $\angle ABC = \angle C = 35°$.

Therefore, the angles of the triangle are $70°, 55°, 55°$ or $110°, 35°, 35°$.

Example 2. In the triangle ABC, $\angle A = 60°$, points M, N, K lie on BC, AC, AB, respectively, and $BK = KM = MN = NC$. If $AN = 2AK$, find the size of $\angle B, \angle C$.

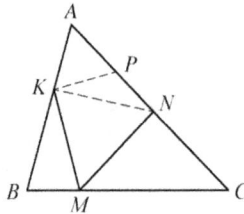

Fig. 18.3

Analysis   As shown in Figure 18.3, since $\triangle BMK$ and $\triangle MCN$ are both isosceles triangles, we have $\angle B = \angle BMK$, $\angle C = \angle CMN$. Meanwhile, from $\angle A = 60°$ and $AN = 2AK$, we may derive the size of $\angle ANM$, and $\angle C$ can be obtained using the exterior angle theorem.

Solution   Let P be the midpoint of AN, and connect KP, KN.

Since $\angle A = 60°, AK = AP$, the triangle APK is equilateral, and $KP = AP = PN, \angle APK = 60°$. Also $\angle APK = 2\angle PNK = 60°$, so $\angle PNK = 30°$.

Since $MN = NC$, we have $\angle C = \angle CMN$, and from $BK = KM$, we get $\angle B = \angle BMK$.

Hence, $\angle KMN = 180° - (\angle BMK + \angle CMN) = 180° - (120° - \angle C) - \angle C = 60°$.

Since $KM = MN$, the triangle MNK is equilateral, and $\angle MNK = 60°$. Therefore, $\angle ANM = 90°, \angle C = \frac{1}{2}\angle ANM = 45°, \angle B = 75°$.

Example 3. In the triangle ABC, $\angle ABC = 60°, \angle ACB = 40°$, and P is the intersection of the angle bisectors of $\angle ABC, \angle ACB$. Prove that $AB = PC$.

Analysis As seen in Figure 18.4, AB and PC are neither sides of the same triangle nor corresponding sides of two congruent triangles. Thus, auxiliary lines are needed, either to put them in the same triangle or to construct congruent triangles, or we may consider proving $AB = m, m = PC$ where m is another object. Due to the special properties of the given angles, we can try to construct equilateral triangles to crack this problem.

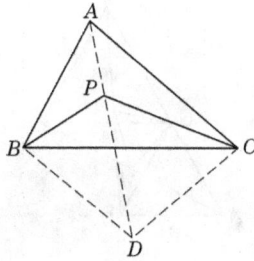

Fig. 18.4

Proof: As in Figure 18.4, we construct an equilateral triangle PDC, where PC is one side, and connect AP. Then, AP bisects $\angle BAC$ (whose reason is explained in the following below), and $\angle PAC = 40°$.

Since $\angle ACP = 20°$, we have $\angle APC = 120°$. Since $\angle CPD = 60°$, we have $\angle APC + \angle CPD = 180°$. Hence, A, P, D are collinear.

On the other hand, $\angle BAC = 80° = \angle ACD, AC = CA, \angle ACB = 40° = \angle CAD$, so we obtain that $\triangle ABC \cong \triangle CDA$, and consequently $AB = PC$.

Remark: As shown in Figure 18.5, let BP, CP be angle bisectors of $\angle ABC, \angle ACB$, respectively, and D, E, F are the projections of P on the three sides. It follows from the property of angle bisectors that

$PD = PF$, $PD = PE$, so that $PE = PF$. This implies that AP is the bisector of $\angle BAC$.

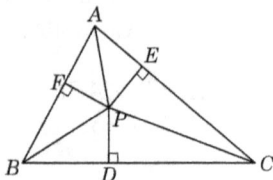

Fig. 18.5

Example 4. As shown in Figure 18.6, $\triangle ABC$ is an equilateral triangle, and E lies on the extension of AC. Choose a point D such that $\triangle CDE$ is equilateral. If M is the midpoint of AD and N is the midpoint of BE, prove that $\triangle CMN$ is an equilateral triangle.

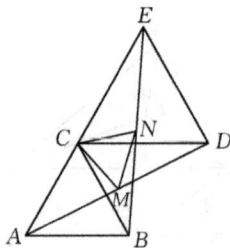

Fig. 18.6

Proof: Since $\triangle ABC, \triangle CDE$ are equilateral, we have $AC = BC, \angle ACD = \angle BCE = 60° + \angle BCD$, $CD = CE$. Thus, it follows that $\triangle ACD \cong \triangle BCE$, and $AD = BE$.

Since M, N are the midpoints of AD, BE, respectively, we have $AM = BN$.

Combining with $\angle CAM = \angle CBN$, we have $\triangle AMC \cong \triangle BNC$, so $CM = CN, \angle ACM = \angle BCN$.

On the other hand, we have $\angle NCM = \angle BCN - \angle BCM, \angle ACB = \angle ACM - \angle BCM = 60°$, which means $\angle NCM = 60°$. Therefore, $\triangle CMN$ is an equilateral triangle.

Remark: In an equilateral triangle, the sides are equal and the internal angles are all 60°. These properties allow us to transform the equality relationships among sides and angles.

Example 5. In $\triangle ABC$, $\angle A = 2\angle B$, CD bisects $\angle C$, $AC = 16, AD = 8$. Find the length of BC.

Analysis   We need to cut BC into two parts to find its relationship with AC, AD.

Solution   Since $\angle A = 2\angle B$, we have $\angle A > \angle B$, and $BC > AC$.

Choose point E on BC such that $EC = AC$, and connect DE, as shown in Figure 18.7.

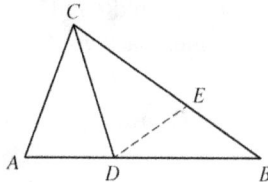

Fig. 18.7

Since CD bisects $\angle ACB$, we have $\angle ACD = \angle ECD$, and combining with $AC = EC$, $CD = CD$, we get $\triangle ACD \cong \triangle ECD$. Thus, $AD = ED, \angle A = \angle CED$.

On the other hand, $\angle CED = \angle B + \angle BDE$, and $\angle A = 2\angle B$, so we have $\angle B = \angle BDE$. This implies $BE = DE, BE = AD$.

Therefore, $BE = AC + AD = 24$.

Remark: Cutting a long segment into two parts is a very common idea. The auxiliary line helps to solve the problem when we can establish its relationship with the given conditions.

Example 6. As shown in Figure 18.8, $\triangle ABC$ is an isosceles triangle with $AB = AC$, and D, E lie on sides BC, AC, respectively. What condition should $\angle BAD, \angle CDE$ satisfy so that $AD = AE$? Give your reasons.

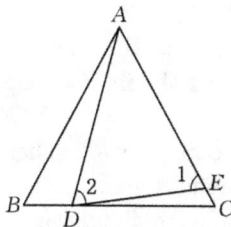

Fig. 18.8

Solution   When $\angle BAD = 2\angle CDE$, it follows that $AB = AC$. The reason is explained as follows.

We have $\angle 1 = \angle C + \angle CDE$, $\angle 2 = \angle ADC - \angle CDE = \angle B + \angle BAD - \angle CDE$.

In order to make $\angle 1 = \angle 2$, we need

$$\angle C + \angle CDE = \angle B + \angle BAD - \angle CDE.$$

Since $AB = AC$, we have $\angle B = \angle C$, so it follows that $\angle BAD = 2\angle CDE$.

Remark: In order to find the required condition, we can first assume the conclusion holds and try to deduce backward.

Reading

Isosceles triangles are a special type of triangle. In addition to the properties and criteria described in the textbook, they also have many other beautiful properties and criteria:

1. (Steiner–Lehmus theorem) If two inner angle bisectors of a triangle are equal, then it is an isosceles triangle.
2. (Viviani's theorem) The sum of the distances from a point in an equilateral triangle to its three sides is a fixed value.
3. For two equilateral triangles, the midpoints of the three segments connecting their corresponding vertices are the vertices of another equilateral triangle.

Students can prove these theorems by themselves, as well as explore and discover more interesting knowledge in isosceles triangles.

Exercises

(I) Multiple-choice questions:

1. As shown in the figure, D, E lie on sides AC, AB of $\triangle ABC$, respectively. Let O be the intersection of BD, CE. We are given four conditions: ①$\angle EBO = \angle DCO$, ②$\angle BEO = \angle CDO$, ③$BE = CD$, ④$OB = OC$. There are six ways to choose two conditions among them, which are ①②, ①③, ①④, ②③, ②④, ③④. How many combinations would imply that $\triangle ABC$ is an isosceles triangle? (   ).

(A) 3   (B) 4   (C) 5   (D) 6

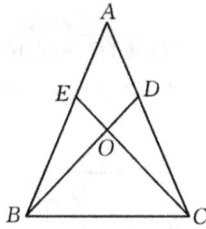

2. In the following figure, $AB = AC, BC = CD, AE = ED = DB$. Then, $\angle A = ($).

    (A) $36°$ (B) $30°$ (C) $\frac{180°}{7}$ (D) $\frac{180°}{7}$

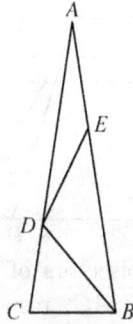

3. As shown in the figure, $\triangle ABC$ is an equilateral triangle, and $\angle ADC = 30°, AD = 3, BD = 5$. Then, $CD = ($).

    (A) 5   (B) 4   (C) 3   (D) $4\frac{1}{2}$

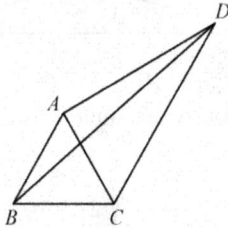

4. Let $\triangle ABC$ be an equilateral triangle and P be a point on the plane such that $\triangle ABP, \triangle BCP, \triangle ACP$ are all isosceles triangles. There are ( ) different choices of P.

    (A) 1   (B) 4   (C) 7   (D) 10

(II) Fill in the blanks:

5. In quadrilateral $ABCD$, $AB = AC = AD, \angle BAC = 25°$, $\angle CAD = 75°$, as shown in the figure. Then, $\angle BDC = $ _____, $\angle DBC = $ _____ .

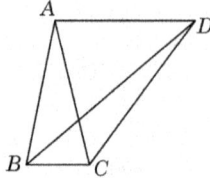

6. In the following figure, $AB = AC$, D lies on BC with $\angle BAD = 50°$, and E lies on AC so that $\angle ADE = \angle AED$. Then, $\angle EDC = $ _____ .

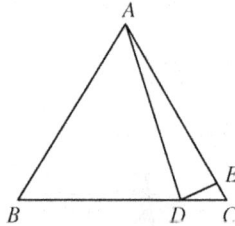

7. Let BI, CI be angle bisectors of $\triangle ABC$, which intersect at I. Points D, E lie on AB, AC, respectively, such that DE passes through I and $DE \parallel BC$. If $AB = 12, AC = 7$, then the perimeter of $\triangle ADE$ is _____ .

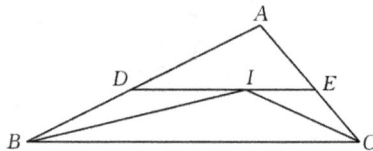

8. In the following figure, $AC = BC, \angle C = 20°$, and M, N lie on sides AC, BC, respectively, such that $\angle BAN = 50°, \angle ABM = 60°$. Then, $\angle NMB = $ _____ .

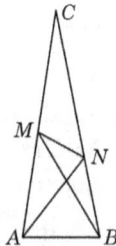

(III) Answer the following questions:

9. Let $ABCD$ be a square and $\triangle PBC, \triangle QCD$ be two equilateral triangles inside it, as shown in the figure. Segments PB, DQ intersect at M, BP, CQ intersect at E, and CP, DQ intersect at F. Prove that $PM = QM$.

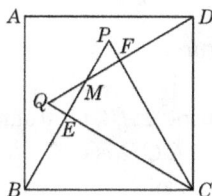

10. As shown in the figure, D is a point on the side BC of $\triangle ABC$, $\angle B = 45°, \angle ADC = 60°, DC = 2BD$. Find the size of $\angle C$.

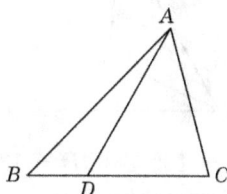

11. In the triangle ABC, D, E lie on the side BC with $AD = AE$, as shown in the figure. Add one condition such that $AB = AC$ can be derived. Write as many conditions as possible, and give the proof for one of them.

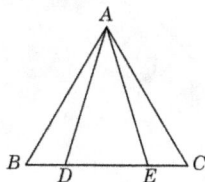

12. As shown in the figure, $\triangle ABC$ is an equilateral triangle, and points D, E, F lie on sides BC, CA, AB, respectively, such that $\triangle DEF$ is also equilateral.

   (1) Besides the given conditions, which segments are also equal? Prove your result.
   (2) For the segments you have proven to be equal, what transformation can we use to transform one segment into another? Specify the details of the transformation.

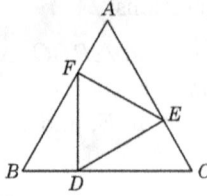

13. As shown in the figure, O is a point inside the equilateral triangle ABC. Suppose $\angle AOB = 110°, \angle BOC = \alpha$, and $\triangle ADC$ can be obtained by rotating $\triangle BOC$ around C in the clockwise direction. It follows that $\triangle COD$ is equilateral. If $\angle AOD$ is an isosceles triangle, find the value of α.

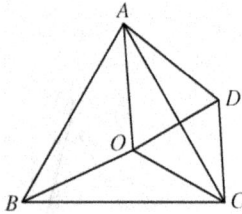

14. In $\triangle ABC$, $\angle ABC = 12°, \angle ACB = 132°$, as shown in the figure. Suppose BM, CN are external angle bisectors, and M, N lie on lines AC, AB, respectively. Prove that $BM = CN$.

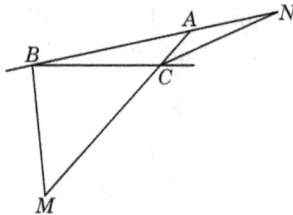

Chapter 19

Right Triangles

In a right triangle, the two acute angles are complementary angles, and the hypotenuse is the longest side. The side opposite a $30°$ angle is half the hypotenuse. Conversely, if a leg equals half the hypotenuse, then its opposite angle equals $30°$. The median on the hypotenuse is equal to half the hypotenuse.

If the two legs are a, b and the hypotenuse is c, then $a^2 + b^2 = c^2$. Conversely, if $a^2 + b^2 = c^2$, then the triangle with side lengths a, b, c is a right triangle, where c is opposite to the right angle.

When solving problems with right triangles, such as proving the equality of sides and angles or perpendicularity, we may use the special properties of right triangles, in addition to methods regarding congruent triangles and isosceles triangles.

For problems where the side lengths are integers, we usually rewrite the equation $a^2 + b^2 = c^2$ and solve using the properties of integers.

Example 1. In Figure 19.1, $\triangle ABC$ is a right triangle with $\angle C = 90°$, $AC = 4, BC = 3$. Concatenate $\triangle ABC$ and another triangle so that the resulting shape is an isosceles triangle. Give at least two ways.

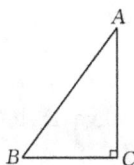

Fig. 19.1

Analysis We note that either one leg or the base of the resulting isosceles triangle coincides exactly with a side of $\triangle ABC$. Thus, every side of $\triangle ABC$ can be a leg or the base of the isosceles triangle, and we can calculate the side lengths of the corresponding triangles.

Solution The different ways of concatenation are illustrated in Figure 19.2.

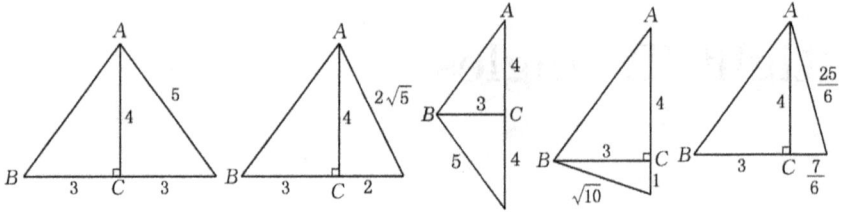

Fig. 19.2

Example 2. In a right triangle ABC, $\angle ACB = 90°$, $AC = BC = \sqrt{2}$, AD is the angle bisector of $\angle BAC$, and E is the projection of B on the line AD. Find the length of AE.

Analysis There are two ways to find AE. One is to write $AD + DE = AE$, and the other is to construct a right triangle which has AE as a side. In the first idea, DE seems hard to find since we need both BD and BE, while in the second idea, only BE is necessary.

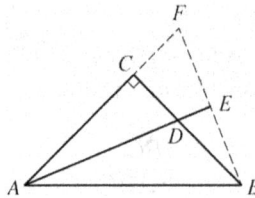

Fig. 19.3

Solution Extend AC, BE, and let them intersect at F, as shown in Figure 19.3.

Since $AC = BC, \angle ACD = \angle BCF = 90°, \angle ADC = \angle BDE$, which implies $\angle CAD = \angle CBF$, we have $\triangle ACD \cong \triangle BCF$, so $CD = CF$, $AD = BF$.

Since AE bisects $\angle BAC$, $AE \perp BF$, we also have $AF = AB, BE = EF = \frac{1}{2}BF$.

By $AC = BC = \sqrt{2}$, we get $AB = \sqrt{(\sqrt{2})^2 + (\sqrt{2})^2} = 2$, so $CD = CF = AF - AC = AB - AC = 2 - \sqrt{2}$.

Further, $AD = \sqrt{AC^2 + CD^2} = \sqrt{(\sqrt{2})^2 + (2 - \sqrt{2})^2} = 2\sqrt{2 - \sqrt{2}}$, so

$$BE = \frac{1}{2}AD = \sqrt{2 - \sqrt{2}}.$$

Therefore, $AE = \sqrt{AB^2 - BE^2} = \sqrt{2^2 - (\sqrt{2 - \sqrt{2}})^2} = \sqrt{2 + \sqrt{2}}.$

Example 3. In Figure 19.4, $AC = BC, \angle ACB = 90°$, and D, E are points on side AB, such that $AD = 3, BE = 4, \angle DCE = 45°$. Find the area of $\triangle ABC$.

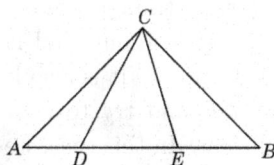

Fig. 19.4

Solution   Rotate $\triangle CEB$ around C by $90°$ in the clockwise direction, and we obtain $\triangle CE'A$, as in Figure 19.5.

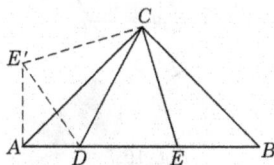

Fig. 19.5

It follows easily that $AE' = BE = 4, \angle E'AD = 90°$, so

$$E'D = \sqrt{AE'^2 + AD^2} = \sqrt{4^2 + 3^2} = 5.$$

On the other hand, $\angle DCE = 45° = \angle DCE'$, so we have $\triangle DCE \cong \triangle DCE'$. Thus, $DE = DE' = 5, AB = 12$. Therefore, $S_{\triangle ABC} = \frac{1}{4}AB^2 = 36$.

Example 4. As shown in Figure 19.6, let P be an arbitrary point in $\triangle ABC$. Construct perpendicular lines to three sides from P, so that $PD \perp BC, PE \perp CA, PF \perp AB$. If $BD = BF, CD = CE$, prove that $AE = AF$.

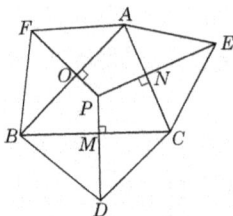

Fig. 19.6

Analysis   There are many right triangles given by the condition, but the equal segments are not in close positions, and it seems hard to find their relationship with the result to prove. It is also hard to find congruent triangles here. If we connect EF and try to prove $\angle AEF = \angle AFE$, the conditions are still not sufficient. In this case, we turn to the Pythagorean theorem for some hope.

Proof: Connect PA, PB, PC as in Figure 19.7.

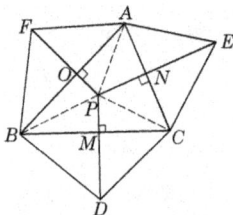

Fig. 19.7

Since $PF \perp AB$, we have

$$AF^2 = AO^2 + FO^2, BP^2 = BO^2 + PO^2,$$
$$AP^2 = AO^2 + PO^2, BF^2 = BO^2 + FO^2.$$

Thus, $AF^2 + BF^2 = AO^2 + BO^2 + FO^2 + PO^2 = AP^2 + BF^2$.

Similarly, $AP^2 + CE^2 = AE^2 + CP^2, BD^2 + CP^2 = BP^2 + CD^2$.

Adding the three equalities, we get $AF^2 + CE^2 + BF^2 = AE^2 + CD^2 + BF^2$.

Since $BD = BF, CD = CE$, we conclude that $AE = AF$.

Remark: The following result comes from the proof above: if the two diagonals of a quadrilateral are perpendicular, then the sum of squares of two pairs of its opposite sides are equal.

Example 5. As shown in Figure 19.8, two snails climb along the sides of a right triangle ABC (where $\angle C$ is the right angle) at the same speed. Both of them start at point A, but one climbs clockwise, while the other climbs counterclockwise. Suppose they meet at point P on BC. Prove that $\frac{2}{AC} + \frac{1}{PC} = \frac{1}{PB}$.

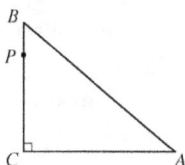

Fig. 19.8

Analysis Given the conditions, we have $AB + BP = AC + CP$ and $AB^2 = AC^2 + BC^2$. We should combine these relations and the desired result to find appropriate transformations.

Proof: It follows that $AB + BP = AC + CP$. Equivalently,

$$AB - AC = PC - PB. \qquad \text{①}$$

Also $AB^2 = AC^2 + BC^2$, so $(AB + AC)(AB - AC) = BC^2$, and

$$AB + AC = \frac{BC^2}{AB - AC} = \frac{BC^2}{PC - PB}. \qquad \text{②}$$

From ①, ②,

$$AC = \frac{1}{2}\left[\frac{BC^2}{PC - PB} - (PC - PB)\right]$$

$$= \frac{1}{2}\left[\frac{(PC + PB)^2 - (PC - PB)^2}{PC - PB}\right] = \frac{2PC \cdot PB}{PC - PB}.$$

Therefore, $\frac{2}{AC} + \frac{1}{PC} = \frac{PC - PB}{PC \cdot PB} + \frac{1}{PC} = \frac{1}{PB}$.

Remark: When transforming geometric expressions, make sure to use the relations among quantities in the given shape, such as $BC = PC + PB$.

Example 6. As shown in Figure 19.9, BD, CE are two altitudes of an acute triangle ABC. The projections of B, C on the line DE are F, G, respectively. Prove that $EF = DG$.

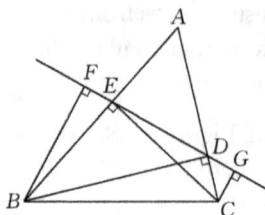

Fig. 19.9

Analysis   Apparently, $\triangle BEF$ and $\triangle CDG$ are not congruent, so in order to prove $EF = DG$, we should either construct congruent triangles or use calculation. In this problem, there are many right angles, so we may try the properties of right triangles. For example, if M is the midpoint of BC and N is the midpoint of FG, then $MN \parallel BF$, and it suffices to prove that N is the midpoint of DE.

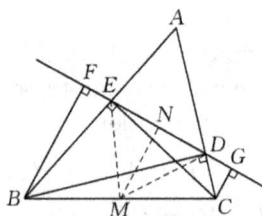

Fig. 19.10

Proof: Choose M, N to be the midpoints of BC, DE, respectively, as in Figure 19.10.

Since $BD \perp AC, CE \perp AB$, we have

$$MD = ME = \frac{1}{2}BC.$$

Since N is the midpoint of ED, we have $MN \perp ED, EN = ND$.

Now, $MN \perp ED, BF \perp ED, CG \perp ED$, so $BF \parallel MN \parallel CG$.

Finally, since M is the midpoint of BC, we have $FN = NG$, and $FN - EN = NG = ND$. The last equality is equivalent to $EF = DG$.

Remark: The theorem that the median on the hypotenuse equals half the hypotenuse helps us build connections between segments in different triangles.

Example 7. Let A be a given positive rational number:

(1) If A is the area of a right triangle with rational side lengths, prove that there exist positive rational numbers x, y, z such that $x^2 - y^2 = y^2 - z^2 = A$.

(2) Prove the converse of (1).

Analysis The numbers x, y, z are not the side lengths of the triangle, but quantities related to the side lengths a, b, c. Hence, we need to find expressions of x, y, z in terms of a, b, c.

Proof: (1) Let a, b, c be the side lengths of the right triangle, then a, b, c are positive rational numbers, and $a^2 + b^2 = c^2$ $\frac{1}{2}ab = A$.

If $a = b$, then $2a^2 = c^2$, $\frac{c}{a} = \sqrt{2}$, which is impossible. Thus, $a \neq b$.

Without loss of generality, assume that $a < b$, and $x = \frac{a+b}{2}, y = \frac{c}{2}$, $z = \frac{b-a}{2}$. Then, x, y, z are positive rational numbers, and

$$x^2 - y^2 = \frac{(a+b)^2 - c^2}{4} = \frac{1}{2}ab = A,$$

$$y^2 - z^2 = \frac{c^2 - (b-a)^2}{4} = \frac{1}{2}ab = A.$$

(2) Suppose three positive rational numbers x, y, z satisfy $x^2 - y^2 = y^2 - z^2 = A$, then $x > y > z$. Let $a = x - z, b = x + z, c = 2y$. Then, a, b, c are positive rational numbers, and

$$a^2 + b^2 = 2(x^2 + z^2) = 4y^2 = c^2,$$

$$\frac{1}{2}ab = \frac{1}{2}(x^2 - z^2) = \frac{1}{2}[(x^2 - y^2) + (y^2 - z^2)] = A.$$

Therefore, the triangle with side lengths a, b, c is a right triangle with area A.

Reading

The Pythagorean theorem can be called an unparalleled problem, and so far there have been more than 300 proofs. For more than 2000 years, people who have participated in the exploration of the proof include professional mathematicians, political dignitaries, and ordinary people. It is legitimate to say that no other mathematical problem has attracted such widespread attention as the Pythagorean theorem. The logo of the 2002 ICM in Beijing was designed after the Pythagorean theorem.

Figures 19.11–19.14 represent some proofs of the Pythagorean theorem for students to appreciate.

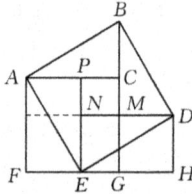

Fig. 19.11

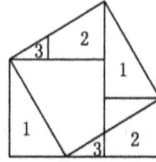

Fig. 19.12

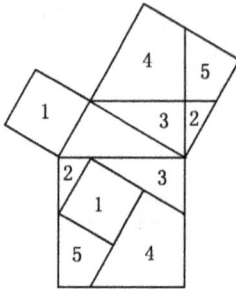

Fig. 19.13

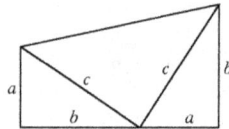

Fig. 19.14

Exercises

(I) Multiple-choice questions:

1. In a right triangle, if the square of the hypotenuse equals twice the product of two legs, then the ratio of its side lengths is (   ).

   (A) $3:4:5$ (B) $1:1:1$ (C) $2:3:4$ (D) $1:1:\sqrt{2}$

2. In the following figure, H is the intersection of altitudes AD, BE of triangle ABC, and $BH = AC$, then $\angle ABC$ equals (   ).

   (A) $30°$ (B) $45°$ (C) $60°$ (D) $75°$

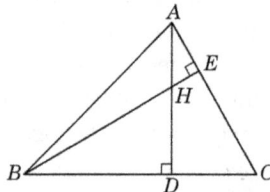

3. In the following figure, $\angle 1 = \angle 2, AD = BD = 4, CE \perp AD$, $2CE = AC$, then the length of CD is ( ).

   (A) 2   (B) 3   (C) 1   (D) 1.5

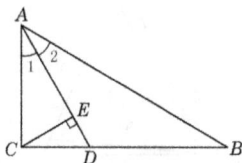

4. As shown in the figure, we concatenate 16 different right triangles to obtain an object that appears like a sea snail. The right angles and the lengths of some segments are marked in the figure. Then, the integer that is closest to the perimeter of this "sea snail" is ( ).

   (A) 19   (B) 20   (C) 21   (D) 22

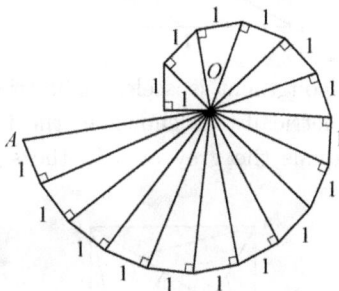

5. What is the number of different right triangles such that the lengths of their legs are integers $a, b (b < 2011)$, and the hypotenuse equals $b + 1$? ( ).

   (A) 28   (B) 29   (C) 30   (D) 31

(II) Fill in the blanks :

6. In a right triangle ABC, $\angle C = 90°, \angle A = 30°$, $AB + BC = 6\sqrt{3}$, then $AC =$ _____.

7. As shown in the figure, $AB \perp AD, AB = 3, BC = 12, CD = 13$, $DA = 4$, then the area of the quadrilateral $ABCD$ is _____.

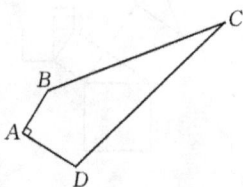

8. In the following figure, $AB = AC, \angle ABN = \angle MBC, BM = MN, BN = a$. Then, the distance from N to BC is _____.

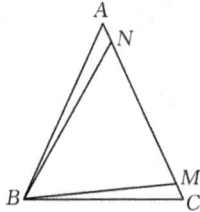

9. In the following figure, AD is a median of $\triangle ABC$ and $\angle ADC = 45°$. We fold the triangle along AD so that $\triangle ACD$ becomes $\triangle AC'D$. If $BC = 4$, then $BC' = $ _____ .

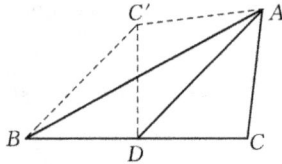

10. We put four congruent isosceles right triangles inside a square $ABCD$ symmetrically, as shown in the following figure. If the white area equals the grey area in the square, then $\angle BAE = $ _____ .

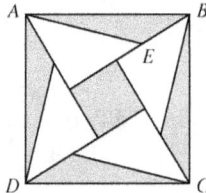

(III) Answer the following questions:

11. In the following figure, all quadrilaterals are squares, and all triangles are right triangles. If the biggest square has a side length of 13 cm, try to find the sum of the areas of the four shaded squares.

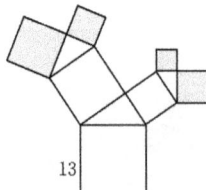

12. In a convex quadrilateral $ABCD$, $\angle ABC = 30°, \angle ADC = 60°$, $AD = DC$, as shown in the following figure. Prove that $BD^2 = AB^2 + BC^2$.

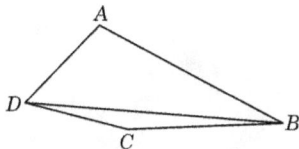

13. In $\triangle ABC$, AD is the median on BC, $AB = \sqrt{2}, AD = \sqrt{6}, AC = \sqrt{26}$. Find the size of $\angle ABC$.

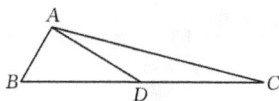

14. In the quadrilateral $ABCD$, $AC = 4, CD = 3, \angle ADB = \angle ABD = \angle ACD = 45°$. Find BC.

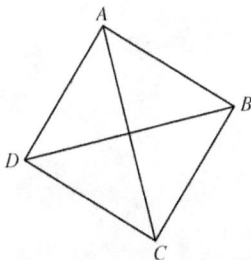

15. Suppose a point P lies inside $\triangle ABC$, and D, E, F are its projections on the three sides, as shown in the following figure. Given $AB = 5, BC = 7, AC = 6, BE - AD = 1$, find $AD + BE + CF$.

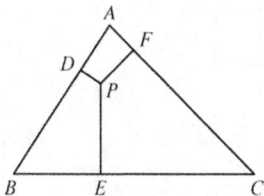

Chapter 20

Parallelograms

Parallelograms are a kind of special quadrilateral. In a parallelogram, each pair of opposite sides is parallel and equal, each pair of opposite angles is equal, and the two diagonals bisect each other.

Rectangles, rhombuses, and squares are special types of parallelograms. In addition to the properties of the parallelogram, these also have their own special properties. In a rectangle, all internal angles are right angles, and the two diagonals are equal. In a rhombus, all sides are equal, and the two diagonals are perpendicular, which also bisect the internal angles. The square has all the properties possessed by the rhombus and the rectangle.

To prove that a quadrilateral is a parallelogram, rectangle, rhombus, or square, we can use a step-by-step method. For example, to prove a quadrilateral is a rectangle, we can first prove it is a parallelogram and then prove that it has a right angle or that its diagonals are equal.

Parallelograms have abundant properties, which bring about much convenience for geometric proofs. For example, we can use parallelograms to prove the equality of segments and angles or to prove that two lines are parallel or perpendicular.

Meanwhile, make sure to use the properties of triangles flexibly as well. In this chapter, an important technique to construct parallelograms to solve problems is described.

Example 1. As shown in Figure 20.1, in the parallelogram $ABCD$, $\angle DAB = 60°$ and E, F lie on the extensions of CD, AB, respectively, with $AE = AD, CF = CB$.

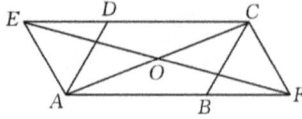

Fig. 20.1

(1) Prove that $AFCE$ is a parallelogram.
(2) If the condition $\angle DAB = 60°$ is omitted, does the same conclusion still hold? Explain your reasons.

Analysis   (1) To prove $AFCE$ is a parallelogram, we already have $AF \parallel CE$, so by characterization of parallelograms, it suffices to show $AF = CE$, or $AE \parallel CF$, or that O is the midpoint of EF.

(2) There are two ways to determine whether the condition $\angle DAB = 60°$ can be omitted. One is drawing another picture where $\angle DAB \neq 60°$ and seeing whether $AFCE$ is still a parallelogram, and the other is trying to prove $AFCE$ is a parallelogram without using this condition.

Solution

(1) Since $ABCD$ is a parallelogram, we have $AB \parallel CD, \angle ADE = \angle DAB = 60°$. Since $AD = AE$, it follows that $\triangle ADE$ is equilateral, so that $\angle DAE = 60°$.
For similar reasons, we have $\angle BFC = 60°$.
Hence, $\angle DAE + \angle DAB + \angle BFC = 180°$, and $AE \parallel CF$. Therefore, $AFCE$ is a parallelogram.

(2) The conclusion still holds.
Since $ABCD$ is a parallelogram, we have $AB \parallel CD, \angle ADE = \angle DAB$.
Since $AD = AE$, we have $\angle ADE = \angle AED, \angle DAE = 180° - 2\angle BAD$.
From $AD \parallel BC$, we get $\angle DAB = \angle CBF$.
Since $BC = BF$, we have $\angle CBF = \angle BFC = \angle DAB$.
Thus, $\angle DAE + \angle DAB + \angle BFC = 180° - 2\angle BAD + \angle BAD + \angle BAD = 180°$, and $AE \parallel FC$. Therefore, $AFCE$ is a parallelogram.

Remark: (2) We can also let $\angle DAB$ be other values, such as $30°$ or $40°$, to find that the conclusion still holds and then try to find a proof.

Example 2. In $\triangle ABC$, D is the midpoint of BC, DE bisects $\angle ADB$, DF bisects $\angle ADC$, $BE \perp DE$, $CF \perp DF$, and P is the intersection of AD and EF, as shown in Figure 20.2. Prove that $EF = 2PD$.

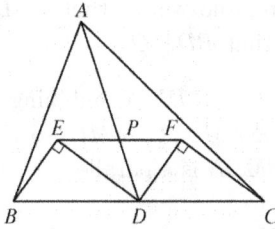

Fig. 20.2

Analysis It suffices to prove that P is the midpoint of EF because $\triangle DEF$ is obviously a right triangle.

Proof: Let $\angle ADB = 2\alpha$, then $\angle ADC = 180° - 2\alpha, \angle ADE = \angle BDE = \alpha, \angle ADF = \angle CDF = 90° - \alpha$.

Since $\triangle BDE$ is a right triangle, we have $\angle BED = 90° - \angle BDE = 90° - \alpha$, so $\angle BED = \angle FDC$.

Also, since D is the midpoint of BC, we have $BD = DC$. Combining with the angles, we get $\triangle BDE \cong \triangle DCF$, and $BE = CF$.

From $\angle EBD = \angle FDC$, we get $EB \parallel FD$, so $BDFE$ is a parallelogram, and $EF \parallel BD$. Thus, $\angle PED = \angle BDE, \angle PFD = \angle CDF$.

Since $\angle BDE = \angle PDE, \angle CDF = \angle PDF$, we have $\angle PED = \angle PDE, \angle PFD = \angle PDF$, which imply $PE = PD = PF$. Therefore, $EF = 2PD$.

Example 3. As shown in Figure 20.3, $ABCD$ and $BFDE$ are rectangles, and $AB = BF$. Prove that $MN \perp CF$.

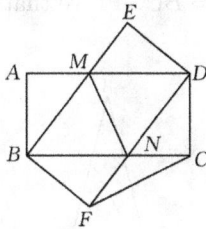

Fig. 20.3

Analysis The figure does not show the intersection of MN and CF, and there are two ways to proceed. One is to construct their intersection, and the other is to prove that CF is parallel to some line that is perpendicular to MN. Observe the figure, and we see that $BNDM$ looks like a rhombus. Thus, it suffices to prove that $BD \parallel CF$.

Proof: We have $AB = BF = DE$. Combining with $\angle A = \angle E = 90°$, $\angle AMB = \angle EMD$, we get $\triangle AMB \cong \triangle EMD$, so $BM = MD$.

It is easily seen that $BMDN$ is a parallelogram, and since $BM = MD$, it is also a rhombus.

As shown in Figure 20.4, we connect BD, then $BD \perp MN, NB = ND$. Hence, $\angle NBD = \angle NDB$.

On the other hand, we can prove that $\triangle BNF \cong \triangle DNC$, so $NF = NC$, and $\angle NFC = \angle NCF$. This implies $\angle NFC = \angle NDB$.

Thus, $BD \parallel CF$ and $MN \perp CF$.

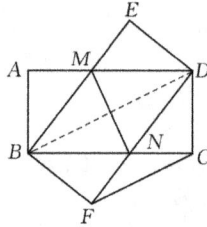

Fig. 20.4

Remark: The property that the two diagonals of a rhombus are perpendicular makes this problem easier to prove.

Example 4. As shown in Figure 20.5, $\triangle ABC$ is an isosceles triangle with vertex angle $\angle A = 30°$. Points P, Q lie on sides AB, AC, respectively, such that $\angle QPC = 45°$ and $PQ = BC$. Prove that $BC = CQ$.

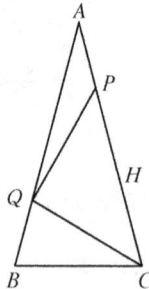

Fig. 20.5

Analysis To prove $BC = CQ$, it suffices to show $\angle CBQ = \angle BQC$, or $PQ = CQ$, since $PQ = BC$ is known. However, neither of them seems easy to prove, and the given condition includes specific angles, so we consider adding auxiliary lines.

Proof: We construct a parallelogram $QPOB$, as shown in Figure 20.6.

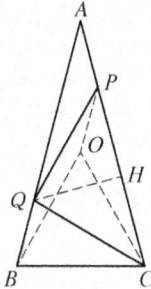

Fig. 20.6

Then, we have $BO = PQ = BC$. Since $\angle QBO = \angle AQP = 45° - 30° = 15°$, it follows that $\angle OBC = (180° - 30°) \div 2 - 15° = 60°$. Hence, $\triangle OBC$ is an equilateral triangle, and $PQ = BC = CO$.

Since $QPOB$ is a parallelogram, we have $\angle OPC = \angle A, \angle PCO = \angle QBO = \angle AQP$, so $\triangle AQP \cong \triangle PCO$ (AAS). Thus, $AQ = CP$.

Let H be the projection of Q on CP, then $QH = HP$.

Since $\angle A = 30°, QH \perp CP$, we have $CP = AQ = 2QH = 2HP$, so H is the midpoint of PC, and QH is the perpendicular bisector of PC. Therefore, $PQ = CQ$ and $BC = PQ = CQ$.

Example 5. In Figure 20.7, $\angle MON = 90°$, and there is a square $AOCD$ inside $\angle MON$. The points A, C lie on rays OM, ON, respectively. Let B_1 be an arbitrary point on ON, and construct a square $AB_1C_1D_1$ inside $\angle MON$.

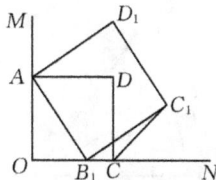

Fig. 20.7

(1) Connect CC_1, guess the size of $\angle C_1CN$, and prove your result.
(2) Choose another point B_2 on ON, and construct a square $AB_2C_2D_2$ inside $\angle MON$. Combining with the result of (1), what conclusion can you draw?

Proof:

(1) $\angle C_1CN = 45°$. The reason is explained as follows.

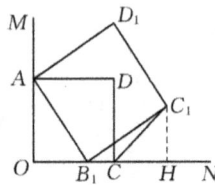

Fig. 20.8

As shown in Figure 20.8, let H be the projection of C_1 on ON. Since $AOCD, AB_1C_1D_1$ are squares, it follows that $\angle AO_1B = \angle C_1HB = 90°, AB_1 = B_1C_1$.

Also $\angle AB_1O + \angle C_1B_1H = 90°, \angle AB_1O + \angle OAB_1 = 90°$, which means $\angle C_1B_1H = \angle OAB_1$, so we have $\triangle OAB_1 \cong \triangle HB_1C_1$, and $B_1H = OA, C_1H = OB_1$.

Finally, by $OA = OC$, we have $OC = B_1H, OB_1 = CH, CH = C_1H$, so that $\angle C_1CN = 45°$.

(2) For the same reason $\angle C_2CN = 45°$. Thus, C, C_1, C_2 lie on the same line.

Remark: This example is in fact a geometric constant problem.

Example 6. In Figure 20.9, M is the midpoint of side AC of triangle ABC, P is a point on AM, and the line through P parallel to AB intersects MB, BC at X, K, respectively. If $PX = 2, XK = 3$, find the length of AB.

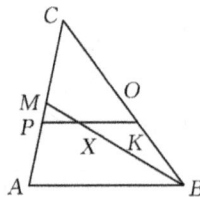

Fig. 20.9

Solution As shown in Figure 20.10, we construct a parallelogram $ABCD$. Extend PK to intersect BD at Q, and let the line through M parallel to AB intersect BC at O and intersect BD at N.

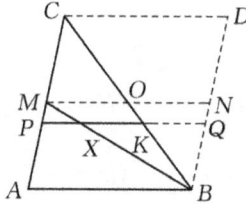

Fig. 20.10

Then, $AB = PQ = MN$. Obviously, $CO = BO$, so O is the center of the parallelogram $ABCD$, and $MO = ON$.

Thus, $KQ = KX = 3, AB = PX + XK + KQ = 2 + 3 + 3 = 8$.

Remark: Constructing parallelograms is a useful method when calculating lengths and angles.

Example 7. As shown in Figure 20.11, a rectangle $ABCD$ has side lengths of $a, b(a < b)$. Now, fold the rectangle once, where PQ is the crease (P lies on BC) so that C is folded to a point C' inside the quadrilateral $APCD$ (see Figure 20.12). The line PC' intersects the line AD at M. Then, we fold the remaining part of the rectangle so that the crease is MN, and A is folded to point A' on the PM (see Figure 20.13).

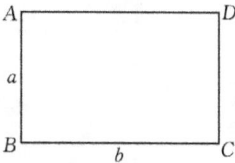

Fig. 20.11

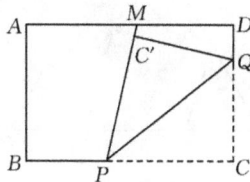

Fig. 20.12

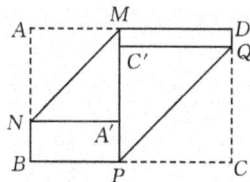

Fig. 20.13

(1) Guess the relationship between the two creases PQ, MN, and prove your result.

(2) If $\angle QPC = 45°$ (as in Figure 20.13), what is the relationship between the perimeters of quadrilaterals $MC'QD, BPA'N$ and a, b? Why?

Solution

(1) $PQ \parallel MN$.

Since $ABCD$ is a rectangle, we have $AD \parallel BC$, and since M lies on AD, it follows that $AM \parallel BC$, so $\angle AMP = \angle MPC$.

By folding, we have

$$\angle MPQ = \angle CPQ = \frac{1}{2}\angle MPC,$$

$$\angle NMP = \angle AMN = \frac{1}{2}\angle AMP.$$

Hence, $\angle MPQ = \angle NMP$, and $PQ \parallel MN$.

(2) If $\angle QPC = 45°$, then $PCQC'$ is a square, and $CQ'DM$ is a rectangle. Since $C'Q = CQ, C'Q+QD = a$, we see that the perimeter of $C'QDM$ is $2a$.

For similar reasons, the perimeter of $BPA'M$ is also $2a$. Therefore, the perimeters of the two quadrilaterals are always $2a$, and b is irrelevant.

Example 8. As shown in Figure 20.14, let $\triangle ABC$ be an equilateral triangle, and choose points A_1, A_2 on side BC, B_1, B_2 on side CA, and C_1, C_2 on side AB so that the hexagon $A_1A_2B_1B_2C_1C_2$ has six equal sides. Prove that A_1B_2, B_1C_2, C_1A_2 are concurrent.

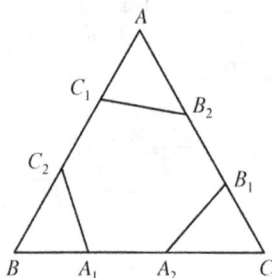

Fig. 20.14

Analysis   There are several common ideas to prove that three lines are concurrent. For example, we can prove that they are the three altitudes/angle-bisectors/medians/perpendicular bisectors of a triangle, or maybe we can prove that each pair of them is a pair diagonals of a certain quadrilateral.

Proof: As shown in Figure 20.15, we construct an equilateral triangle PA_1A_2 inside $\triangle ABC$ and connect PB_1, PB_2, PC_1, PC_2, A_1B_1, B_1C_1, C_1A_1.

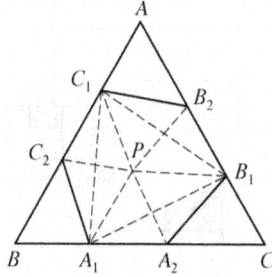

Fig. 20.15

Since $C_1C_2 \parallel PA_2, C_1C_2 = PA_1 = A_1C_2, B_1B_2 \parallel A_2P, B_1B_2 = A_2P = A_2B_1$, it follows that $PA_1C_2C_1, PB_1B_2C_1$ are both rhombuses, and $\triangle PB_2C_1$ is an equilateral triangle.

Note that

$$\angle C_1B_2B_1 = 60° + \angle PB_2B_1 = 60° + \angle PA_2B_1 = \angle A_1A_2B_1,$$

$$\angle A_1C_2C_1 + \angle A_2B_1B_2 = \angle A_1PC_1 + \angle A_2PB_2 = 360° - 60° - 60° = 240°,$$

$$\angle A_1A_2B_1 + \angle A_2B_1B_2 = (\angle A_2B_1C + \angle C) + (\angle B_1A_2C + \angle C)$$

$$= (\angle A_2B_1C + \angle C + \angle B_1A_2C) + 60°$$

$$= 180° + 60° = 240°,$$

so we have $\angle A_1C_2C_1 = \angle A_1A_2B_1$. Hence, $\angle A_1C_2C_1 = \angle A_1A_2B_1 = \angle C_1B_2B_1$.

It is easy to see that $\triangle B_1A_2A_1 \cong \triangle C_1B_2B_1 \cong \triangle A_1C_2C_1$, so $A_1B_1 = B_1C_1 = C_1A_1$.

Therefore, A_1B_2, B_1C_2, C_1A_2 are the perpendicular bisectors of the three sides of $\triangle A_1B_1C_1$, which must be concurrent.

Reading

Can we concatenate squares of different sizes to obtain a big square? Don't think this is a simple graphic puzzle problem since it is actually difficult to do. If it is possible, then the big square is called a perfect square. In other words, a perfect square refers to a big square composed of smaller squares whose sizes are different. In search of perfect squares, mathematicians have made unremitting efforts.

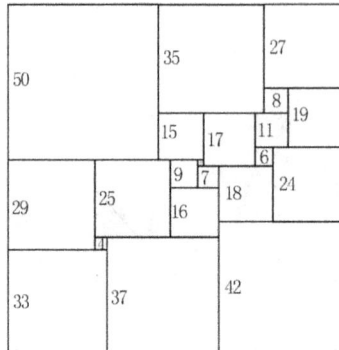

Fig. 20.16

In 1930, Soviet mathematician Luzin believed that perfect squares do not exist.

In 1939, Sprague discovered the first perfect square, which consists of 55 small squares with a side length of 4205 units.

In 1939, Brooks, Smith, Stone, and Tutte, four students from Cambridge University in the United Kingdom, were addicted to this problem and spent a long time on it. Finally, with theoretical guidance, they found a perfect square composed of 28 small squares, whose side length is 1015 units.

In 1946, Willcocks discovered a perfect square composed of 24 small squares, whose side length is 175 units.

In 1978, Dutch mathematician Duijvestijn discovered a perfect square composed of 21 small squares with the help of a computer, as shown in Figure 20.16. It is proven to be a perfect square composed of the least number of small squares.

Exercises

(I) Multiple-choice questions:

1. Which of the following conditions imply that a convex quadrilateral $ABCD$ is a parallelogram? ( ).

   ① $AB \parallel CD, AD \parallel BC$; ② $\angle A = \angle C, \angle B = \angle D$;
   ③ $AB = CD, AD \parallel BC$; ④ $AD \parallel BC, \angle A = \angle C$.

   (A) ①,②,③ (B) ②,③,④ (C) ①,②,④ (D) ①,③,④

2. As shown in the figure, the rectangle has a size of 8×5, then the area of the shaded region is ( ).

   (A) $4\frac{3}{5}$ (B) $5\frac{1}{4}$ (C) $6\frac{1}{2}$ (D) 5

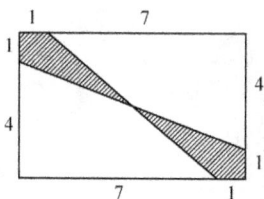

3. In the figure below, a parallelogram $ABCD$ has $\angle ABC = 72°$, $AF \perp BC$ where F is the foot, and AF intersects BD at E. If $DE = 2AB$, then $\angle AED = $ ( ).

   (A) $66°$ (B) $60°$ (C) $56°$ (D) $50°$

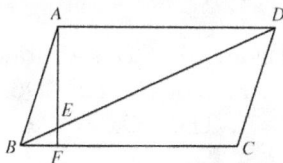

4. As shown in the figure, we divide a rectangle $ABCD$ into 15 identical squares, and E, F, G, H lie on sides AD, AB, BC, CD, respectively, where each of them is a vertex of some square. If the area of $EFGH$ is 1, then the area of $ABCD$ is ( ).

   (A) 2 (B) $\frac{4}{3}$ (C) $\frac{3}{2}$ (D) $\frac{5}{3}$

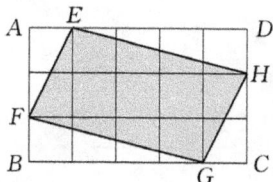

5. In the following figure below, the diagonals of a rectangle $ABCD$ intersect at O, AE bisects $\angle BAD$ and intersects BC at E, and $\angle CAE = 15°$. Then, $\angle BOE = (\quad)$.

   (A) 60°   (B) 45°   (C) 50°   (D) 75°

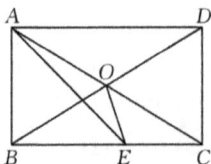

6. In the following figure, $ABCD$ is a rhombus, $AB = AE$, points E, F lie on sides BC, CD, respectively, and $\triangle AEF$ is equilateral. Then, the size of $\angle C$ is $(\quad)$.

   (A) 100°   (B) 120°   (C) 80°   (D) 60°

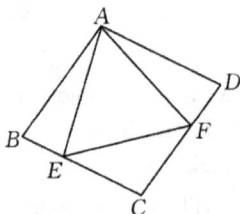

(II) Fill in the Blanks

7. In the following figure, $ABCD$ is a parallelogram, $\angle ABC = 3\angle A$, F lies on the extension of CB, $EF \perp DC$ where E is the foot, and $CF = CD, EF = 1$. Then, $DE = $ _____.

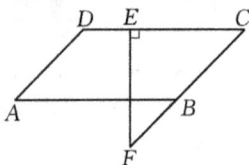

8. In the following figure, $ABCD$ is a parallelogram, where $AE \perp BC, AF \perp DC, BC : CD = 3 : 2, AB = EC$. Then, $\angle EAF = $ _____.

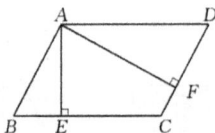

9. As shown in the following figure, five identical squares (shaded) are placed inside a big square without overlapping, and the midpoint of each side of the middle square is a vertex of another square. If the big square has a side length of 1 and the small square has a side length of $\frac{a-\sqrt{2}}{b}$, where a, b are positive integers, then $a + b =$ _____.

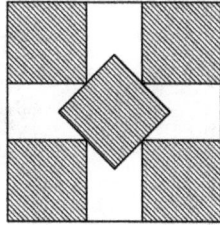

10. In the following figure, $BM = 6$ and points A, C, D lie on segments MB, NB, MN, respectively, such that $ABCD$ is a parallelogram. If $\angle NDC = \angle MDA$, then the perimeter of $ABCD$ is

_____.

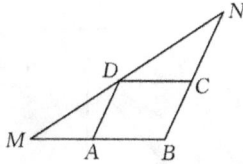

(III) Answer the following questions:

11. As shown in the figure, AK, CS, BJ, DL are the internal angle bisectors of parallelogram $ABCD$. Let E, F, G, H be their intersections. If $AB = 10, AD = 6$, find the length of EG.

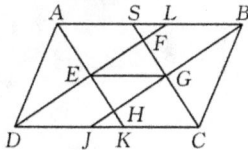

12. In the following figure, $ABCD$ is a parallelogram, E, F are midpoints of AB, CD, respectively, and $AG \parallel DB$, with G lying on the extension of CB. If $BEDF$ is a rhombus, what kind of quadrilateral is $AGBD$? Prove your result.

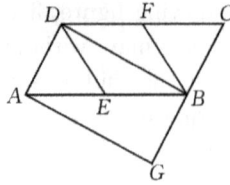

13. In the following figure, $ABCD$ is a square, Q is the midpoint of CD, and P lies on CD with $\angle BAP = 2\angle QAD$. Prove that $AP = CP + BC$.

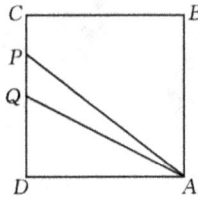

14. In the following figure, $ABCD$ is a rhombus, $AB = 4a$, E lies on BC such that $EC = 2a$, and $\angle BAD = 120°$. If P is a point on BD, find the minimum of $PE + PC$.

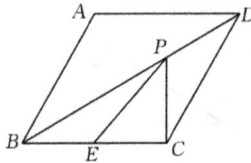

15. In a rectangle $ABCD$, points M, N, P, Q lie on sides AB, BC, CD, DA, respectively. Suppose $S_{\triangle AQM} = S_{\triangle BMN} = S_{\triangle CNP} = S_{\triangle DPQ}$. Prove that $MNPQ$ is a parallelogram.

Chapter 21

Trapezoids

A trapezoid is a quadrilateral where one pair of opposite sides is parallel while the other pair is not. Be sure to pay attention to the restriction that "the other pair of sides is not parallel"; otherwise, parallelograms could be mistaken as a special type of trapezoids.

A trapezoid is an isosceles trapezoid if its two legs are equal, two internal angles on the same base are equal, or the two diagonals are equal.

When solving a problem with trapezoids, auxiliary lines are often needed to reduce it to a problem of triangles and parallelograms. Figures 21.1–21.6 show six common ways to add auxiliary lines.

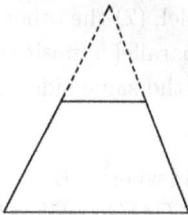

Fig. 21.1

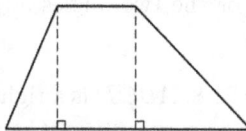

Fig. 21.2

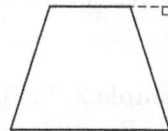

Fig. 21.3

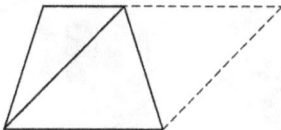

Fig. 21.4

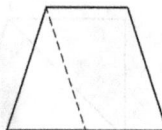

Fig. 21.5

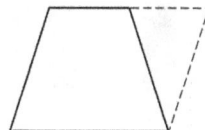

Fig. 21.6

Example 1. In Figure 21.7, $\triangle ABC$ is a right triangle, and E is the midpoint of the hypotenuse AB. D lies on AC such that $DE \parallel BC$, and F

lies on the extension of BC such that $DF \parallel EC$. Prove that $EBFD$ is an isosceles trapezoid.

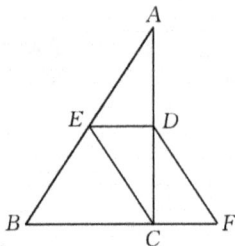

Fig. 21.7

Proof: Since $DE \parallel CF$ and $DF \parallel EC$, the quadrilateral $ECFD$ is a parallelogram, so $DF = EC$.

Since EC is the median on the hypotenuse of the right triangle ABC, we have $EC = \frac{1}{2}AB = EB$, so that $DF = EC = EB$.

Now, $DF \parallel EC$, and EC intersects EB at E. This means EB is not parallel to DF because, otherwise, there are two lines through E and parallel to DF, which is impossible.

Therefore, $EBFD$ is an isosceles trapezoid.

Remark: To prove that a quadrilateral is an isosceles trapezoid, we need three conditions: (1) a pair of opposite sides are parallel; (2) the other pair of opposite sides are not parallel; (3) the pair of non-parallel opposite sides have the same length, or the two angles adjacent to the same side in the parallel pair are equal.

Example 2. In Figure 21.8, $ABCD$ is a right trapezoid where $\angle B = \angle C = 90°$, $AB = BC$. Let M be a point on BC such that $\angle DMC = 45°$. Prove that $AD = AM$.

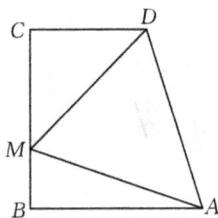

Fig. 21.8

Proof: Let E be the projection of A on the line CD, as in Figure 21.9.

Since $\angle B = \angle C = \angle E = 90°$ and $AB = BC$, we see that $ABCE$ is a square, so $AE = AB = CE$.

Since $\angle DMC = 45°$, we have $CM = CD$, so that $BM = DE$. Hence, $\text{Rt}\triangle ABM \cong \text{Rt}\triangle AED$, and $AM = AD$.

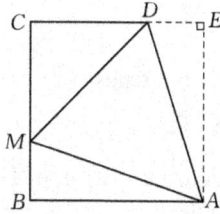

Fig. 21.9

Remark: We can also construct the projection of D on AB as H and prove $\text{Rt}\triangle ABM \cong \text{Rt}\triangle DHA$.

Example 3. In Figure 21.10, $ABCD$ is a trapezoid with $AD \parallel BC$. Suppose $AD + BC = 3, AC = \sqrt{3}, BD = \sqrt{6}$. Find the area of $ABCD$.

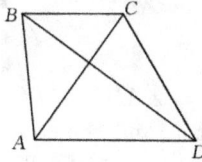

Fig. 21.10

Solution Extend AD to E so that $DE = BC$, and connect CE, as shown in Figure 21.11. Then, $CE = BD$.

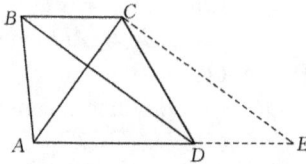

Fig. 21.11

Since $AC = \sqrt{3}$, $CE = BD = \sqrt{6}$, $AD + BC = 3$, we have $AE^2 = 9 = AC^2 + CE^2$, and thus $\angle ACE = 90°$. In other words, $AC \perp CE$ and $AC \perp BD$.

Therefore, the area of the trapezoid is equal to

$$\frac{1}{2}AC \times BD = \frac{1}{2} \times \sqrt{3} \times \sqrt{6} = \frac{3\sqrt{2}}{2}.$$

Remark: The area of a trapezoid whose diagonals are perpendicular equals half the product of two diagonals.

Example 4. As shown in Figure 21.12, $ABCD$ is a trapezoid, where $BC \parallel AD$, CA bisects $\angle BCD$, O is the intersection of the diagonals, $CD = AO$, $BC = OD$. Find the size of $\angle ABC$.

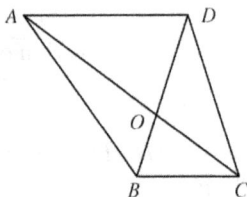

Fig. 21.12

Analysis We have $\angle ABC = \angle ABD + \angle CBD$. The conditions include parallel lines, an angle bisector, and the equality of segments. These give us the relationship among angles, which allows us to establish equations and solve for the desired angle.

Solution Let $\angle OCD = \alpha$, $\angle ADO = \beta$.

Since CA bisects $\angle BCD$, we have $\angle OCD = \angle OCB = \alpha$.

By $BC \parallel AD$, we have $\angle ADO = \angle OBC = \beta$, $\angle DAO = \angle OCB = \alpha$.

Thus, $\angle OCD = \angle DAO = \alpha$, and $AD = CD$.

Also, we have $CD = AO$, so $AD = AO$, and $\angle ADO = \angle AOD = \angle BOC = \angle OBC = \beta$. Hence, $OC = BC$.

As $BC = OD$, we get $OC = OD$, which means $\angle ODC = \angle OCD = \alpha$.

Since $\angle BOC = \angle ODC + \angle OCD$, $\angle BOC + \angle OBC + \angle OCB = 180°$, we have $\begin{cases} \beta = 2\alpha, \\ \alpha + 2\beta = 180°, \end{cases}$ which implies $\alpha = 36°$, $\beta = 72°$.

Thus, $\angle DBC = \angle BCD = 72°, BD = CD = AD$, and
$$\angle ABD = \angle BAD = \frac{180° - \beta}{2} = 54°.$$
Therefore, $\angle ABC = \angle ABD + \angle DBC = 126°$.

Example 5. In Figure 21.13, $AB \parallel CD, AD = BC, BD = DC, AC \perp BD$ where M is the foot of perpendicularity. Prove that $CM = \frac{1}{2}(AB + CD)$.

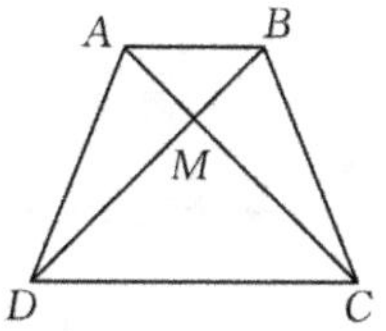

Fig. 21.13

Proof: Let E be a point on the extension of DC such that $BE \parallel AC$, and let F be the projection of B on DC, as shown in Figure 21.14.

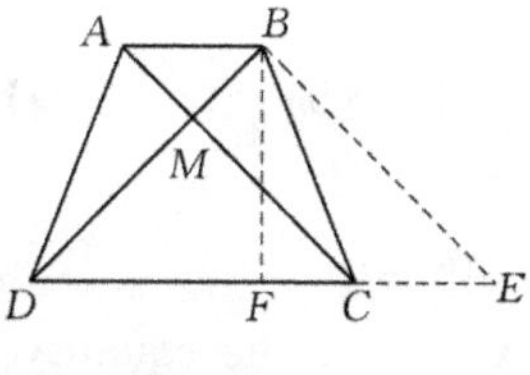

Fig. 21.14

Since $AB \parallel CD$, the quadrilateral $ABEC$ is a parallelogram, and $AB = CE, BE = AC$.

Since $AD = BC$, we have $BD = AC = BE$, and consequently $DF = FE$.

Since $AC \perp BD, BE \parallel AC$, we have $BD \parallel BE$, and
$$BF = \frac{1}{2}DE = \frac{1}{2}(DC + CE) = \frac{1}{2}(AB + DC).$$
Note that in $\triangle DBF$ and $\triangle DCM$, $\angle BFD = \angle CMD = 90°$, $\angle BFD$ is a shared angle, and $BD = DC$, so it follows that $\triangle DBF \cong \triangle DCM$. Therefore, $CM = BF = \frac{1}{2}(AB + DC)$.

Example 6. In Figure 21.15, $ABCD$ is a trapezoid with $AB \parallel CD$. Let $S_{ABCD} = S, S_{\triangle AOB} = S_1, S_{\triangle DOC} = S_2$. Try to determine the relationship between $\sqrt{S_1} + \sqrt{S_2}$ and $\sqrt{S}$, and explain.

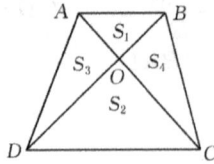

Fig. 21.15

Solution Let $S_{\triangle AOD} = S_3, S_{\triangle BOC} = S_4$, then

$$S = S_1 + S_2 + S_3 + S_4.$$

Since $\frac{S_1}{S_3} = \frac{OB}{OD}, \frac{S_4}{S_2} = \frac{OB}{OD}$, we have $\frac{S_1}{S_3} = \frac{S_4}{S_2}$, or, equivalently, $S_1 S_2 = S_3 S_4$.

On the other hand, since $AB \parallel CD$, we have $S_{\triangle ACD} = S_{\triangle BCD}$, which means $S_2 + S_3 = S_2 + S_4$, and $S_3 = S_4$. Thus,

$$(\sqrt{S_1} + \sqrt{S_2})^2 - (\sqrt{S})^2 = S_1 + S_2 + 2\sqrt{S_1 S_2} - (S_1 + S_2 + S_3 + S_4)$$
$$= 2\sqrt{S_1 S_2} - (S_3 + S_4),$$

where $\sqrt{S_1 S_2} = S_3 = S_4$, so $2\sqrt{S_1 S_2} - (S_3 + S_4) = 0$.

Therefore, $\sqrt{S_1} + \sqrt{S_2} = \sqrt{S}$.

Remark: In this problem if $AB \parallel CD$ is omitted, then the conclusion would be $\sqrt{S_1} + \sqrt{S_2} \le \sqrt{S}$, and the equality is attained if and only if $AB \parallel CD$.

Reading

Qin Jiushao, an ancient Chinese mathematician, raised the following question in his book, *The Mathematical Treatise in Nine Sections*.

Three brothers together own a trapezoid field. The south base of the field is 34 steps long, the north base is 52 steps long, and the distance between them is 150 steps. It is required to divide this field into three equal parts. Then, what are the sizes of each part?

Using the area formula of the trapezoid, we can find that the area of this field is $(34 + 52) \times 150 \div 2 = 6450\,(\text{step}^2)$, so everyone owns $6450 \div 3 = 2150\,(\text{step}^2)$.

As shown in Figure 21.16, we can find the lengths of the relevant segments by solving this trapezoid, but the calculation is very complicated. Interested students may try to calculate with the help of a calculator, and

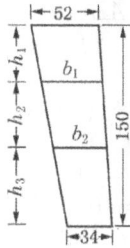

Fig. 21.16

the answers are

$$h_1 = 43\frac{448886027045}{843370901905}, \quad b_1 = 46\frac{6567454825283}{8482689572651},$$

$$h_2 = 49\frac{26276319}{412406309}, \quad b_2 = 40\frac{52284}{58709}, \quad h_3 = 57\frac{853}{2043}.$$

Exercises

(I) Multiple-choice questions:

1. As shown in the figure, the quadrilateral $ABCD$ can be divided into four identical isosceles trapezoids. Then, the size of $\angle A$ is ( ).

(A) 30°   (B) 40°   (C) 50°   (D) 60°

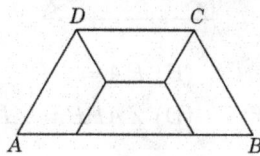

2. In the following figure, $ABCD$ is a trapezoid with $AD \parallel BC$, $AB = 3, BC = 4, CD = 2, AD = 1$. Then, the area of the trapezoid is ( ).

(A) $\frac{10\sqrt{2}}{3}$ (B) $\frac{10\sqrt{3}}{3}$ (C) $3\sqrt{2}$ (D) $3\sqrt{3}$

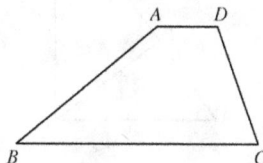

3. In a right trapezoid $ABCD$, the bases $AB = 13, CD = 8$, $AD \perp AB$, and $AD = 12$. Then, the distance from A to BC is ( ).

    (A) 12   (B) 13   (C) $\frac{12 \times 21}{13}$ (D) 10.5

4. In the following figure, $ABCD$ is an isosceles trapezoid with $AB \parallel CD, \angle ABC = 60°$, and $AB = 20, AC \perp BD$. Then, the perimeter of $ABCD$ is ( ).

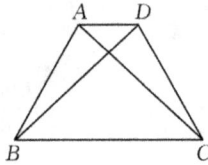

    (A) 100   (B) $50\sqrt{3}$ (C) $40 + 20\sqrt{3}$ (D) $60\sqrt{3}$

5. In the following figure, $ABCD$ is a right trapezoid with $AB \parallel CD, AD \perp AB$. If P is a point on AD, then the minimum of $PB + PC$ is attained when ( ).

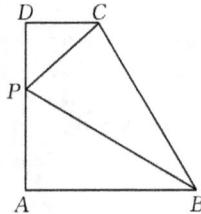

    (A) $PB = PC$ (B) $PA = PD$
    (C) $\angle BPC = 60°$ (D) $\angle APB = \angle DPC$

(II) Fill in the blanks:

6. As shown in the figure, $ABCD$ is a trapezoid, $\angle DCB = 90°, AB \parallel CD, AB = 25, BC = 24$. Now, fold the trapezoid along the crease BE so that A coincides with D, then the length of AD is _____.

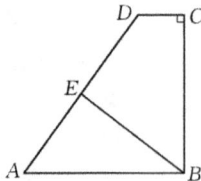

7. In the following figure, $ABCD$ is a trapezoid, $AD \parallel BC, AD = 2, AC = 4, BC = 6, BD = 8$. Then, the area of $ABCD$ is _____ .

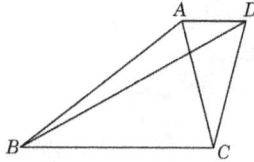

8. We have a rectangle aluminum plate of size $2.1\,\text{m} \times 1.6\,\text{m}$, from which we would like to cut several congruent right trapezoids whose size is shown in the following figure (the unit is mm). Then, we can cut at most _____ copies.

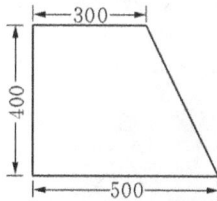

9. In the following figure, $ABCD$ is a trapezoid with $AD \parallel BC$, and AC, BD intersect at M. Suppose $AB = AC, AB \perp AC, BC = BD$, then the size of $\angle AMB$ is _____ .

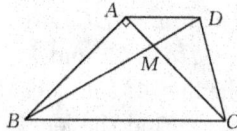

10. In the following figure, $ABCD$ is a trapezoid with $AD \parallel BC, \angle A = 90°$. If E is a point on AB with $AE = 42, BE = 28, BC = 70, \angle DCE = 45°$, then $DE = $ _____ .

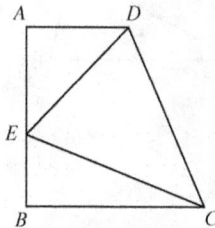

(III) Answer the following questions:

11. In the following figure, $ABCD$ is a trapezoid with $AD \parallel BC, BC = DC$, CF bisects $\angle BCD$, $DF \parallel AB$, and E is the intersection of BF and the extension of DC. Prove that
    (1) $\triangle BFC \cong \triangle DFC$;   (2) $AD = DE$.

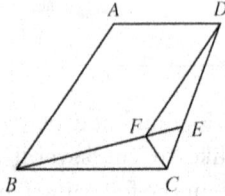

12. In the following figure, $ABCD$ is a trapezoid, where $AB \parallel CD$, $AC \perp BD$. Prove:

$$AC^2 + BD^2 = (AB + DC)^2.$$

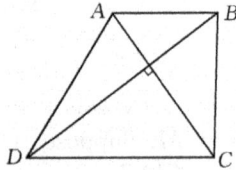

13. In the following figure, $ABCD$ is a trapezoid with $AB \parallel CD, AB = a, CD = b\,(a > b)$, and M is a point on the extension of DC such that the line AM divides the trapezoid into two parts with equal perimeter. Find the length of CM.

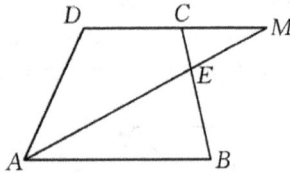

14. In the following figure, $ABCD$ is a trapezoid, where $\angle A + \angle B = 90°$, $AB \parallel CD$, and M, N are the midpoints of AB, CD, respectively. Prove that $MN = \frac{1}{2}(AB - CD)$.

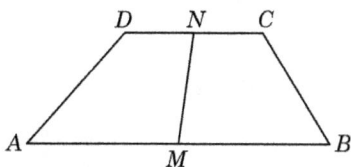

Chapter 22

The Angles and Diagonals of a Polygon

Polygons include convex and concave polygons. However, in middle school, we usually consider only convex polygons.

The sum of the internal angles of an n-gon is $(n-2) \times 180°$, and the sum of its external angles is $360°$. If all the internal angles of an n-gon are equal, then they are all equal to $\frac{(n-2) \times 180°}{n}$, in which case every external angle is $\frac{360°}{n}$.

In an n-gon, the segment connecting two non-adjacent vertices is called a diagonal. An n-gon has $\frac{n(n-3)}{2}$ diagonals.

Common problems include finding the internal (or external) angles, the number of sides, or the number of diagonals of a polygon, and in these problems, we can use the sum of the internal angles of a triangle as well as knowledge regarding equations and inequalities.

Example 1. As shown in Figure 22.1, if $\angle A + \angle B + \angle C + \angle D + \angle E + \angle F + \angle G = n \times 90°$, then $n =$ _____.

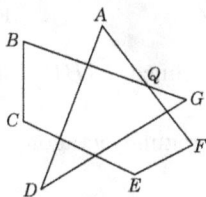

Fig. 22.1

Solution   Let Q be the intersection of AF, BG, then

$$\angle AQG = \angle A + \angle D + \angle G.$$

Thus,

$$\angle A + \angle B + \angle C + \angle D + \angle E + \angle F + \angle G$$

$$= \angle B + \angle C + \angle E + \angle F + \angle AQG$$

$$= \angle B + \angle C + \angle E + \angle F + \angle BQF = 540° = 6 \times 90°.$$

Therefore, $n = 6$.

Remark: Figure 22.1 is an example of a star polygon. In such problems, the common idea is to reduce the desired angles to the sum of the internal angles of triangles or quadrilaterals. Finding correct triangles/ quadrilaterals will bring more convenience.

Example 2. In Figure 22, $ABCDEFGH$ is a regular octagon and $ABIJ$ is a square. Suppose $AB = 1$. Find the length of CJ.

Solution   As shown in Figure 22, connect JB and consider $\triangle JBC$.
Since $AB = 1$, we have $BC = 1$.
Since $ABCD$ is a square, it follows that

$$JB = \sqrt{AJ^2 + AB^2} = \sqrt{2}.$$

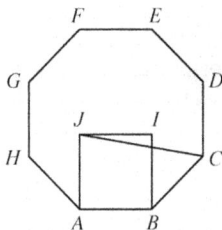 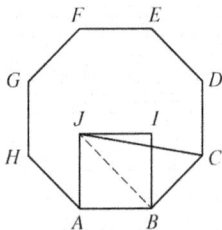

Fig. 22.2   Fig. 22.3

We prove that $JB \perp BC$. Since $\angle JBI = 45°$, it suffices to show that $\angle IBC = 45°$.

Since $ABCDEFGH$ is a regular octagon, the sum of its internal angles is $(8 - 2) \times 180° = 1080°$.

Thus, $\angle ABC = \frac{1080°}{8} = 135°$, and $\angle IBC = 135° - 90° = 45°$. Hence, $\angle JBC = 90°$, and $\triangle JBC$ is a right triangle. Therefore, $JC = \sqrt{JB^2 + BC^2} = \sqrt{3}$.

Example 3. Suppose a regular n-gon has $n + 3$ diagonals. Its perimeter is x, and the sum of its diagonals is y. Find the value of $\frac{y}{x}$.

Solution We have $\frac{n(n-3)}{2} = n + 3$ so that $n^2 - 5n - 6 = 0$, and $n = 6$ since $n > 0$.

Let the regular hexagon have a side length of 1, as shown in Figure 22.4.

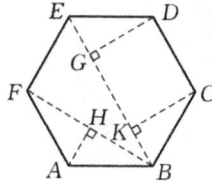

Fig. 22.4

The diagonals of this polygon can have two lengths, either BE or BF.

Let H be the projection of A on BF, then $\triangle ABF$ is an isosceles triangle with a vertex angle of 120°. Thus, $\angle ABH = 30°, AH = \frac{1}{2}AB = \frac{1}{2}$, and $BH = \sqrt{AB^2 - AH^2} = \sqrt{1^2 - \left(\frac{1}{2}\right)^2} = \frac{\sqrt{3}}{2}, BF = \sqrt{3}$.

It is easy to see that $BCDE$ is an isosceles trapezoid, so we have $BK = EG = \frac{1}{2}, BE = 2 \times \frac{1}{2} + 1 = 2$.

In the hexagon, six diagonals have the same length as BF, while three diagonals have the same length as BE. Therefore, $\frac{y}{x} = \frac{6\sqrt{3}+6}{6} = 1 + \sqrt{3}$.

Example 4. Suppose $ABCDE$ is a regular pentagon and O is a point on the plane such that $\triangle DOE$ is an equilateral triangle. Find the size of $\angle AOC$.

Solution If O lies inside the pentagon, as shown in Figure 22.5, then since $\triangle DOE$ is equilateral, we have $\angle AEO = 108° - 60° = 48°$, and $OE = DE = AE$, so $\angle 2 = \frac{180° - 48°}{2} = 66°$.

Similarly, $\angle 1 = 66°$. Combining with $\angle DOE = 60°$, we have

$$\angle AOC = 360° - (\angle 2 + \angle EOD + \angle 1) = 360° - (66° + 60° + 66°) = 168°.$$

If O lies outside the pentagon, as shown in Figure 22.6, then we have

$$\angle COD = \angle AOE = \frac{1}{2}[180° - (108° + 60°)] = 6°.$$

Thus, $\angle AOC = \angle DOE - \angle AOE - \angle COD = 60° - 6° - 6° = 48°$.

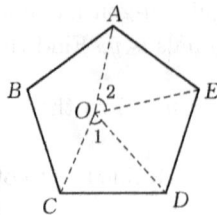

Fig. 22.5

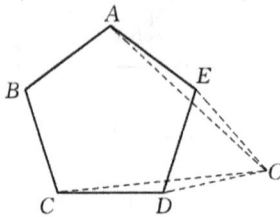

Fig. 22.6

Example 5. In Figure 22.7, we translate the lines of each side of a pentagon $ABCDE$ outward (in the direction perpendicular to that side) by 4 units, and obtain a new pentagon $A'B'C'D'E'$.

(1) There are five shaded regions $AHA'G, BFB'P, COC'N, DMD'L$, $EKE'I$. Do they compose a pentagon by concatenation? Explain.
(2) Prove that the perimeter of $A'B'C'D'E'$ is greater than the perimeter of $ABCDE$ by at least 25 units.

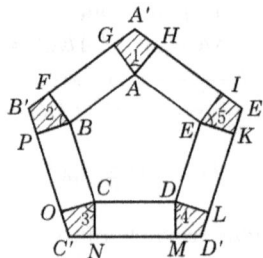

Fig. 22.7

Analysis If we can concatenate the five shaded regions to obtain a pentagon, then the following conditions need to be satisfied: the five tuples $A'GB', B'PC', C'ND', D'LE', E'IA'$ are all collinear, and $\angle 1 + \angle 2 + \angle 3 + \angle 4 + \angle 5 = 360°$.

(2) The increased perimeter is exactly $A'H + A'G + B'F + B'P + C'O + C'N + D'M + D'L + E'K + E'I$. We estimate this length using the circumference of a circle.

Solution (1) The five shaded regions compose a pentagon.

We have $BF = AG = AH = EI = EK = DL = DM = CN = CO = BP = 4$, and

$$\angle BFB' = \angle AGA' = \angle COC' = \angle BPB' = \angle DMD' = \angle CNC'$$

$$= \angle EKE' = \angle DLD' = \angle AHA' = \angle EIE' = 90°.$$

Moreover, $(A' + \angle B' + \angle C' + \angle D' + \angle E') + (\angle 1 + \angle 2 + \angle 3 + \angle 4 + \angle 5) = 5 \times 180°$, while $A' + \angle B' + \angle C' + \angle D' + \angle E' = (5-2) \times 180° = 3 \times 180°$, so $\angle 1 + \angle 2 + \angle 3 + \angle 4 + \angle 5 = 360°$. Therefore, the five pieces can be put together to obtain a pentagon.

(2) The increased perimeter equals $A'H + A'G + B'F + B'P + C'O + C'N + D'M + D'L + E'K + E'I$, which is exactly the perimeter of the pentagon obtained by concatenating the five shaded regions. This pentagon has an incircle whose radius is 4, so its perimeter is greater than the circumference of the incircle, which is 8π. Therefore,

$$S > 8\pi > 8 \times 3.14 = 25.12 > 25.$$

Example 6.

(1) Find the number of diagonals of a convex decagon as well as the number of triangles whose vertices are all vertices of the given decagon.

(2) We write a natural number at each vertex of a convex decagon, and for each triangle discussed in (1), we call it an odd triangle if the sum of numbers on its vertices is odd or an even triangle if the sum of numbers on its vertices is even. Is the number of odd triangles odd or even? Prove your result.

Solution (1) A convex decagon has $10 \times (10 - 3) \div 2 = 35$ diagonals. Since the total number of sides and diagonals is 45, each belonging to 8 triangles, while every triangle has 3 sides, we conclude that the number of such triangles is $\frac{45 \times 8}{3} = 120$.

(2) The number of odd triangles is even.

Note that every vertex of the decagon belongs to 36 triangles, so the number on every vertex is counted 36 times in the total sum of sums in all the triangles. Hence, calculating the total sum would give 36 times the sum of numbers written on all vertices, which must be even.

If the number of odd triangles is odd, then the sum of sums in these triangles is still odd (since an odd sum of odd numbers is still odd), while the sum of sums in even triangles is always even. Since odd + even = odd, the total sum should also be odd, which results in a contradiction.

Reading

How many different ways are there to divide a convex n-gon with unequal sides into triangles with its diagonals? This question was proposed by the great mathematician Euler in 1751.

Of course, it is quite difficult to solve this problem. As shown in Figure 22.8, we can first see how many divisions there are when $n = 3, 4, 5$. (Here, E_n is the number of different ways to divide.)

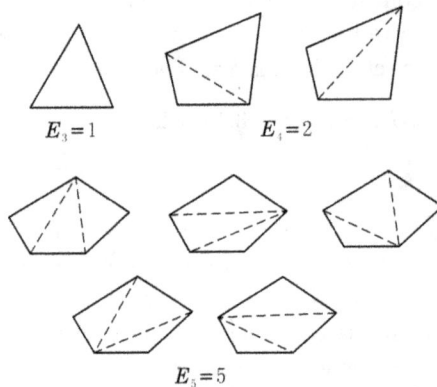

$E_3 = 1$ $E_4 = 2$

$E_5 = 5$

Fig. 22.8

Euler himself found the answer to some special cases, for example: $E_3 = 1, E_4 = 2, E_5 = 5, E_6 = 14, E_7 = 42, E_8 = 132, E_9 = 429$.

In 1758, the following formula was found as a method to calculate E_n:

$$E_n = E_2 E_{n-1} + E_3 E_{n-2} + \cdots + E_{n-2} E_3 + E_{n-1} E_2 (n \geq 3),$$

where $E_2 = 1$.

For example, we can calculate the number of ways to divide a convex decagon with unequal sides into triangles using the formula above:

$$E_{10} = E_2 E_9 + E_3 E_8 + E_4 E_7 + E_5 E_6 + E_6 E_5 + E_7 E_4 + E_8 E_3 + E_9 E_2$$

$$= 1 \times 429 + 1 \times 132 + 2 \times 42 + 5 \times 14 + 14 \times 5 + 42 \times 2$$

$$+ 132 \times 1 + 429 \times 1 = 1430.$$

Exercises

(I) Multiple-choice questions:

1. In the following figure, $\angle A + \angle B + \angle C + \angle D + \angle E + \angle F + \angle G$ equals (   ).

   (A) 360°   (B) 450°   (C) 540°   (D) 720°

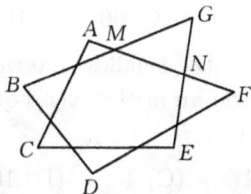

2. As shown in the figure, we fold a hexagon $ABCDEF$, so that A, B are folded to $A'B'$, which lie inside the hexagon $CDEFGH$. Let $\angle C + \angle D + \angle E + \angle F = \alpha$ then which of the following is true? (   ).

   (A) $\angle 1 + \angle 2 = 900° - 2\alpha$ (B) $\angle 1 + \angle 2 = 1080° - 2\alpha$
   (C) $\angle 1 + \angle 2 = 720° - \alpha$ (D) $\angle 1 + \angle 2 = 360° - \frac{1}{2}\alpha$

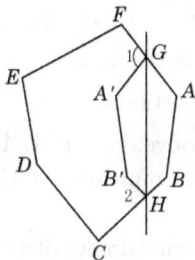

3. We put a square, a regular pentagon, and a regular hexagon of equal side length together, as shown in the following figure. Then, $\angle 1 = (\quad)$.

(A) $32°$ 　　 (B) $36°$ 　　 (C) $40°$ 　　 (D) $42°$

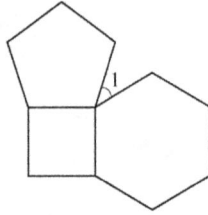

4. If the sum of all internal angles and external angles of a polygon is $2160°$, then the number of diagonals of this polygon is (　).

(A) 54 　　 (B) 65 　　 (C) 60 　　 (D) 55

5. In a convex n-gon, the smallest internal angle is $95°$ and all its internal angles form an arithmetic sequence with a common difference of $10°$. Then, $n = (\quad)$.

(A) 6 　　 (B) 12 　　 (C) 4 　　 (D) 10

(II) Fill in the blanks:

6. If the sum of internal angles of an $(n+x)$-gon is greater than the sum of internal angles of an n-gon by $720°$, then $x = $ _____ .

7. A convex 2000-gon has at most _____ acute internal angles.

8. In a convex n-gon, if the sum of all internal angles except for one equals $8940°$, then the excluded angle equals _____ .

9. In a convex hexagon $ABCDEF$, $\angle A = \angle B = \angle C = \angle D = \angle E = \angle F$, $EF = 1, ED = 5$, then $AB + BC = $ _____ .

10. If a convex n-gon has exactly four obtuse internal angles, then the maximum of n is _____ .

(III) Answer the following questions:

11. Suppose that in a convex n-gon $A_1 A_2 \ldots A_n (n > 4)$, all the internal angles are integer multiples of $15°$, and $\angle A_1 + \angle A_2 + \angle A_3 = 285°$, find n.

12. Suppose that all internal angles of a convex n-gon are equal, which is an odd number of degrees. How many polygons are there (up to the value of n)? Prove your result.

13. We put black dots on the sides of regular polygons, as shown in the following figure. With the same pattern, how many dots do we put on the nth polygon?

 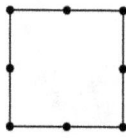  

Fig. E13.1   Fig. E13.2   Fig. E13.3   Fig. E13.4

14. The following figures are obtained by connecting $1, 2, n$ squares, respectively. In Figure E14.1, $x = 70°$. In Figure E14.2, $y = 28°$. By calculation and analogy with the first two cases, give the relationship between $a + b + c + \cdots + d$ and n in Figure E14.3.

_____ .

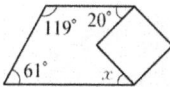

Fig. E14.1

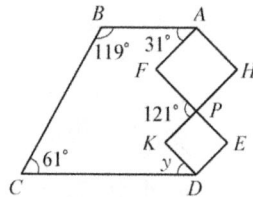

Fig. E14.2

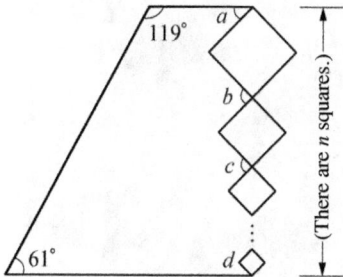

Fig. E14.3

Chapter 23

Proportion of Segments

In this chapter, we deal with algebraic proportions and proportions of segments, where the latter is based on the former, as well as the arithmetic relationship of segments given by geometric structures.

If $\frac{a}{b} = \frac{c}{d}$, then $ad = bc$. This is the fundamental property of proportions, and it is the foundation of all other properties and transformations of proportions.

From $\frac{a}{b} = \frac{c}{d}$, it follows that $\frac{a \pm b}{b} = \frac{c \pm d}{d}, \frac{a}{b \pm a} = \frac{c}{d \pm c}$, which is called the property of sum and difference proportions.

If $\frac{a}{b} = \frac{c}{d} = \frac{e}{f} = \cdots$, then $\frac{a+c+e+\cdots}{b+d+f+\cdots} = \frac{a}{b}$, which is called the property of equal proportions.

In Figure 23.1, if $l_1 \parallel l_2 \parallel l_3$, then $\frac{AB}{BC} = \frac{DE}{EF}$. Combining with the property of proportions, we can also obtain many other proportional relations.

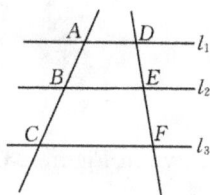
Fig. 23.1

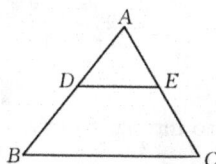
Fig. 23.2

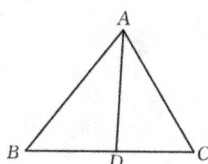
Fig. 23.3

Figure 23.2 can be regarded as a special case of Figure 23.1. In this case, $DE \parallel BC$, and we have

$$\frac{AD}{AB} = \frac{AE}{AC} = \frac{DE}{BC}.$$

In Figure 23.3, if AD is an angle bisector of $\triangle ABC$, then $\frac{BD}{CD} = \frac{AB}{AC}$.

For the three propositions above, their converses are also true.

Problems with proportions often involve solving equations, while for geometric proportions, it is common to construct parallel lines.

Example 1. Suppose x, y, z are real numbers such that $\frac{2}{x} = \frac{3}{y-z} = \frac{5}{x+5}$, then $\frac{5x-y}{y+2z} = ($ $)$.

(A) 1   (B) $\frac{1}{3}$ (C) $-\frac{1}{3}$ (D) $\frac{1}{2}$

Solution From $\frac{2}{x} = \frac{3}{y-z} = \frac{5}{x+5}$, we can derive that $y = 3x, z = \frac{3}{2}x$. Thus, $\frac{5x-y}{y+2z} = \frac{5x-3x}{3x+3x} = \frac{1}{3}$, and the answer is (B).

Remark: The given condition is a continued equality, so we can use the property of equal proportions to establish equations and find the relationship among variables.

Example 2. In Figure 23.4, $ABCD$ is a parallelogram, E, F are points on AB, AD, respectively, and EF intersects AC at P. If $\frac{AE}{EB} = \frac{a}{b}, \frac{AF}{FD} = \frac{m}{n}$ (a, b, m, n are all positive), then $\frac{AP}{PC} = ($ $)$.

(A) $\frac{am}{an+bm}$ (B) $\frac{bn}{an+bm}$ (C) $\frac{am}{am+an+bm}$ (D) $\frac{bn}{an+bm+bn}$

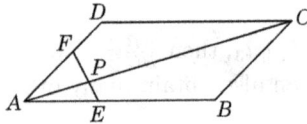

Fig. 23.4

Solution Extend CB, FE to intersect at G, as shown in Figure 23.5.

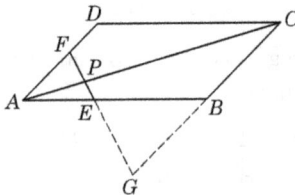

Fig. 23.5

Since $AD \parallel CG$, we have $\frac{BG}{AF} = \frac{EB}{AE} = \frac{b}{a}$, so

$$\frac{CG - BC}{AF} = \frac{b}{a}, \quad \frac{CG}{AF} - \frac{BC}{AF} = \frac{b}{a}.$$

On the other hand, $\frac{AF}{FD} = \frac{m}{n}$, so $\frac{FD}{AF} = \frac{n}{m}$. Thus, $\frac{AD}{AF} = \frac{m+n}{m}$.

Since $BC = AD$, we have $\frac{BC}{AF} = \frac{m+n}{m}$. Therefore, $\frac{CG}{AF} = \frac{m+n}{m} + \frac{b}{a} = \frac{am+an+bm}{am}$.

Since $AD \parallel CG$, it follows that $\frac{AP}{PC} = \frac{AF}{GC}$, so $\frac{AP}{PC} = \frac{am}{am+an+bm}$, and the answer is (C).

Remark: When using the theorem that parallel lines cut lines into proportional segments, we need to proficiently make use of the property of sum/difference proportion to obtain the desired quantity or proportions based on the figure and the objective of the problem.

Example 3. In Figure 23.6, $AB = AC, \angle BAC = 90°$, BD is the median of $\triangle ABC$ on side AC, $AE \perp BD$, where E is their intersection. Prove that $BE = 2EC$.

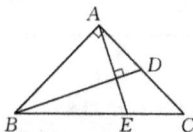

Fig. 23.6

Proof: Extend BA to F such that $AF = AD$, and connect FC, as shown in Figure 23.7.

We have $AB = AC, \angle BAD = \angle CAF = 90°, AF = AD$, so $\triangle ABD \cong \triangle ACF$.

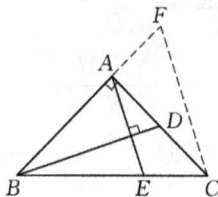

Fig. 23.7

Thus, $\angle ABD = \angle ACF$. Since $\angle BAC = 90°$, $AE \perp BD$, we have $\angle ABD = \angle DAE$.

This implies $\angle ACF = \angle EAC$, and $AE \parallel FC$. Hence, $\frac{BE}{EC} = \frac{BA}{AF}$.

Since BD is a median, it follows that $AB = 2AD = 2AF$, so $BE = 2EC$.

Remark: Constructing parallel lines is a very common tool to prove the relationship between segments.

Example 4. In Figure 23.8, P is a point on the midsegment of $\triangle ABC$ parallel to BC. The lines BP, CP intersect sides AC, AB at D, E, respectively. Prove that

$$\frac{AD}{DC} + \frac{AE}{EB} = 1.$$

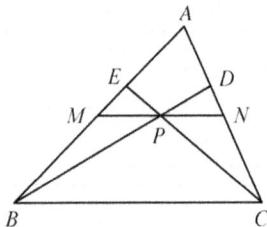

Fig. 23.8

Proof: We have

$$\frac{AD}{DC} = \frac{2CN - DC}{DC} = \frac{2CN}{DC} - 1 = \frac{2(DC - DN)}{DC} - 1$$

$$= 1 - \frac{2DN}{DC} = 1 - \frac{2PN}{BC},$$

$$\frac{AE}{EB} = \frac{2BM - EB}{EB} = \frac{2BM}{EB} - 1 = \frac{2(EB - BM)}{EB} - 1$$

$$= 1 - \frac{2EM}{EB} = 1 - \frac{2PM}{BC}.$$

Thus, $\frac{AD}{DC} + \frac{AE}{EB} = \left(1 - \frac{2PN}{BC}\right) + \left(1 - \frac{2PM}{BC}\right) = 2 - 2\left(\frac{PN}{BC} + \frac{PM}{BC}\right) = 2 - \frac{2MN}{BC}$.

Since MN is a midsegment, we have $MN = \frac{1}{2}BC$. Therefore,

$$\frac{AD}{DC} + \frac{AE}{EB} = 2 - 2 \times \frac{1}{2} = 1.$$

Example 5. In Figure 23.9, AD is a median of $\triangle ABC$. Let F be any point on CD, and EG is the line through F parallel to AB, such that G lies on AC and E lies on the extension of AD.

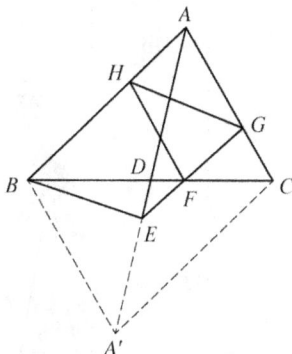

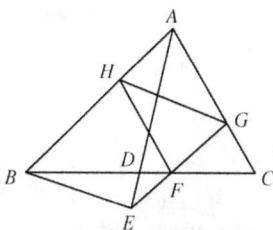

Fig. 23.9   Fig. 23.10

Let $FH \parallel AC$ and intersect AB at H. Prove that $HG = BE$.

Proof: Extend AD to A' such that $DA' = AD$, and connect $A'B$, $A'C$, as shown in Figure 23.10.

Since $BD = CD$, $ABA'C$ is a parallelogram, so $A'C \parallel AB$, $A'C = AB$. Since $EG \parallel AB$, $EG \parallel A'C$, we have

$$\frac{EG}{A'C} = \frac{AG}{AC}.$$

Also, from $EG \parallel AB$, $FH \parallel A'C$, we get

$$\frac{AG}{AC} = \frac{BF}{BC}, \quad \frac{BF}{BC} = \frac{BH}{BA}.$$

These imply that $\frac{EG}{A'C} = \frac{BH}{BA}$. Thus, $EG \parallel BH$ and $EG = BH$.
    Therefore, $BEGH$ is a parallelogram, and $HG = BE$.

Example 6. In $\triangle ABC$, D is the midpoint of BC, E is an arbitrary point on AC, and BE intersects AD at O, as shown in Figure 23.11. Someone discovers the following facts:

① If $\frac{AE}{AC} = \frac{1}{2} = \frac{1}{1+1}$, then $\frac{AO}{AD} = \frac{2}{3} = \frac{2}{2+1}$.

② If $\frac{AE}{AC} = \frac{1}{3} = \frac{1}{1+2}$, then $\frac{AO}{AD} = \frac{2}{4} = \frac{2}{2+2}$.

③ If $\frac{AE}{AC} = \frac{1}{4} = \frac{1}{1+3}$, then $\frac{AO}{AD} = \frac{2}{5} = \frac{2}{2+3}$.

Please guess the general result when $\frac{AE}{AC} = \frac{1}{1+n}$, and prove your result. (Here, n is a positive integer.)

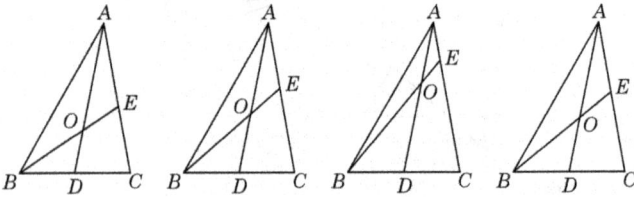

Fig. 23.11

Analysis   The given condition concerns the proportion of segments, so we may construct parallel lines. Consider the line through D and parallel to BE, which intersects AC at F. Then, $\frac{AO}{AD} = \frac{AE}{AF}$. Since $\frac{AE}{AC} = \frac{1}{1+n}$, by the property of parallel lines and the property of proportions, we can obtain $\frac{AE}{AF} = \frac{2}{n+2}$. This proves $\frac{AO}{AD} = \frac{2}{n+2}$.

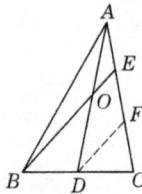

Fig. 23.12

Solution   Guess: $\frac{AO}{AD} = \frac{2}{n+2}$.

As in Figure 23.12, let F be a point on AC such that $DF \parallel BE$, then $\frac{AO}{AD} = \frac{AE}{AF}$. Since D is the midpoint of BC, we have $CF = EF = \frac{1}{2}EC$. Since $\frac{AE}{AC} = \frac{1}{n+1}$, we have $\frac{AE}{AE+2EF} = \frac{1}{1+n}, \frac{AE}{EF} = \frac{2}{n}, \frac{AE}{AF} = \frac{2}{2+n}$. Therefore, $\frac{AO}{AD} = \frac{2}{n+2}$.

Reading

There are two famous theorems regarding segment proportion in triangles: Menelaus's theorem and Ceva's theorem. Menelaus was a Greek mathematician in the 2nd century, and Ceva was an Italian mathematician in the 18th century. These two theorems have wide applications in proving segment proportions as well as the collinearity of three points.

Menelaus's Theorem. If a line DF intersects the lines AB, BC, CA at D, F, E, respectively, then

$$\frac{AD}{DB} \cdot \frac{BF}{FC} \cdot \frac{CE}{EA} = 1.$$

Proof: Construct $CH \parallel DF$ which intersects AB at H, as shown in Figure 23.13. Then,

$$\frac{AD}{DH} = \frac{AE}{EC}, \quad \frac{BF}{CF} = \frac{BD}{HD}.$$

Hence,

$$\frac{AD}{DB} \cdot \frac{BF}{FC} \cdot \frac{CE}{EA} = \frac{AD}{DB} \cdot \frac{BD}{HD} \cdot \frac{DH}{AD} = 1.$$

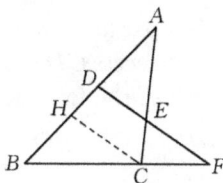
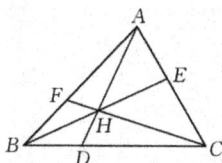

Fig. 23.13   Fig. 23.14

Ceva's Theorem. As shown in Figure 23.14, if H is a point in $\triangle ABC$ and the extensions of AH, BH, CH intersect BC, CA, AB at D, E, F, respectively, then $\frac{BD}{DC} \cdot \frac{CE}{EA} \cdot \frac{AF}{FB} = 1.$

Proof: We apply Menelaus's theorem as follows:

The line FC intersects the sides of $\triangle ABD$, so $\frac{AF}{FB} \cdot \frac{BC}{CD} \cdot \frac{DH}{HA} = 1$.

The line BE intersects the sides of $\triangle ADC$, so $\frac{AH}{HD} \cdot \frac{DB}{BC} \cdot \frac{CE}{EA} = 1$.

Multiplying the above results, we have $\frac{BD}{DC} \cdot \frac{CE}{EA} \cdot \frac{AF}{FB} = 1$.

The inverse propositions of these two theorems also hold and are frequently used to prove that three points are collinear or that three lines are concurrent.

Exercises

(I) Multiple-choice questions:

1. If $\frac{a}{b} = \frac{3}{4}, \frac{b}{c} = \frac{3}{2}, \frac{c}{d} = \frac{4}{5}$, then $\frac{ac}{b^2+d^2}$ equals (   ).

   (A) $\frac{19}{60}$ (B) $\frac{18}{61}$ (C) $\frac{7}{54}$ (D) $\frac{11}{54}$

2. If a, b, c are three different positive numbers, and $\frac{a-c}{b} = \frac{c}{a+b} = \frac{b}{a}$, then (   ).

   (A) $3b = 2c$ (B) $3a = 2b$ (C) $2b = c$ (D) $2a = b$

3. As shown in the figure, $l_1 \parallel l_2$, $AF : FB = 2 : 5$, $BC : CD = 4 : 1$, then $AE : EC = \underline{\hspace{2cm}}$.

4. In the following figure, $ABCD$ is a parallelogram, the angle bisector $\angle BAD$ intersects BD at E, intersects CD at F, and intersects the extension of BC at G. Which of the following is true?

   (A) $AE^2 = EF \cdot FG$ (B) $AE^2 = EF \cdot EG$
   (C) $AE^2 = EG \cdot FG$ (D) $AE^2 = EF \cdot AG$

5. In the following figure, $\frac{BD}{DC} = \frac{2}{3}, \frac{AE}{EC} = \frac{3}{4}$, and AD, BE intersect at F. Then, $\frac{AF}{FD} \cdot \frac{BF}{FE}$ equals ( ).

   (A) $\frac{7}{3}$ (B) $\frac{14}{9}$ (C) $\frac{35}{12}$ (D) $\frac{56}{15}$

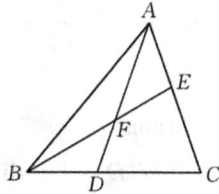

(II) Fill in the blanks:

6. If $\frac{3}{x+y} = \frac{4}{y+z} = \frac{5}{z+x}$, then $\frac{x^2+y^2+z^2}{xy+yz+zx} = $ _____.

7. In the following figure, CD is the bisector of $\angle ACB$ and $DE \parallel BC$. If $BC = a, AC = b$, then $DE = $ _____.

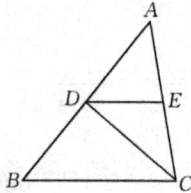

8. In a trapezoid $ABCD, AB \parallel CD$ and AC intersects BD at O. The line through O and parallel to AB intersects AD, BC at M, N, respectively. If $MN = 1$, then $\frac{1}{AB} + \frac{1}{CD} = $ _____.

9. In the following figure, $AB \parallel EF \parallel CD$. Suppose $AC + BD = 240, BC = 100, EC + ED = 192$, then $CF = $ _____.

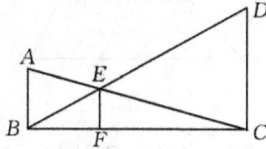

10. In $\triangle ABC$, D, F are points on sides AB, AC, respectively, and $AD : DB = CF : FA = 2 : 3$. Let E be the intersection of lines DF, BC, then $EF : DF = $ _____.

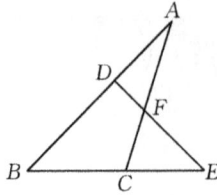

(III) Answer the following questions:

11. In a quadrilateral $ABCD$, the diagonals AC, BD intersect at O. A line $l \parallel BD$ and intersects the extensions of AB, DC, BC, AD, AC at M, N, R, S, P, respectively (as shown in the figure). Prove that $PM \cdot PN = PR \cdot PS$.

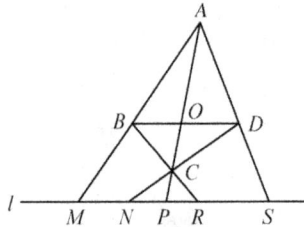

12. Let O be a point inside $\triangle ABC$ and P, Q, R be points on sides AB, BC, CA, respectively, such that $OP \parallel BC, OQ \parallel CA, OR \parallel AB$ (as shown in the figure). Suppose that $OP = OQ = OR = x$, and let $BC = a, CA = b$. Find x in terms of a, b.

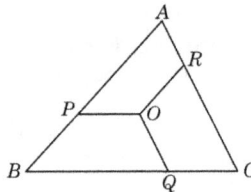

13. In trapezoid $ABCD$, $AD \parallel BC (AD < BC)$, AC intersects BD at M. The line through M and parallel to AD intersects AB, CD at E, F, respectively. Let N be the intersection of EC and FB,

and the line through N and parallel to AD intersects AB, CD at G, H, respectively. Prove that

$$\frac{1}{AD} + \frac{2}{BC} = \frac{1}{EF} + \frac{2}{GH}.$$

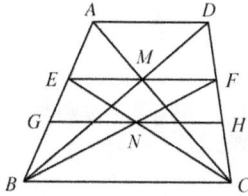

Chapter 24

Similar Triangles

There are several theorems to determine whether two triangles are similar, as follows: two triangles are similar if their two pairs of corresponding angles are equal; if two pairs of corresponding sides are proportional and the angle between them are equal; or if the three pairs of corresponding sides are all proportional.

If two triangles are both right triangles, another criterion is that the ratio of one leg and the hypotenuse is the same in two triangles.

The intersection of the three medians of a triangle is its centroid. The centroid divides each median into two parts with a ratio of 2:1, where the distance from the centroid to the vertex is twice its distance to the corresponding midpoint.

The similarity ratio of two similar triangles equals the ratio of their perimeters, corresponding medians, corresponding altitudes, and corresponding angle bisectors. The square of the similarity ratio equals the ratio of their areas.

Two polygons are similar if and only if their corresponding angles are equal and their corresponding sides are proportional. Sometimes, we divide the polygons into triangles, and if all pairs of corresponding triangles are similar, then the two polygons are also similar.

In many problems, we are required to prove the proportions of segments or the equality of angles to prove that two lines are parallel or perpendicular or to find areas with the help of similar triangles. When solving a problem, we should carefully analyze the characteristics of the conditions and conclusions in the problem, find out what kind of auxiliary line should be added, and reduce the problem to proving the similarity of triangles.

Example 1. In Figure 24.1, $AD \parallel BC$, $\angle D = 90°$, $DC = 7$, $AD = 2$, $BC = 4$. How many points P are there on side DC such that $\triangle PAD$ and $\triangle PBC$ are similar triangles? Prove your result.

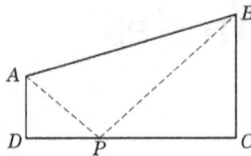

Fig. 24.1

Solution   There are three desired points.

Since $\triangle ADP$ and $\triangle PCB$ are similar, and $\angle D = \angle C = 90°$, where a triangle can have only one right angle, they must be corresponding angles. Thus, there are two cases of segment proportions:

$$① \frac{AD}{BC} = \frac{DP}{CP}; \quad ② \frac{AD}{PC} = \frac{DP}{BC}.$$

For ①, we have $\frac{DP}{CP} = \frac{2}{4} = \frac{1}{2}$. Since $DP + CP = 7$, it follows that $DP = \frac{7}{3}$, and in this case, P lies on segment DC.

For ②, we have $\frac{2}{PC} = \frac{PD}{4}$, so that $PC \cdot PD = 8$. This means $(7 - PD) \cdot PD = 8$, and solving the equation, we have $PD = \frac{7 \pm \sqrt{17}}{2}$. Note that both satisfy $0 < PD < 7$, so there are two such points.

In summary, there are three points that satisfy the condition.

Remark: This problem seeks to find the condition for two triangles to be similar, where the condition already implies a pair of right angles. However, there are two cases where the legs are proportional, so make sure to consider both of them.

Example 2. In Figure 24.2, D is a point on AB, and E, F lie on AC, BC, respectively, such that $DE \parallel BC$, $DF \parallel AC$. If the areas of $\triangle ADE$, $\triangle DBF$ are m, n, respectively, find the area of $DECF$.

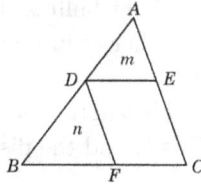

Fig. 24.2

Solution Let S denote the area of $\triangle ABC$. Since $\triangle ADE \sim \triangle ABC$, we have

$$\frac{AD}{AB} = \sqrt{\frac{S_{\triangle ADE}}{S}} = \sqrt{\frac{m}{S}}.$$

Similarly, since $\triangle BDF \sim \triangle BAC$, we have $\frac{BD}{AB} = \sqrt{\frac{S_{\triangle BDF}}{S}} = \sqrt{\frac{n}{S}}$.

Note that $\frac{AD}{AB} + \frac{BD}{AB} = 1$, so we have $\sqrt{\frac{m}{S}} + \sqrt{\frac{n}{S}} = 1$, which means $S = (\sqrt{m} + \sqrt{n})^2$.

Therefore, the area of $DECF$ is $S - m - n = (\sqrt{m} + \sqrt{n})^2 - m - n = 2\sqrt{mn}$.

Example 3. Several students have observed that for the same object, the size of its shadow depends on the position of the light source. Based on this phenomenon, they are trying to determine the position of the light source according to the shadow. The following figure shows their attempts.

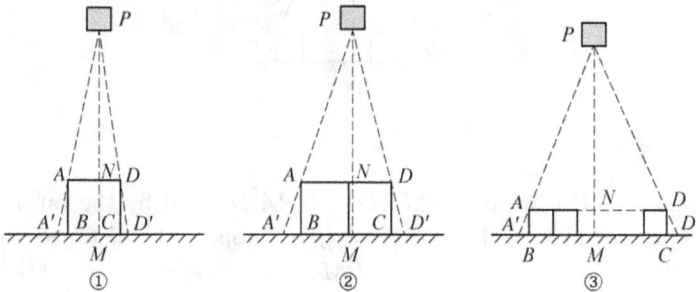

Fig. 24.3

(1) In Figure 24.3①, $ABCD$ is a square on the floor with a side length of $AB = 30\,\text{cm}$, and there is a light bulb above the square. Suppose the shadow of the square is $A'D'$, and the sum of lengths of $A'B, D'C$ is $6\,\text{cm}$. Then, the distance from the light bulb to the floor is _____ .

(2) Suppose the altitude of the light bulb is the same as in Figure 24.3①, and we put two squares together as in Figure 24.3②. What is the sum of $A'B, D'C$, in this case?

(3) If we put n squares with a side length of a together, as in Figure 24.3③, and the sum of $A'B, D'C$ is b, find the distance from the light bulb to the floor.

Solution (1) Assume that the distance from the light bulb to the floor is x cm.

Since $AD \parallel A'D'$, we have $\angle PAD = \angle PA'D'$, $\angle PDA = \angle PD'A'$. Thus, $\triangle PAD \sim \triangle PA'D'$.

Since the ratio between corresponding altitudes of similar triangles equals the similarity ratio, we have $\frac{AD}{A'D'} = \frac{PN}{PM}$. Hence, $\frac{30}{36} = \frac{x-30}{x}$, so that $x = 180$.

(2) Assume that the sum of the shadow lengths $A'B, D'C$ is y cm.

Then, similarly to (1), we have $\frac{60}{60+y} = \frac{150}{180}$, so $y = 12$.

(4) Let P be the position of the light bulb, as shown in Figure 24.4.

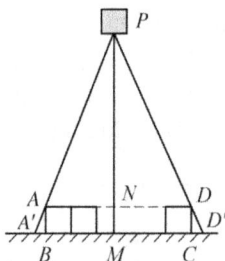

Fig. 24.4

Similarly to (1), we have $\triangle PAD \sim \triangle PA'D'$, and by the ratio of corresponding altitudes, we have $\frac{AD}{A'D'} = \frac{PN}{PM}$. Suppose the distance from the light bulb to the floor is x, then $PM = x$, $PN = x - a$, $AD = na$, $A'D' = na + b$. Thus,

$$\frac{na}{na+b} = \frac{x-a}{x} = 1 - \frac{a}{x},$$

$$\frac{a}{x} = 1 - \frac{na}{na+b}, \qquad x = \frac{na^2 + ab}{b}.$$

Example 4. In Figure 24.5, $\triangle ABC$ is a right triangle with $\angle ACB = 90°$, CD is the angle bisector, E lies on AC with $DE \parallel BC$, and F lies on BC

with $DF \parallel AC$. Prove that

$$CD^2 = 2AE \cdot BF.$$

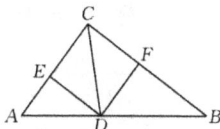

Fig. 24.5

Proof: Since $DE \parallel BC$, we have $\angle ADE = \angle B$. Thus, $\triangle AED \sim \triangle DFB$, and $\frac{AE}{DF} = \frac{DE}{BF}$, or, equivalently, $DE \cdot DF = AE \cdot BF$.

Since $\angle ACB = 90°$, $DE \parallel BC$, $DF \parallel AC$, we have $DE \perp AC$, $DF \perp BC$, so that $\angle ECF = \angle DEC = \angle DFC = 90°$. Also, CD bisects $\angle ACB$, so $DE = DF$.

Hence, $CEDF$ is a square, and $CD = \sqrt{2}DE = \sqrt{2}DF$. Therefore,

$$CD^2 = \sqrt{2}DE \cdot \sqrt{2}DF = 2DE \cdot DF = 2AE \cdot BF.$$

Example 5. In Figure 24.6, $AB = AC$, $AD \perp BC$ where D lies on BC. E, G are the midpoints of AD, AC, respectively, and F is the projection of D on BE. Prove that $FG = DG$.

Proof: Connect AE, CF, as in Figure 24.7.

Since $AD \perp BC$, $DF \perp BE$, we find that $\angle FDB = \angle FED$ and $\triangle BFD \sim \triangle DFE$. Thus, $\frac{DF}{EF} = \frac{BD}{ED}$.

It follows by condition that $CD = BD$, $DE = EA$, so $\frac{DF}{EF} = \frac{DC}{EA}$. Since $\angle FDC = \angle FEA$, we also have $\triangle DFC \sim \triangle EFA$, so $\angle DFC = \angle EFA$.

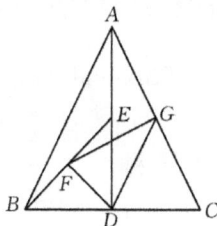

Fig. 24.6

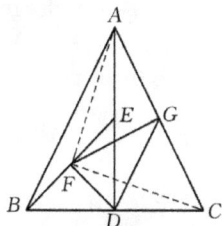

Fig. 24.7

Since $\angle DFE = 90°$, we have $\angle AFC = 90°$, and $FG = \frac{1}{2}AC$.

Comparing with $DG = \frac{1}{2}AC$, we conclude that $FG = DG$.

Remark: If we try to prove $\angle GDF = \angle GFD$, then it will be hard to use the given conditions. Thus, we use another segment AC as a bridge and reduce the problem to proving similarity between triangles.

Example 6. In Figure 24.8, $ABCD$ is a parallelogram whose diagonals intersect at O. A line through O intersects BC at E and intersects the extension of AB at F. If $AB = a$, $BC = b$, $BF = c$, find BE.

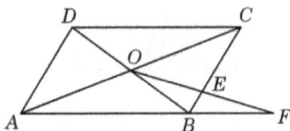

Fig. 24.8

Analysis   It seems hard to use the equality relations in the parallelogram since the segments AB, BC, BF are not in very good positions. Thus, we need to add auxiliary lines to put them together in a familiar structure. Let G be a point on AB such that $OG \parallel BC$, then $\triangle FEB \sim \triangle FOG$, and we can proceed to solve the problem.

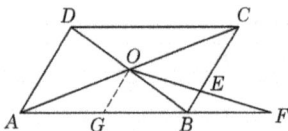

Fig. 24.9

Solution   Let G be a point on AB such that $OG \parallel BC$, as shown in Figure 24.9. Then, OG is a midsegment of $\triangle ABC$, so $OG = \frac{1}{2}BC = \frac{1}{2}b$, $GB = \frac{1}{2}AB = \frac{1}{2}a$.

Since $GO \parallel EB$, we have $\triangle FOG \sim \triangle FEB$, and $\frac{BE}{OG} = \frac{FB}{FG}$. Thus,

$$BE = \frac{BF}{FG} \cdot OG = \frac{c}{c + \frac{a}{2}} \cdot \frac{b}{2} = \frac{bc}{a + 2c}.$$

Example 7. In $\triangle ABC$, $\angle A : \angle B : \angle C = 1 : 2 : 4$. Prove that

$$\frac{1}{AB} + \frac{1}{AC} = \frac{1}{BC}.$$

Analysis   It is equivalent to proving $\frac{AB+AC}{AB \cdot AC} = \frac{1}{BC}$, or $\frac{AB+AC}{AB} = \frac{AC}{BC}$.

If we can construct a pair of similar triangles, where four sides have lengths of $AB, BC, CA, AB + AC$, then the equality can be proven.

Note that the original triangle already contains three of the above four lengths, so we can try to add auxiliary lines to this triangle and construct a new triangle that is similar to $\triangle ABC$.

Proof: As shown in Figure 24.10, we extend AB to D, such that $CD = AC$. Thus, $AD = AB + AC$. Also, extend BC to E, such that $AE = AC$, and connect ED. Then, we prove that $\triangle ABC \sim \triangle DAE$.

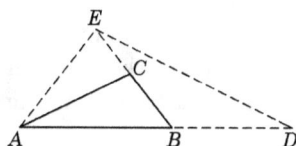

Fig. 24.10

Let $\angle A = \alpha$, $\angle B = 2\alpha$, $\angle C = 4\alpha$, then

$$\angle A + \angle B + \angle C = 7\alpha = 180°.$$

Since $\angle ACB$ is an external angle of the isosceles triangle ACE, we have

$$\angle CAE = 180° - 3\alpha - 3\alpha = 7\alpha - 6\alpha = \alpha, \quad \angle ACE = 180° - 4\alpha = 3\alpha.$$

Thus, $\angle EAB = 2\alpha = \angle EBA$, $AE = BE$.

Also, by construction, we have $AE = AC$, $AE = BD$, so $BE = BD$.

Since $\triangle BDE$ is an isosceles triangle, we have $\angle D = \angle BED = \alpha = \angle CAB$.

Therefore, $\triangle ABC \sim \triangle DAE$, and $\frac{AD}{AE} = \frac{AB}{BC}$. Equivalently, $\frac{AB+AC}{AC} = \frac{AB}{BC}$, and $\frac{1}{AB} + \frac{1}{AC} = \frac{1}{BC}$.

Reading

The use of the knowledge of similar triangles for measurement appeared in ancient Chinese mathematical works, such as *The Nine Chapters on the Mathematical Art* and *The Sea Island Mathematical Manual*. Here, we choose two examples for students to read:

1. *The Nine Chapters on the Mathematical Art*: A rectangular city is 7 li long from east to west and 9 li long from south to north. There is a gate in the middle of each side of the city, and there is a tree located 15 li east of the east gate. (1 km equals 2 li.) Question: how many steps do you need to walk south from the south gate to see the tree? (1 li equals 300 steps.)

2. *The Sea Island Mathematical Manual*: Someone is looking at an island from the coast for measurement. He sets up two benchmarks, each of height 3 zhang (1 zhang equals $\frac{10}{3}$ m), where the distance between the front benchmark and the rear benchmark is 1000 steps, so that the two

benchmarks and the peak of the island lie on the same plane. Walk back 123 steps from the front benchmark and look at the peak of the island with eyes on the ground. Then, the peak coincides with the top of the front benchmark. Meanwhile, walk back 127 steps from the rear benchmark and look at the peak of the island with eyes on the ground. Then, the peak coincides with the top of the rear benchmark. Question: what is the distance between the front benchmark and the island peak, or the island, respectively?

Exercises

(I) Multiple-choice questions:

1. Suppose $ABCD$ is a parallelogram, where the angle bisector of $\angle BAD$ intersects BD at E, intersects CD at F, and intersects the extension of BC at G. Which of the following is true? ( ).

   (A) $AE^2 = EF \cdot FG$ (B) $AE^2 = EF \cdot EG$
   (C) $AE^2 = EG \cdot FG$ (D) $AE^2 = EF \cdot AG$

2. In the following figure, $\triangle ABC$ is a right triangle with $\angle C = 90°$, $ED \perp AB$ where D lies on AB, $AD = BD$, $AB = 20$, $AC = 12$. Then, the length of DE is ( ).

   (A) 10   (B) 8.5   (C) 9.5   (D) 7.5

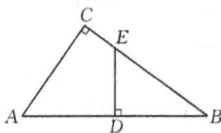

3. In a right trapezoid $ABCD$, the bases $AD = \sqrt{3}$, $BC = 3\sqrt{3}$, and the leg $AB = 6$, which is perpendicular to the bases. Choose a point P on side AB such that $\triangle PAD$ and $\triangle PBC$ are similar. Then, there are ( ) such points.

   (A) 1   (B) 2   (C) 3   (D) 0

4. Suppose the area of a right triangle ABC is 120, where $\angle BAC = 90°$. Let D be the midpoint of the hypotenuse BC and E be the projection of D on AB. Let F be the intersection of CE, AD. Then, the area of $\triangle AEF$ is ( ).

   (A) 18   (B) 20   (C) 22   (D) 24

5. If a, b, c are the side lengths of $\triangle ABC$ such that $\frac{a}{b} = \frac{a+b}{a+b+c}$, then the relationship between the internal angles $\angle A, \angle B$ is ( ).

(A) $\angle B > 2\angle A$ (B) $\angle B = 2\angle A$
(C) $\angle B < 2\angle A$ (D) uncertain

(II) Fill in the blanks:

6. In the following figure, $ABCD$ is a square, and E lies on the side CD with $DE = 2$. If the distance from B to AE is $BF = 3$, then the side length of $ABCD$ is _____ .

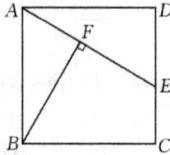

7. In the following figure, $\angle A = 2\angle B$, $AC = 4$, $AB = 5$. Then, $BC = $ _____ .

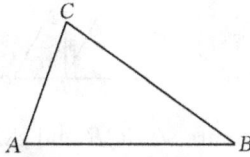

8. Let $\triangle ABC$ be an isosceles right triangle with $\angle C = 90°$, $AC = BC$, $BE = ED = CF$, as shown in the following figure. Then, $\angle CEF + \angle CAD = $ _____ .

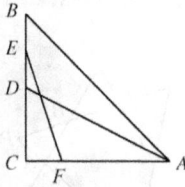

9. In the following figure, $ABCD$ is a square with a side length of $5\,\text{cm}$, and $EF = FG, FD = DG$. Then, the area of $\triangle ECG$ is

_____ .

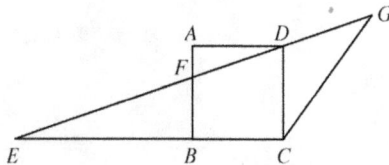

(III) Answer the following questions:

10. In the following figure, $ABCD$ is a square, and M, N are points on sides AB, BC, respectively, such that $BM = BN$. Let P be the projection of B on MC. Prove that $PD \perp PN$.

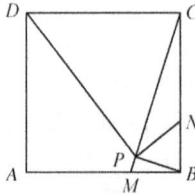

11. In the following figure, D lies on BC, E lies on AB, and $\angle 1 = \angle 2 = \angle 3$. If the perimeters of $\triangle ABC$, $\triangle EBD$, $\triangle ADC$ are m, m_1, m_2, respectively, and $BD = 6$, $CD = 2$, find the value of $\frac{m_1 + m_2}{m}$.

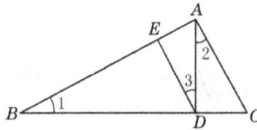

12. As shown in the figure, $\triangle PQR$ and $\triangle P'Q'R'$ are two congruent equilateral triangles whose overlapping is a hexagon $ABCDEF$. If the side lengths of the hexagon are $AB = a_1$, $BC = b_1$, $CD = a_2$, $DE = b_2$, $EF = a_3$, $FA = b_3$, prove that

$$a_1^2 + a_2^2 + a_3^2 = b_1^2 + b_2^2 + b_3^2.$$

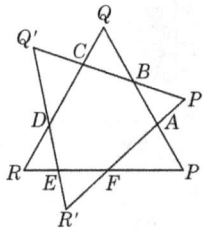

Chapter 25

The Midsegment

A midsegment of a triangle is the segment connecting the midpoints of two sides. It is parallel to the third side, and its length equals half of the third side.

The midsegment of a trapezoid is the segment connecting the midpoints of its two legs. It is parallel to both bases, and its length equals the arithmetic mean of the bases.

There are two conditions and two assertions regarding the midsegment of both triangles and trapezoids. By swapping any condition with any assertion, we still get a correct statement. This is frequently used in problems.

When there is a midpoint in the conditions, one possible idea is to construct a midsegment. This happens in many problems in this chapter.

Example 1. In Figure 25.1, D, E are points on AB, AC, respectively. If $AD = DB$, $AE = 2EC$, and F is the intersection of BE, CD, prove that $EF = \frac{1}{2}BE$.

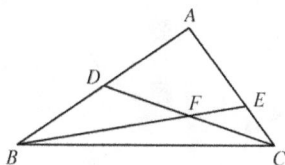

Fig. 25.1

Analysis In this problem, D is the midpoint of AB, so we may consider constructing midsegments. Let M be the midpoint of AE, then $EF = \frac{1}{2}DM = \frac{1}{4}BE$.

Solution Let M be the midpoint of AE, and connect DM, as shown in Figure 25.2.

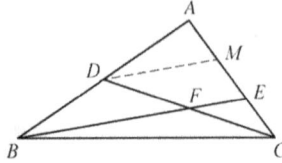

Fig. 25.2

Since $AD = DB$, $AM = ME$, we have $DM \parallel BE$ and $DM = \frac{1}{2}BE$.

Since $AE = 2EC$, $AE = 2ME$, we have $ME = EC$.

Combining with $DM \parallel EF$, we see that EF is a midsegment of $\triangle DMC$, and $DF = FC$, $EF = \frac{1}{2}DM$, so $EF = \frac{1}{4}BE$.

Remark: When constructing midsegments, we usually choose one side whose midpoint is given, and the second side based on other considerations. Another method is to construct the midpoint of BE as N and connect DN. Details are left for the reader.

Example 2. In Figure 25.3, BE is the angle bisector of $\angle ABC$, D is the midpoint of BC, and $AD \perp BE$. Suppose $BE = AD = 4$. Find the three side lengths of $\triangle ABC$.

Solution Let the line through D and parallel to BE intersect AC at F, as shown in Figure 25.4.

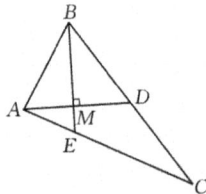
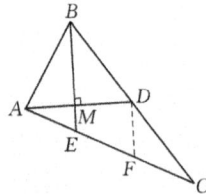

Fig. 25.3   Fig. 25.4

Since BE bisects $\angle ABC$, $BE \perp AD$, we see that M is the midpoint of AD. Since D is the midpoint of BC and $DF \parallel BE$, the point F is also the midpoint of EC. Thus, DF is a midsegment of $\triangle BEC$, and ME is a midsegment of $\triangle ADF$.

Since $BE = 4$, we have $DF = 2$, $ME = 1$. Thus, $MB = 3$, $MD = 2$, and

$$BD = \sqrt{MB^2 + MD^2} = \sqrt{3^2 + 2^2} = \sqrt{13}, \quad BC = 2\sqrt{13}.$$

Since $\triangle ABD$ is an isosceles triangle, we have $AB = BD = \sqrt{13}$.

In the right triangle AME,

$$AE = \sqrt{MA^2 + ME^2} = \sqrt{2^2 + 1^2} = \sqrt{5}.$$

In the right triangle ADF,

$$AF = \sqrt{AD^2 + DF^2} = \sqrt{4^2 + 2^2} = 2\sqrt{5}.$$

Therefore, $AB = \sqrt{13}$, $AC = 3\sqrt{5}$, $BC = 2\sqrt{13}$.

Remark: Since D is the midpoint of BC, we construct a line parallel to BE through F, and the resulting intersection F is also a midpoint. This construction helps us to make the best use of the given conditions.

Example 3. In Figure 25.5, AD is an angle bisector of $\triangle ABC$, $AB < AC$. Let E be a point on AC such that $CE = AB$, and let M, N be the midpoints of BC, AE, respectively. Prove that $MN \parallel AD$.

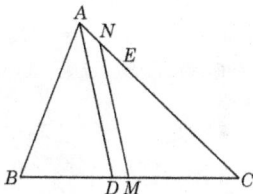

Fig. 25.5

Proof: Connect BE, and let F be the midpoint of BE, then connect FN, FM, as shown in Figure 25.6.

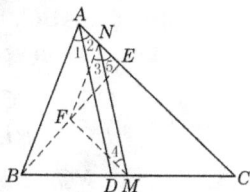

Fig. 25.6

Since FN is a midsegment of $\triangle EAB$, we have $FN = \frac{1}{2}AB$, $FN \parallel AB$.
Since FM is a midsegment of $\triangle BCE$, we have $FM = \frac{1}{2}CE$, $FM \parallel CE$.
From $CE = AB$, we get $FM = FN$, so $\angle 3 = \angle 4$.

Also, $\angle 4 = \angle 5$, so $\angle 3 = \angle 5$. Since $\angle 1 + \angle 2 = \angle 3 + \angle 5 = \angle BAC$ while $\angle 1 = \angle 2$, it follows that $\angle 2 = \angle 5$, and $MN \parallel AD$.

Example 4. In the quadrilateral $ABCD$ in Figure 25.7, E, F are the midpoints of AB, CD, respectively. P is a point on the extension of AC, and the lines PF, PE intersect AD, BC at M, N, respectively. Let K be the intersection of EF, MN. Prove that K is the midpoint of MN.

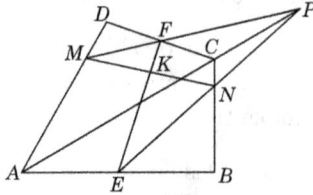

Fig. 25.7

Proof: Choose a point G on PF such that $GF = FM$, and let L be the midpoint of AC. Connect GC, GN, LE, LF, as shown in Figure 25.8.

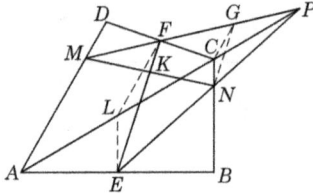

Fig. 25.8

Here, LE, LF are midsegments of $\triangle ABC$, $\triangle ACD$, respectively, and $CG \parallel DM$, $LF \parallel AD$, $LE \parallel CB$. Thus, we have

$$\angle GCN = \angle FLE, \quad \frac{CG}{LF} = \frac{PC}{PL} = \frac{CN}{LE}.$$

These show that $\triangle CNG \sim \triangle LEF$, and $NG \parallel EF$.

Consequently, FK is a midsegment of $\triangle MNG$, and K is the midpoint of MN.

Example 5. In Figure 25.9, $\triangle ABC$ is an arbitrary triangle, E, F are the midpoints of sides AB, AC, respectively, and M, N are the trisection points of AC. Let D be the intersection of lines EM, FN. Prove that $ABCD$ is a parallelogram.

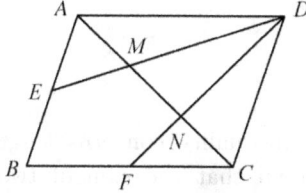

Fig. 25.9

Analysis The conditions contain many midpoints, so we need to find a good way to add midsegments. Here, connecting BN is better than connecting EF, and combining with BM, we may prove that $BMDN$ is a parallelogram, which leads to the conclusion.

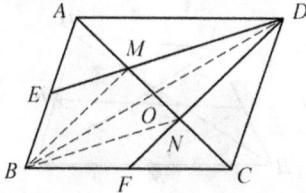

Fig. 25.10

Proof: Connect BM, BN, and let BD intersect AC at O, as shown in Figure 25.10.

Since $AE = EB$, $AM = MN$, we have $EM \parallel BN$.

Similarly, $BM \parallel FN$, so $BMDN$ is a parallelogram, and $OB = OD$, $OM = ON$.

Combining with $AM = NC$, it follows that $OA = OC$, so $ABCD$ is a parallelogram.

Remark: When there are many midpoints in the given conditions, we can add midsegments in multiple triangles.

Example 6. In Figure 25.11, $ABCD$ is a quadrilateral and E, F are the midpoints of sides AB, CD, respectively. Prove that $EF < \frac{1}{2}(AC + BD)$.

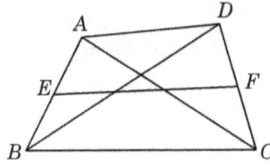

Fig. 25.11

Analysis The desired inequality concerns lengths of segments, which reminds us of the theorem that the sum of two sides in a triangle is greater than the third side. Thus, we construct auxiliary lines to put $EF, \frac{1}{2}AC, \frac{1}{2}BD$ in one triangle.

Proof: As shown in Figure 25.12, let M be the midpoint of BC, and connect EM, FM. Since E is the midpoint of AB, and F is the midpoint of DC, it follows that $EM = \frac{1}{2}AC$, $FM = \frac{1}{2}BD$. Hence, $EF < EM+FM$, which means $EF < \frac{1}{2}(AC + BD)$.

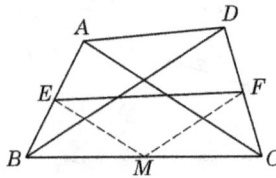

Fig. 25.12

Think about it The midsegments of a triangle divide it into triangles that are similar to the original triangle. Are the two trapezoids divided by the midsegment of a trapezoid similar to the original trapezoid? If this is not true, what conditions of the trapezoid should be satisfied so that the two parts are similar to the original trapezoid?

Exercises

(I) Multiple-choice questions:

    1. Let $ABCD$ be an arbitrary quadrilateral, and the midpoints of its sides are E, F, G, H, as shown in the following figure. If the

diagonals AC and BD both have a length of $20\,\mathrm{cm}$, then the perimeter of $EFGH$ is ( ).

(A) $80\,\mathrm{cm}$ (B) $40\,\mathrm{cm}$ (C) $20\,\mathrm{cm}$ (D) $10\,\mathrm{cm}$

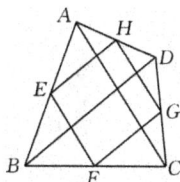

2. In the following figure, $ABCD$ is a quadrilateral and E, F are the midpoints of AC, BD, respectively. Then, the relationship between EF and $AB + CD$ is ( ).

(A) $EF = \frac{1}{2}(AB + CD)$ (B) $EF > \frac{1}{2}(AB + CD)$
(C) $EF < \frac{1}{2}(AB + CD)$ (D) $EF \geq \frac{1}{2}(AB + CD)$

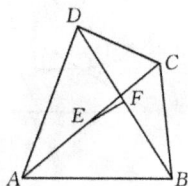

3. As shown in the figure, M, P lie on AB, AC, respectively, and $AM = BM$, $AP = 2CP$. Let N be the intersection of BP, CM, and suppose $PN = 1$, then the length of PB is ( ).

(A) 2   (B) 3   (C) 4   (D) 5

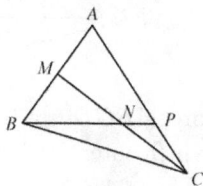

4. In the following figure, E is the midpoint of AC, D lies on BC such that $CD = 2BD$, F is the intersection of AD, BE, and the area of $\triangle BDF$ is 1. Then, the area of $\triangle ABC$ is (   ).

   (A) 6   (B) 12   (C) 5   (D) 10

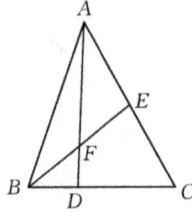

5. Let $ABCD$ be a trapezoid with $AD \parallel BC$, $\angle B = 30°$, $\angle C = 60°$. Let E, M, F, N be the midpoints of AB, BC, CD, DA, respectively, as shown in the following figure. Suppose $BC = 7$, $MN = 3$, then the length of EF is (   ).

   (A) 4   (B) 4.5   (C) 5   (D) 6

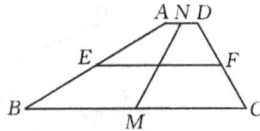

(II) Fill in the blanks:

6. As shown in the figure, M, P lie on AB, AC, respectively, and $AM = BM$, $AP = 2CP$. Let N be the intersection of BP, CM, and suppose $PN = 1$. Then, the length of PB is _____.

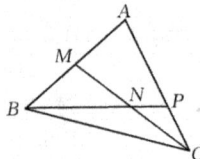

7. In the following figure, $ABCD$ is a trapezoid with $AB \parallel CD$, EF is a midsegment, which intersects AC at G, and $EF = 16\,\text{cm}$, $EG - GF = 4\,\text{cm}$. Then, $AB =$ _____.

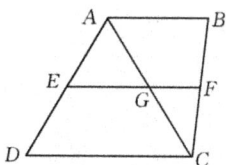

8. In $\triangle ABC$, AD bisects $\angle BAC$, $BD \perp AD$, $DE \parallel AC$ and E lies on AB. If $AB = 5$, then $DE = $ _____.

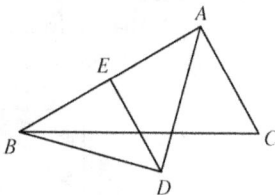

9. In the following figure, $ABCD$ is a square, E is the midpoint of CD, and F is the midpoint of AD. Connect BE, CF, which intersect at P. If $\angle CBP = 27°$, then $\angle BAP = $ _____.

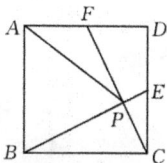

10. In the following figure, $\triangle ABC$ is a given triangle, and l is a line passing through A. Let E, F be the projections of B, C on l, respectively, and P be the midpoint of BC. If the area of $\triangle PEF$ is $\sqrt{2}$, then the area of $BCFE$ is _____.

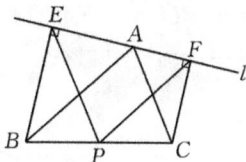

(III) Answer the following questions:

11. In the following figure, D, E are points on AB, AC, respectively, such that $BD = CE$. Let M, N be the midpoints of BE, CD, respectively, and the line MN intersects AB, AC at P, Q, respectively. Prove that $AP = AQ$.

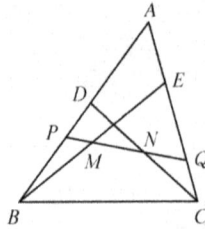

12. In the following figure, $ABCD$ is a trapezoid with $AD \parallel BC$, $\angle A = 90°$, and E is the midpoint of AB, with $\angle CED = 90°$. Prove that the distance from E to CD is equal to EA.

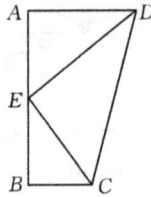

13. In the following figure, $ABCD$ is a square, whose diagonals AC, BD intersect at O. The angle bisector of $\angle CAB$ intersects BD at F and intersects BC at G. Prove that $CG = 2OF$.

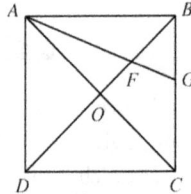

14. In $\triangle ABC$, $AB > AC$, AD is the angle bisector of $\angle BAC$, M is the midpoint of BC, and E is the projection of B on the line AD. Prove that $EM = \frac{1}{2}(AB - AC)$.

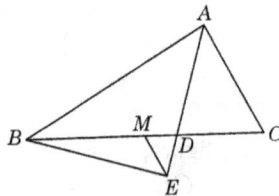

Chapter 26

Translation and Symmetry

Translation and symmetry are common methods of transformation in plane geometry.

Translation refers to moving a figure by a certain distance in a certain direction in a plane. By definition, the shape, size, and orientation of the figure do not change after translation (so that the two figures before and after translation are congruent):

(1) The corresponding segments are parallel (or collinear) and equal, and the corresponding angles are equal. The segment between two points in one figure is parallel and equal to the segment between their corresponding points in the second figure.
(2) The segments connecting corresponding points before and after translation are parallel (or collinear) and equal.
(3) The composition of multiple translations is still a translation.
(4) A translation is determined by direction and distance.

These are key properties that we may use to solve problems with translation.

Symmetry is another geometric transformation. Commonly, we know central symmetry and axial symmetry, whose properties are listed in Table 26.1.

Table 26.1

Transformation definition and property	Central symmetry	Axial symmetry
Definition	Two figures are centrally symmetric with respect to a point if we can rotate one figure around the point by 180° so that it coincides with the second figure. The point is called the center of symmetry.	Two figures are axially symmetric with respect to a line if we can reflect one figure across the line so that it coincides with the second figure. The line is called the axis of symmetry.
Property	Two centrally symmetric figures are congruent, and the center of symmetry is the midpoint of the segment connecting each pair of corresponding points.	Two axially symmetric figures are congruent, and the segment between corresponding points is perpendicular to and bisected by the axis of symmetry. If two corresponding segments (or lines) intersect, then their intersection lies on the axis of symmetry.

Example 1. In Figure 26.1, $\triangle ABC$ is a right triangle with $\angle C = 90°$, $MN \parallel AB$, and P, Q are the midpoints of MN, AB, respectively. Prove that $PQ = \frac{1}{2}(AB - MN)$.

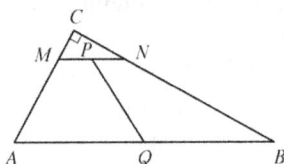

Fig. 26.1

Analysis   We want to prove $2PQ = AB - MN$, but we observe that the three segments are separated, and their relationship is hard to find. Thus, we consider translation to put them together into one figure.

As shown in Figure 26.2, we translate MA to PX and translate NB to PY. Then, X, Y lie on $AB - MN = AB - (MP + NP) = AB - (AX + BY) = XY$. Then, it suffices to prove that $2PQ = XY$.

Proof: We translate MA to PX and translate NB to PY, as shown in Figure 26.2.

Since $MN \parallel AB$, the points X, Y both lie on AB, and $MAXP$, $NBYP$ are parallelograms.

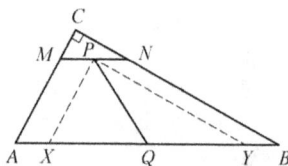

Fig. 26.2

Since $AX = MP$, $BY = NP$, $\angle XPY = \angle C = 90°$, we have

$$AB - MN = AB - (MP + NP) = AB - (AX + BY) = XY.$$

Also, $\angle XPY = \angle ACB = 90°$, so PQ is the median on the hypotenuse of the right triangle PXY. This means $PQ = \frac{1}{2}XY = \frac{1}{2}(AB - MN)$.

Example 2. A rectangular piece of paper $ABCD$ of size 3×4 is folded along BD, and C becomes C'. Let G be the intersection of BC', AD, as shown in Figure 26.3. Then, fold once more to make A and D coincide, where the crease is EN, and let M be the intersection of EN, AD, as shown in Figure 26.4. Find the length of ME.

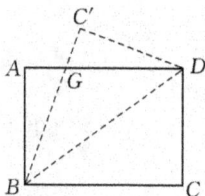
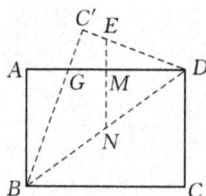

Fig. 26.3   Fig. 26.4

Solution Apparently, $\triangle ABG \cong \triangle C'DG$, so $C'G = AG$.

Let $C'G = x$, then $CD = 4 - x$, and in right triangle $C'DG$,

$$GD^2 = C'G^2 + C'D^2.$$

Thus, $(4 - x)^2 + x^2 + 3^2$, and solving the equation, we have $x = \frac{7}{8}$, so $C'G = \frac{7}{8}$.

Since EN is the perpendicular bisector of AD, we have $MD = \frac{1}{2} AD = 2$.

Also, $\triangle EMD \sim \triangle GC'D$, so $\frac{ME}{C'G} = \frac{MD}{C'D}$, and we conclude that $ME = \frac{7}{12}$.

Remark: Problems concerning reflection across a line can be solved using the properties of axial symmetry.

Example 3. In Figure 26.5, $ABCD$ is a square, whose area is S.

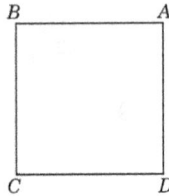

Fig. 26.5

(1) Construct a quadrilateral $A'B'C'D'$, such that A' and A are symmetric with respect to B, B and B' are symmetric with respect to C, C and C' are symmetric with respect to D, and D and D' are symmetric with respect to A. (Draw only the figure; explanation is not required.)

(2) If the area of the figure constructed in (1) is S', express S' in terms of S.

(3) If $ABCD$ is an arbitrary quadrilateral instead of a square, whose area is still S. Construct the quadrilateral $A'B'C'D'$ in the same way as (1), and let S'' be the area of $A'B'C'D'$. Is S'' equal to S'? Why?

Solution (1) The quadrilateral is shown in Figure 26.6.

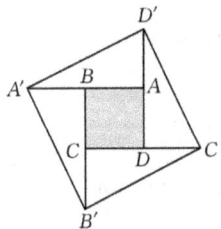

Fig. 26.6

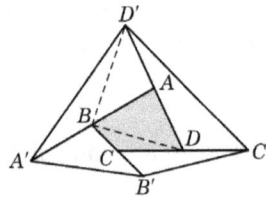

Fig. 26.7

(2) Let a be the side length of $ABCD$, then $AA' = 2a$, $S_{\triangle AA'D'} = \frac{1}{2}AA' \cdot AD' = a^2$.

Similarly, $S_{\triangle BB'A'} = S_{\triangle CC'B'} = S_{\triangle DD'C'} = a^2$.

Thus, $S' = S_{\triangle AA'D'} + S_{\triangle BB'A'} + S_{\triangle CC'B'} + S_{\triangle DD'C'} + S_{ABCD} = 5a^2 = 5S$.

(3) $S' = S''$. The proof is as follows. Connect BD, BD', as in Figure 26.7. Since AB is a median of $\triangle BDD'$, we have $S_{\triangle ABD'} = S_{\triangle ABD}$.

Also, BD' is a median of $\triangle AA'D'$, so $S_{\triangle ABD'} = S_{\triangle A'BD'}$.

Thus, $S_{\triangle AA'D'} = 2S_{\triangle ABD}$. Similarly, $S_{\triangle CC'B'} = 2S_{\triangle CBD}$, so we obtain $S_{\triangle AA'D'} + S_{\triangle CC'B'} = 2S$.

For similar reasons, $S_{\triangle BB'A'} + S_{\triangle DD'C'} = 2S$. Therefore,

$$S'' = S_{\triangle AA'D'} + S_{\triangle BB'A'} + S_{\triangle CC'B'} + S_{\triangle DD'C'} + S_{ABCD} = 5S.$$

From (2), we have $S' = 5S$, so $S' = S''$.

Remark: Central symmetry is a special case of the rotation transformation, which is also called point reflection. Please note its similarities and differences compared to axial symmetry.

Example 4. In Figure 26.8, $ABCD$ is a square with a side length of 3, E lies on BC, such that $BE = 2$, and P is a moving point on BD. What is the minimum of $PE + PC$?

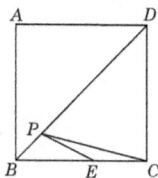

Fig. 26.8

Analysis Note that since P is a moving point, both PE and PC are varying, so the minimum is not easy to find in a straightforward way. However, we can try to put them in one triangle (or segment) that is easier to deal with. Such conversion is realized by transformations such as axial symmetry.

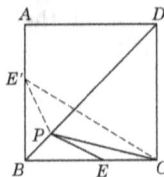

Fig. 26.9

Solution   As shown in Figure 26.9, let E' be the reflection of E across BD. Since $ABCD$ is a square, it follows that E' lies on AB. Thus, $PE = PE'$, and the minimum of $PE + PC$ is the same as the minimum of $PE' + PC$. Note that $PE' + PC$ is minimal when P lies on the segment between E' and C, so the minimum of $PE + PC$ is exactly the length of CE', which is

$$CE' = \sqrt{BC^2 + BE^2} = \sqrt{3^2 + 2^2} = \sqrt{13}.$$

Example 5. Suppose the lengths of the three medians of $\triangle ABC$ are 3, 4, and 5 cm. Find the area of the triangle.

Analysis   Assume that $AE = 3\,\mathrm{cm}$, $BF = 4\,\mathrm{cm}$, $CG = 5\,\mathrm{cm}$, which are the three medians. We try to put them in the same triangle through translation, so that they compose a triangle of fixed shape. Thus, we translate CG to FH, and naturally we have $HB = AE$. Then, $\triangle BFH$ is a triangle whose sides equal the medians AE, BF, CG. If we find the relationship between the area of $\triangle ABC$ and the area of $\triangle BFH$. Then, the desired area can be calculated.

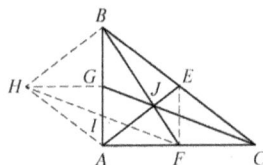

Fig. 26.10

Solution   We translate CG to FH as in Figure 26.10, then $HFCG$ is a parallelogram. Connect HB, HA.

Since $HG \parallel FC$, $HG = FC$, where F is the midpoint of AC, we have $HG \parallel AF$, $HG = AF$, and consequently $HA \parallel GF$, $HA = GF$ (where GF is not indicated in the figure). Also, $GF \parallel BE$, $GF = BE$, so $HA \parallel BE$, $HA = BE$, and $HB \parallel AE$, $HB = AE$.

Hence, $HB = AE = 3\,\mathrm{cm}$, $BF = 4\,\mathrm{cm}$, $HF = GC = 5\,\mathrm{cm}$, and $\triangle BFH$ is a right triangle.

Its area is $S_{\triangle BFH} = \frac{1}{2} \times BF \times BH = \frac{1}{2} \times 4 \times 3 = 6(\mathrm{cm}^2)$.

Note that in Figure 26.10, $\triangle BIH$ is the overlapping of $\triangle ABC$ and $\triangle BFH$.

Since $GHAF$ is a parallelogram, we have $S_{\triangle IAF} = S_{\triangle IGH}$. Connect EF.

Also, $S_{\triangle EFC} = S_{\triangle BGH}$, and $S_{\triangle FBE} = \frac{1}{2}S_{\triangle FBC} = \frac{1}{4}S_{\triangle ABC}$.

Thus, $S_{\triangle BIF} + S_{\triangle IAF} + S_{\triangle EFC} + S_{\triangle FBE} = (S_{\triangle BIF} + S_{\triangle IGH} + S_{\triangle BGH}) + \frac{1}{4}S_{\triangle ABC}$.

In other words, $S_{\triangle ABC} = S_{\triangle BFH} + \frac{1}{4}S_{\triangle ABC}$, and $S_{\triangle BFH} = \frac{3}{4}S_{\triangle ABC}$.

Therefore, $S_{\triangle ABC} = \frac{4}{3}S_{\triangle BFH} = 8(\mathrm{cm}^2)$.

Remark: In problems (computations, proofs, or constructions) concerning the medians of a triangle, it is common to use translation.

Reading: Finding the Treasure

Once, an adventurous young man who found a piece of parchment in his great-grandfather's relic, indicating a treasure. It said as follows:

On the north coast of a desert island somewhere, there is an oak tree, a pine tree, and a gallows on a large meadow. Go from the gallows to the oak tree and remember the number of steps. When you reach the oak tree, turn right at a right angle, walk the same number of steps, and insert a small pile there. Then, go back to the gallows and walk toward the pine tree while remembering the number of steps. When you reach the pine tree, turn left at a right angle, walk the same number of steps, and then insert another pile. Dig in the middle of the two piles, and you can find the treasure.

However, when the young man reached the island, although the oak tree and the pine tree were found, the gallows disappeared. The place was too big to dig at all, and the young man had no choice but to return with empty hands.

Can we apply the mathematical knowledge we have learned to find the location of the treasure?

As shown in Figure 26.11, we use A, B, C, D, X to denote the oak tree, the pine tree, the first pile, the second pile, and the gallows, respectively. E be the midpoint of CD, that is, the location of the treasure. Let M, N be the midpoints of CX, DX, respectively, and P be the intersection of AM, DX.

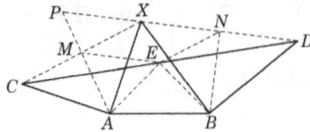

Fig. 26.11

From the property of the midsegment, it follows easily that $MENX$ is a parallelogram, so $ME = XN = BN$, $EN = MX = AM$, $\angle AME = 90° - \angle XME = 90° - \angle XNE = \angle ENB$. Thus, $\triangle AME \cong \triangle ENB$, and $EA = EB$.

Since the bottom angles of an isosceles right triangle are 45°, and the alternate interior angles are equal when a transversal intersects parallel lines, we can derive that $\angle AXC = \angle BXD$. On the other hand, isosceles right triangles are always congruent or similar, so $\frac{AM}{AX} = \frac{BN}{BX}$, that is, $\frac{AM}{AX} = \frac{BM}{BX}$. Hence, $\triangle AME$ and $\triangle AXB$ are similar, which yields $\angle EAB = \angle XAM = 45°$, so $\angle EBA = \angle EAB = 45°$.

Therefore, $\triangle AEB$ is also an isosceles right triangle whose hypotenuse AB is already known, so the location of E (the treasure) is not hard to find.

Exercises

1. If we translate the right triangle ABC along the hypotenuse AB to the right by 5 cm, we obtain $\triangle DEF$. Suppose $AB = 10\,\text{cm}$, $BC = 8\,\text{cm}$. Find the perimeter of the shaded region in the figure.

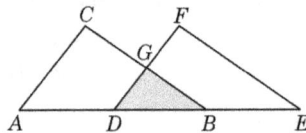

2. As shown in the figure, D, E are points on AB, AC, respectively. If we fold the triangle along DE so that A coincides with a point A' inside the quadrilateral $BCED$, then there is a fixed quantitative relationship between $\angle A'$ and $\angle 1 + \angle 2$. Find this relationship.

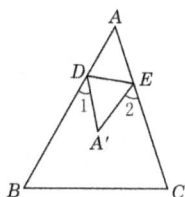

3. In the following figure, $ABCD$ is a rectangle whose diagonal has a length of 2. Suppose $\angle 1 = \angle 2 = \angle 3 = \angle 4$, Find the perimeter of $EFGH$.

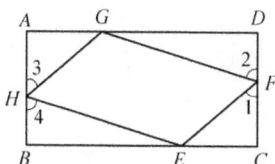

4. Suppose $\triangle ABC$ has a perimeter of 1999 cm, and a squirrel is located at the point P on AB (endpoints not included). The squirrel first runs along a direction parallel to BC until it reaches P_1 on AC, and then it runs along a direction parallel to AB until it reaches P_2 on BC. Next, it runs along a direction parallel to AC until it reaches P_3 on AB, and so on. In this process, will the squirrel return to P eventually? If so, what is the minimal distance it has traveled?

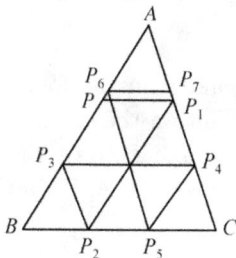

5. In the following figure, $ABCD$ is a square. Fold the square along BE so that A coincides with a point A' on BD, and connect $A'C$. Find the size of $\angle BA'C$.

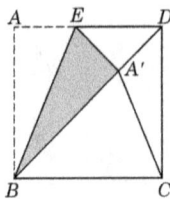

6. In the following figure, $\angle ABC = 46°$, D is a point on BC such that $DC = AB$, $\angle DAB = 21°$. Find the size of $\angle CAD$.

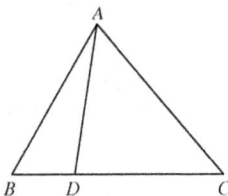

7. In the following figure, $ABCD$ is a rectangle where $AB = 8, AD = 6$. Fold the rectangle so that AD coincides with a part of AB, and let AE be the crease. Then, fold $\triangle AED$ along DE to the right, and let F be the intersection of AE and BC after folding. Find the area of $\triangle CEF$.

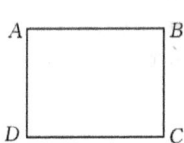

Fig. E7.1

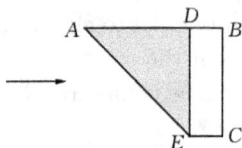

Fig. E7.2

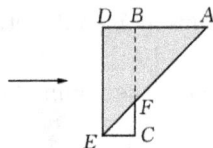

Fig. E7.3

8. Experiment and reasoning: in the following two figures, $ABCD$ is a square, and M is a point on the extension of AB. Suppose one leg of a set square passes through D, its right-angled vertex E slides along AB (E does not coincide with A, B), and the other leg intersects the bisector of $\angle BCM$ at F.

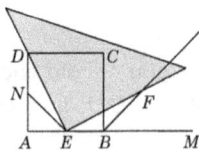

Fig. E8.1

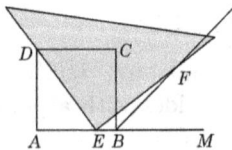

Fig. E8.2

(1) In Figure E8.1, if E is the midpoint of AB, guess the relationship between DE, EF by measuring their lengths. Let N be the midpoint of AD and connect EN. Also, guess the relationship between NE and BF. Prove your hypotheses.

(2) In Figure E8.2, E is in arbitrary location on AB. Try to find a point N on AD such that $NE = BF$, and guess the relationship between DE, EF in this case.

9. The following figure shows a windmill-shaped object, where $AA' = BB' = CC' = 2$, $\angle AOB' = \angle BOC' = \angle COA' = 60°$. Prove that $S_{\triangle AOB'} + S_{\triangle BOC'} + S_{\triangle COA'} < \sqrt{3}$.

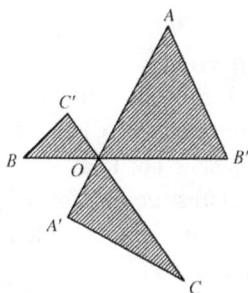

10. Let $ABCD$ be a square with a side length of 1. Let there be a moving set square whose right-angled vertex P moves on AC, and one leg passes through the vertex B, while the other leg intersects the ray DC at Q (as shown in Figures E10.1 and E10.2). Assume the distance between A, P is x.

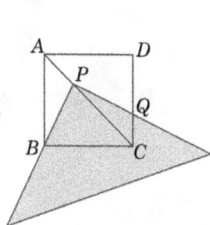

Fig. E10.1

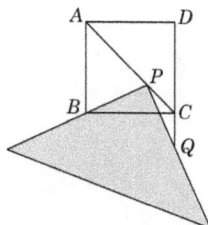

Fig. E10.2

(1) If Q lies on the segment DC, what is the relationship between PQ and PB? Try to prove your result.

(2) When P is sliding on AC, is it possible that $\triangle PCQ$ is an isosceles triangle? If possible, find all values of x such that $\triangle PCQ$ is isosceles, and if not, explain the reason.

11. In the following figure, $ABCD$ is a rhombus with $AB = 2$, $\angle A = 120°$. Points P, Q, K lie on segments BC, CD, BD, respectively. Explain when $PK + QK$ is minimal.

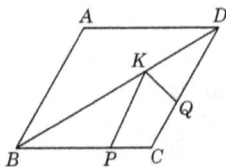

12. In the following figure, $\triangle PMN$ is a right triangle with $\angle P = 90°$, $PM = PN$, $MN = 8$ cm. A rectangle $ABCD$ has side lengths of 8 and 2 cm, where C coincides with M and BC lie on the same line as MN, as shown in Figure E12.1. Suppose $\triangle PMN$ is fixed, and the rectangle is moving along the line MN at a speed of 1 cm/s (as shown in Figure E12.2) until C coincides with N. After moving for x s, the area of the overlapping region of the triangle and the rectangle is y cm^2. Find the functional relationship between y and x.

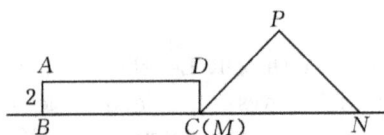

Fig. E12.1

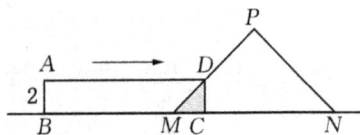

Fig. E12.2

Chapter 27

The Area

The area is an important attribute of a geometric object. Solving a problem with knowledge regarding the area is called the area method. The area method deals with not only problems where the area appears but also those where the area does not appear. The basic idea of this method is to try to express the area of a plane object in different ways or establish a relationship between areas of different figures in order to find information about geometric quantities, such as lengths and angles.

Solving problems using the area method requires the following fundamental formulas and theorems:

1. The area formulas:

$$S_{\text{triangle}} = \frac{1}{2}ah_a = \frac{1}{2}bh_b = \frac{1}{2}ch_c;$$

$$S_{\text{parallelogram}} = ah_a;$$

$$S_{\text{trapezoid}} = \frac{1}{2}(a+b)h;$$

$$S_{\text{rhombus } ABCD} = \frac{AC \times BD}{2}.$$

2. Theorems of equal area:

   (1) Two congruent objects have the same area.
   (2) Two triangles with equal bases and equal altitudes have the same area.
   (3) The area of an object equals the sum of the areas of its different parts.

3. Theorems of area ratio

  (1) The area ratio of two triangles equals the ratio of their base–altitude products.

  (2) The area ratio of two triangles with equal bases is the ratio of their altitudes.

  (3) The area ratio of two triangles with equal altitudes is the ratio of their bases.

  (4) The area ratio of two similar triangles equals the square of their similarity ratio.

4. Important theorems about area:

  (1) If $ABCD$ is a quadrilateral whose diagonals intersect at O, then

$$\frac{S_{\triangle DAC}}{S_{\triangle BAC}} = \frac{OD}{OB}, \quad S_{\triangle OAD} \cdot S_{\triangle OBC} = S_{\triangle OAB} \cdot S_{\triangle OCD}.$$

  (2) If P, Q do not lie on segment AB, then

$$S_{\triangle PAB} = S_{\triangle QAB} \Leftrightarrow \begin{cases} AB \parallel PQ, & P, Q \text{ lie on the same side of } AB; \\ AB \text{ bisects } PQ, & P, Q \text{ lie on different sides of } AB. \end{cases}$$

When using the area method, we need to apply the above knowledge and equal-area transformations adeptly. Area–area and area–length conversions are important tools in such problems.

Example 1. In Figure 27.1, P is an arbitrary point in $\triangle ABC$. Let h_a, h_b, h_c be the altitudes of $\triangle ABC$ on sides a, b, c, respectively, and let t_a, t_b, t_c be the distances from P to the three sides, respectively. Prove that

$$\frac{t_a}{h_a} + \frac{t_b}{h_b} + \frac{t_c}{h_c} = 1.$$

Proof:　Connect PA, PB, PC, as in Figure 27.2. Since $S_{\triangle ABC} = S_{\triangle PAB} + S_{\triangle PBC} + S_{\triangle PCA}$, we have

$$\frac{S_{\triangle PAB}}{S_{\triangle ABC}} + \frac{S_{\triangle PBC}}{S_{\triangle ABC}} + \frac{S_{\triangle PCA}}{S_{\triangle ABC}} = 1.$$

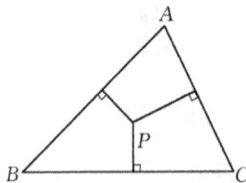

Fig. 27.1

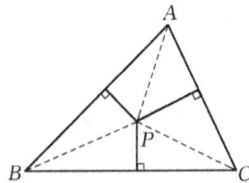

Fig. 27.2

Since $S_{\triangle PAB} = \frac{1}{2}ct_c$, $S_{\triangle PBC} = \frac{1}{2}at_a$, $S_{\triangle PCA} = \frac{1}{2}bt_b$, while $S_{\triangle ABC} = \frac{1}{2}ah_a = \frac{1}{2}bh_b = \frac{1}{2}ch_c$, it follows that

$$\frac{S_{\triangle PAB}}{S_{\triangle ABC}} = \frac{t_b}{h_b}, \quad \frac{S_{\triangle PBC}}{S_{\triangle ABC}} = \frac{t_b}{h_b}, \quad \frac{S_{\triangle PCA}}{S_{\triangle ABC}} = \frac{t_a}{h_a}.$$

Therefore, $\frac{t_a}{h_a} + \frac{t_b}{h_b} + \frac{t_c}{h_c} = 1$.

Think about it This example implies the following two results. Can you prove them? Give them a try:

(1) $at_a + bt_b + ct_c = 2S_{\triangle ABC}$;
(2) if P is an arbitrary point in $\triangle ABC$, and the lines AP, BP, CP intersect their opposite sides at E, F, G, respectively, then

$$\frac{PE}{AE} + \frac{PF}{BF} + \frac{PG}{CG} = 1, \quad \frac{AP}{AE} + \frac{BP}{BF} + \frac{CP}{CG} = 2.$$

Example 2. In Figure 27.3, $OABC$ is a square whose vertices are $O(0,0)$, $A(100,0)$, $B(100,100)$, $C(0,100)$. If there is a grid point P (whose x and y coordinates are both integers) inside the square (excluding the boundary) such that $S_{\triangle POA} \cdot S_{\triangle PBC} = S_{\triangle PAB} \cdot S_{\triangle POC}$, then we call P a "good point." Find the number of good points inside $ABCD$.

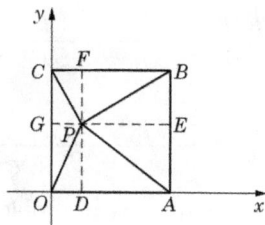

Fig. 27.3

Solution Let PD, PE, PF, PG be the perpendicular segments from P to sides OA, AB, BC, OC, respectively. Then, we easily have $PF + PD = PE + PG = 100$.

If $S_{\triangle POA} \cdot S_{\triangle PBC} = S_{\triangle PAB} \cdot S_{\triangle POC}$, then $PD \cdot PF = PE \cdot PG$, or, equivalently,

$$PD(100 - PD) = PG(100 - PG).$$

Thus, $PD = PG$ or $PG = PF$.

This means P lies on either OB or AC, and all such points satisfy $S_{\triangle POA} \cdot S_{\triangle PBC} = S_{\triangle PAB} \cdot S_{\triangle POC}$. Hence, P is a good point if and only if P is a grid point on either OB or AC.

It is easy to see that both OB and AC contain 99 good points, and their intersection is also a good point, which is counted twice. Therefore, the number of good points is $99 + 99 - 1 = 197$.

Example 3. In Figure 27.4, P is an arbitrary point inside $\triangle ABC$, and the lines AP, BP, CP intersect the opposite sides at D, E, F, respectively. Prove that

$$\frac{PD}{AD} + \frac{PE}{BE} + \frac{PF}{CF} = 1.$$

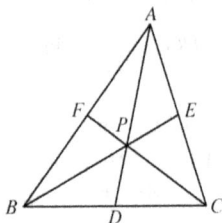

Fig. 27.4

Proof: Since $\dfrac{S_{\triangle PBC}}{S_{\triangle ABC}} = \dfrac{PD}{AD}$, $\dfrac{S_{\triangle PCA}}{S_{\triangle ABC}} = \dfrac{PE}{BE}$, $\dfrac{S_{\triangle PAB}}{S_{\triangle ABC}} = \dfrac{PF}{CF}$, it follows that

$$\frac{PD}{AD} + \frac{PE}{BE} + \frac{PF}{CF} = \frac{S_{\triangle PBC} + S_{\triangle PCA} + S_{\triangle PAB}}{S_{\triangle ABC}} = 1.$$

Remark: This result is used frequently in problems where three segments in a triangle are concurrent.

Example 4. In Figure 27.5, $ABCD$ is a parallelogram, P is a point inside $\triangle BAD$, such that $\triangle PAB$ has an area of 2, $\triangle PCB$ has an area of 5. Find the area of $\triangle PBD$.

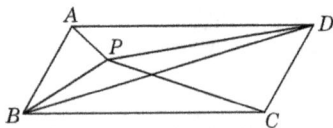

Fig. 27.5

Solution   It is easy to see that $S_{\triangle ABP} + S_{\triangle CDP} = S_{\triangle ADP} + S_{\triangle BCP}$, so $S_{\triangle CDP} - S_{\triangle ADP} = 3$.

Let E, F, G be the projections of A, B, C on the line DP, respectively, and let H be the projection of B on CG, as shown in Figure 27.6.

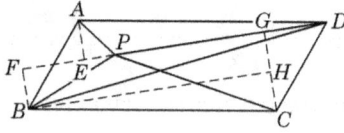

Fig. 27.6

Then, $BFGH$ is a rectangle, so $BF = GH$. Also, $\triangle ADE \cong \triangle CBH$, so $AE = CH$.

Thus, $CG - AE = BF$. Since

$$S_{\triangle ADP} = \frac{1}{2}PD \cdot AE, \quad S_{\triangle CDP} = \frac{1}{2}PD \cdot CG, \quad S_{\triangle PBD} = \frac{1}{2}PD \cdot BF,$$

we conclude that $S_{\triangle PBD} = S_{\triangle CDP} - S_{\triangle ADP} = 3$.

Example 5. Let $ABCD$ be a rectangle, and P be an arbitrary point.

(1) If P is located as in Figure 27.7(1), prove that $S_{\triangle PBC} = S_{\triangle PAC} + S_{\triangle PCD}$.
(2) If P is located as in Figure 27.7(2), guess the quantitative relationship of $S_{\triangle PBC}$, $S_{\triangle PAC}$, $S_{\triangle PCD}$, and prove your result.

Proof: (1) Let EF be a segment passing through P and perpendicular to BC, such that E, F lie on AD, BC, respectively. Since

$$S_{\triangle PBC} + S_{\triangle PAD} = \frac{1}{2}BC \cdot PF + \frac{1}{2}AD \cdot PE = \frac{1}{2}BC \cdot (PE + PF)$$

$$= \frac{1}{2}BC \cdot EF = \frac{1}{2}S_{ABCD}$$

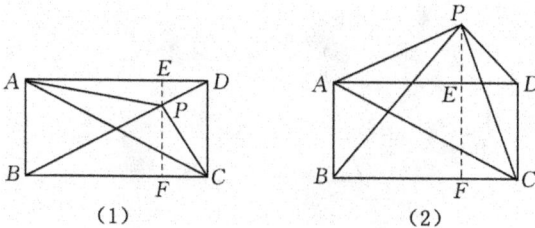

(1)   (2)

Fig. 27.7

and $S_{\triangle PAC} + S_{\triangle PCD} + S_{\triangle PAD} = S_{\triangle ACD} = \frac{1}{2}S_{ABCD}$, we see that

$$S_{\triangle PBC} + S_{\triangle PAD} = S_{\triangle PAC} + S_{\triangle PCD} + S_{\triangle PAD}.$$

Therefore, $S_{\triangle PBC} = S_{\triangle PAC} + S_{\triangle PCD}$.

(2) Guess $S_{\triangle PBC} = S_{\triangle PAC} + S_{\triangle PCD}$.

Let the line through P and perpendicular to BC intersect AD, BC at E, F, respectively. Since

$$S_{\triangle PBC} = \frac{1}{2}BC \cdot PF = \frac{1}{2}BC \cdot PE + \frac{1}{2}BC \cdot EF$$

$$= \frac{1}{2}AD \cdot PE + \frac{1}{2}BC \cdot EF = S_{\triangle PAD} + \frac{1}{2}S_{ABCD}$$

and $S_{\triangle PAC} + S_{\triangle PCD} = S_{\triangle PAD} + S_{\triangle PCD} = S_{\triangle PAD} + \frac{1}{2}S_{ABCD}$, it follows that

$$S_{\triangle PBC} = S_{\triangle PAC} + S_{\triangle PCD}.$$

Think about it

If P is located as in Figure 27.8, what is the relationship of $S_{\triangle PBC}$, $S_{\triangle PAC}$, $S_{\triangle PCD}$ then?

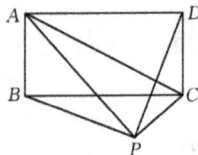

Fig. 27.8

Example 6. In Figure 27.9, $ABCD$ is a trapezoid with $AD \parallel BC$, and the extensions of BA, CD intersect at E. F lies on the extension of CB such that $EF \parallel DB$, and G lies on the extension of BC such that $CG = BF$. Prove that $EG \parallel AC$.

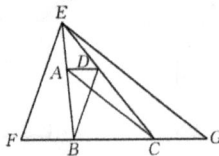

Fig. 27.9

Proof: Connect AG, DF as in Figure 27.10.

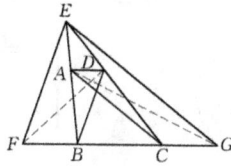

Fig. 27.10

Since $AD \parallel BC$, we have $S_{\triangle ABD} = S_{\triangle ACD}$, which implies $S_{\triangle EBD} = S_{\triangle EAC}$.

Since $EF \parallel DB$, we have $S_{\triangle EBD} = S_{\triangle FBD}$, which implies $S_{\triangle EAC} = S_{\triangle FBD}$.

Since $AD \parallel BC$ and $BF = CG$, it follows that $S_{\triangle FBD} = S_{\triangle GAC}$. Thus, $S_{\triangle EAC} = S_{\triangle GAC}$, and we conclude that $EG \parallel AC$.

Example 7. Suppose the area of $\triangle ABC$ is $1\,\text{cm}^2$, and $AD = DE = EC$, $BG = GF = FC$, as shown in Figure 27.11. Find the area of the shaded quadrilateral.

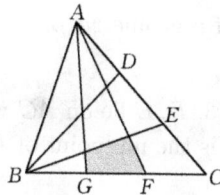

Fig. 27.11

Solution Let AG, BE intersect at N, let AF, BE intersect at P, and connect NC, PC, as shown in Figure 27.12. Assume that $S_{\triangle NGB} = x\,\text{cm}^2$, $S_{\triangle NCE} = y\,\text{cm}^2$, then $S_{\triangle NCG} = 2x\,\text{cm}^2$, $S_{\triangle NEA} = 2y\,\text{cm}^2$.

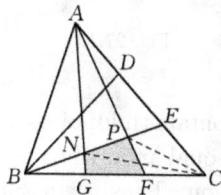

Fig. 27.12

Since $S_{\triangle ABC} = 1 \,\text{cm}^2$ and $AD = DE = EC$, $BG = GF = FC$, we see that $S_{\triangle BCE} = S_{\triangle ACF} = \frac{1}{3} \,\text{cm}^2$, $S_{\triangle ACG} = \frac{2}{3} \,\text{cm}^2$, and

$$\begin{cases} 3x + y = \dfrac{1}{3}, \\ 2x + 3y = \dfrac{2}{3}, \end{cases} \Longrightarrow \begin{cases} x = \dfrac{1}{21}, \\ y = \dfrac{4}{21}. \end{cases}$$

Hence, $S_{\triangle NGB} = \frac{1}{21} \,\text{cm}^2$.

Let $S_{\triangle PCF} = u \,\text{cm}^2$, $S_{\triangle PCE} = v \,\text{cm}^2$, then $S_{\triangle PBF} = 2u \,\text{cm}^2$, $S_{\triangle PAE} = 2v \,\text{cm}^2$. We have

$$\begin{cases} 3u + v = \dfrac{1}{3}, \\ u + 3v = \dfrac{1}{3}, \end{cases}$$

which implies $4u + 4v = \frac{2}{3}$, and $u + v = \frac{1}{6}$. Thus, $S_{PECF} = \frac{1}{6} \,\text{cm}^2$.

Therefore, $S_{\text{shade}} = \frac{1}{3} - \frac{1}{21} - \frac{1}{6} = \frac{5}{42} \,(\text{cm}^2)$.

Remark: (1) When calculating the area of an irregular figure, we can divide it into several simpler parts and compute the sum of their areas, or subtract some parts from a larger object. The method we use depends on the specific problem.

(2) Reducing the area of a geometric object to solve equations can be an effective method.

Example 8. In Figure 27.13, D, E lie on AC with $AD = DE = EC$, F is the midpoint of BC, and G is the midpoint of FC. If $S_{\triangle ABC} = 2004$, find the area of the shaded region.

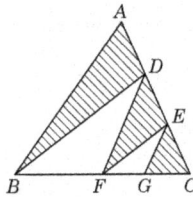

Fig. 27.13

Analysis The condition contains midpoints and trisection points, so some triangles in the figure have equal area, while some triangles have areas that are in integer-multiple relation. Thus, we assume one triangle has an area of x and try to find its relationship with other triangles.

Solution Let $S_{\triangle EGC} = x$. Since $FG = GC$, we have $S_{\triangle EFG} = S_{\triangle EGC} = x$, so

$$S_{\triangle EFC} = S_{\triangle EFG} + S_{\triangle EGC} = 2x.$$

Since $DE = EC$, we have $S_{\triangle DEF} = S_{\triangle EFC} = 2x$, $S_{\triangle DFC} = S_{\triangle DEF} + S_{\triangle EFC} = 4x$.

Since $BF = FC$, we have $S_{\triangle DBF} = S_{\triangle DFC} = 4x$, so

$$S_{\triangle DBC} = S_{\triangle DBF} + S_{\triangle DFC} = 8x.$$

Since $AD : DC = 1 : 2$, it follows that $S_{\triangle ABD} = \frac{1}{2}S_{\triangle DBC} = 4x$. Thus,

$$S_{\triangle ABC} = S_{\triangle ABD} + S_{\triangle DBC} = 4x + 8x = 12x = 2004,$$

so $x = 167$. Therefore,

$$S_{\text{shade}} = 4x + 2x + x = 7x = 7 \times 167 = 1169.$$

Example 9. In Figure 27.14, $ABCD$ is a parallelogram, and F, E lie on sides AD, CD, respectively, such that $AE = CF$. Let P be the intersection of AE, CF. Prove that BP bisects $\angle APC$.

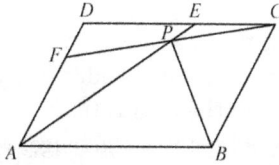

Fig. 27.14

Proof: Connect AC, BE, BF, as in Figure 27.15.

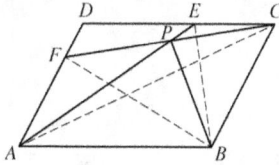

Fig. 27.15

Since $ABCD$ is a parallelogram, we have $S_{\triangle ABE} = S_{\triangle ABC}$, $S_{\triangle BFC} = S_{\triangle ABC}$, so $S_{\triangle ABE} = S_{\triangle BFC}$.

Since $AE = CF$ and $S_{\triangle ABE} = S_{\triangle BFC}$, the corresponding altitudes on AE and CF must be equal. In other words, the distance from B to PA equals the distance from B to PC. Therefore, B lies on the angle bisector of $\angle APC$, or, equivalently, PB bisects $\angle APC$.

Reading

A patient of Hippocrates (an ancient Greek physician and the founder of Western medicine) learned that Hippocrates was good at geometry, so he asked him a question (more of "testing" than asking for advice): what is the simplest way to divide a square into three pieces with an area ratio of 3:4:5?

This is not difficult to solve. For example, for a square of side length 12, Figure 27.16 shows a rather simple way to divide.

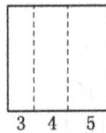

Fig. 27.16

"This requires dividing the sides equally," said the patient, who shook his head after seeing this method and said, "It is required that you do not use a ruler or a compass, and the area ratio is accurate."

This question was undoubtedly a challenge for Hippocrates, who had always been used to ruler-compass construction! However, Hippocrates thought calmly and skillfully applied his proficient geometric knowledge to solve the problem quickly. The method he used was paper folding, and the steps are shown in Figure 27.17.

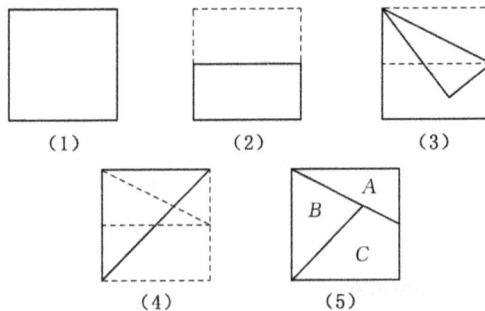

Fig. 27.17

Several folds create three creases on the square paper, thus dividing it into three parts A, B, C, with an area ratio of 3:4:5, as shown in Figure 27.17(5).

Can you prove that this is correct?

Exercises

(I) Multiple-choice questions:

1. In the following figure, $ABCD$ is a square with a side length of 2, E lies on AB, and $EFGB$ is also a square. Let S be the area of $\triangle AFC$, then (   ).

   (A) $S = 2$ (B) $S = 2.4$ (C) $S = 4$ (D) S depends on BE

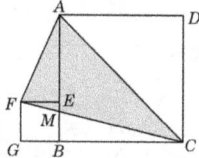

2. In quadrilateral $ABCD$, M, N are the midpoints of AB, CD, respectively. The segments AN, BN, DM, CM divide the quadrilateral into seven parts, whose areas are $S_1, S_2, S_3, S_4, S_5, S_6, S_7$, as shown in the following figure. Then, it is always true that (   ).

   (A) $S_2 + S_6 = S_4$ (B) $S_1 + S_7 = S_4$
   (C) $S_2 + S_3 = S_4$ (D) $S_1 + S_6 = S_4$

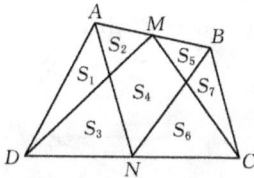

3. In the following figure, $AHSP, BCPQ, DERQ, FGSR$ are all squares, whose areas are $144, 48, 121, 25$, respectively. Suppose $PR = 13$, then the area of the octagon $ABCDEFGH$ is (   ).

   (A) $428 + 66\sqrt{3}$ (B) $520 + \sqrt{2}$ (C) 256 (D) $72\sqrt{6}$

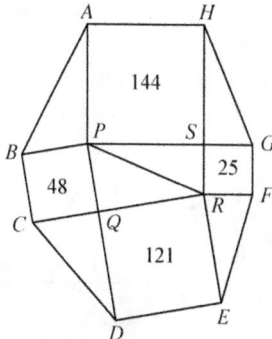

4. In the following figure, $ABCD$ is a rectangle, E, F are the midpoints of BC, CD, respectively, and segments AE, AF intersect BD at G, H, respectively. Let S be the area of the rectangle, then how many of the following assertions are correct? (   ).
   (1) $AG{:}GE = 2{:}1$; (2) $BG{:}GH{:}HD = 1{:}1{:}1$; (3) $S_1{+}S_2{+}S_3 = \frac{1}{3}S$;
   (4) $S_2{:}S_4{:}S_6 = 1{:}2{:}4$.

   (A) 1   (B) 2   (C) 3   (D) 4

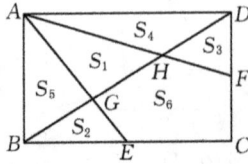

5. In $\triangle ABC$, the side lengths are $a = 3$, $b = 4$, $c = 6$, and h_a, h_b, h_c denote the altitudes on a, b, c, respectively. Then, $(h_a + h_b + h_c)(\frac{1}{h_a} + \frac{1}{h_b} + \frac{1}{h_c})$ equals (   ).
   (A) 4   (B) $\frac{39}{4}$ (C) 8   (D) $\frac{39}{2}$

(II) Fill in the blanks:

6. Let $ABCD$ be a square with a side length 2 and E, F be the midpoints of AB, AD, respectively. G is a point on CF, such that $3CG = 2GF$. Then, the area of $\triangle BEG$ is _____.

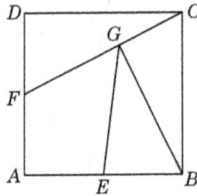

7. Let $ABCD$ be a convex quadrilateral whose diagonals intersect at O, and the areas of $\triangle ABC$, $\triangle ACD$, $\triangle ABD$ are $S_1 = 5$, $S_2 = 10$, $S_3 = 6$, respectively. Then, the area of $\triangle ABO$ is _____.

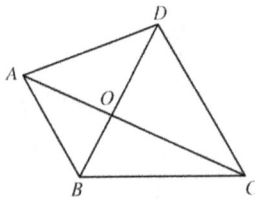

8. In the following figure, $ABCD$ is a parallelogram whose area is $30\,\mathrm{cm}^2$, E is a point on the extension of AD, and EB intersects CD at F. Suppose the area of $\triangle FBC$ is greater than the area of $\triangle DEF$ by $9\,\mathrm{cm}^2$, and $AD = 5\,\mathrm{cm}$, then $DE = $ _____.

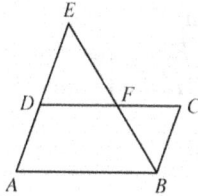

(III) Answer the following questions:

9. A quadrilateral $ABCD$ is divided by its diagonals AC, BD into four triangles, denoted as X, Y, Z, T. Suppose $BE = 80\,\mathrm{cm}$, $CE = 60\,\mathrm{cm}$, $DE = 40\,\mathrm{cm}$, $AE = 30\,\mathrm{cm}$, then what is the ratio of the area of $Z + T$ to the area of $X + Y$?

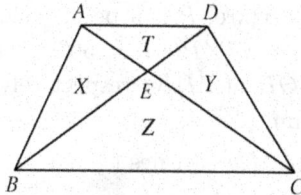

10. Let P be an arbitrary point inside or on the boundary of $\triangle ABC$, and the distances from P to sides a, b, c are x, y, z, respectively. Prove that $ax + by + cz$ is a constant.

11. In the following figure, $\triangle ABC$ is an equilateral triangle with an area of 1. Let P be a point on the plane such that $\triangle PAB$, $\triangle PBC$, $\triangle PCA$ all have the same area. Construct all such points P, and find the corresponding area of $\triangle PAB$.

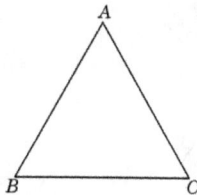

12. Prove that the area of any triangle whose vertices lie inside or on the boundary of a rectangle of area S is not greater than $\frac{1}{2}S$.

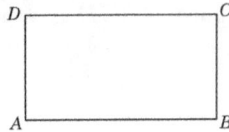

13. In the following figure, $\triangle ABC$ has an area of 1. Extend AB, BC, CA to D, E, F, respectively, such that $AB = BD$, $BC = CE$, $CA = AF$, and connect DE, EF, FD. Find the area of $\triangle DEF$.

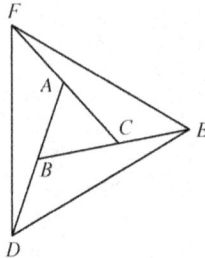

14. In the following figure, P is a point inside a right angle $\angle AOB$, such that $OP = a$, $\angle POA = 30°$. A line passing through P intersects OA, OB at M, N, respectively, such that the area of $\triangle MON$ is minimal.

(1) In this case, MN satisfies ( ).

   (A) $MN \perp OP$ (B) $OM = NM$ (C) $OM = 2ON$
   (D) $PM = PN$

(2) In this case, the area of $\triangle MON$ is _____.

(3) If $\angle AOB$ is an acute angle instead of a right angle, P is a fixed point inside the angle, and a line passing through P intersects OA, OB at M, N, respectively, how would you characterize the line that makes the area of $\triangle MON$ minimal? Prove your result.

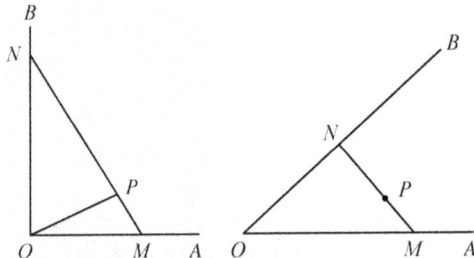

Solution 1

Linear Equations with Absolute Values

1. a can be 0, so none of (A), (B), and (C) is necessarily true. The answer is (D).
2. Either $3990x + 1995 = 1995$ or $3990x + 1995 = -1995$. Thus, $x = 0$ or $x = -1$, and the answer is (B).
3. It follows that $2010x + 2010 = 20 \times 2010$ or $2010x + 2010 = -20 \times 2010$, so $x = 19$ or $x = 21$, and the answer is (D).
4. We have $|x + 1| + |x - 1| = \begin{cases} 2x, & x \geq 1 \\ 2, & -1 < x < 1 \\ -2x, & x \leq -1 \end{cases}$. Since the equation has solutions, it follows that $a \geq 2$, so the answer is (D).
5. $x = -2, y = -2, x + y = -4$. The answer is (A).
6. By $|x| = x + 2$, we have $x = x + 2$ or $-x = x + 2$, and apparently only $-x = x+2$ yields a solution, which is $x = -1$. Hence, $19x^{94} + 3x + 27 = 19 - 3 + 27 = 43$.
7. It follows that $10|x| = 5x$, so $x = 2|x|$. Hence, $x = 0$.
8. It follows that $5 - 4x = 8$ or $5 - 4x = -8$, so $x = -\frac{3}{4}$ or $x = \frac{13}{4}$.
9. Since $|x - |x|| = -2x$, we have $x \leq 0$, so

$$|x - |x|| = |x + x| = |2x|,$$

and $|2x| = -2x$. Therefore, all $x \leq 0$ solve the equation.
10. If $x > 0$, then the equation reduces to $\left(\frac{a}{1997} - 1\right) x = 1997$. If $a = 1997$, then the equation has no solution. If $a \neq 1997$, then $x = \frac{1997^2}{a - 1997}$, and in this case, we have $a > 1997$.

If $x < 0$, then the equation reduces to $\left(\frac{a}{1997} + 1\right) x = -1997$. If $a = -1997$, then the equation has no solution. If $a \neq -1997$, then $x = -\frac{1997^2}{a+1997}$, and in this case, $a > -1997$.

In summary, $-1997 < a \leq 1997$.

11. Remove the absolute value symbol step by step, and note that the solutions are always different. The total number of solutions is 46.

12. It follows that $5x + 6 = 6x - 5$, in which case $x = 11$, or $5x + 6 = 5 - 6x$, in which case $x = -\frac{1}{11}$. However, since $6x - 5 < 0$, $x = -\frac{1}{11}$ is not valid, so the solution is $x = 11$.

13. The original equation reduces to $|x| - 4 = 5$ or $|x| - 4 = -5$, so $|x| = 9$ or $|x| = -1$ (which has no solution). Hence, $x = 9$ or $x = -9$.

14. The original equation reduces to $5x - 3 = \pm(2 + 3x)$ and $2 + 3x \geq 0$. If $5x - 3 = 2 + 3x$, then $x = \frac{5}{2}$, while if $5x - 3 = -(2 + 3x)$, then $x = \frac{1}{8}$. Note that both solutions satisfy $2 + 3x \geq 0$, so the solutions are $x = \frac{5}{2}$ and $x = \frac{1}{8}$.

15. The original equation implies $||x - 2| + 1| = 9$ or $||x - 2| + 1| = -15$, among which the latter has no solution. Further, remove the absolute value symbol, and we have $|x - 2| = 8$ or $|x - 2| = -10$, among which the latter has no solution. Therefore, $x - 2 = 8$ or $x - 2 = -8$, and the solutions are $x = 10$ and $x = -6$.

16. The original equation reduces to $|4x + 8| - 3x = \pm 5$.

If $|4x + 8| = 5 + 3x$, then the solutions are $x = -\frac{13}{7}$ and $x = -3$, but both make $5 + 3x < 0$, so there is no solution in this case.

If $|4x + 8| = -5 + 3x$, then $x = -13$ or $x = -\frac{3}{7}$, but both make $-5 + 3x < 0$, so there is still no solution in this case.

Therefore, the original equation has no solution.

17. If $x \leq 2$, then the equation reduces to $2 - x + 3 - x = 1$, so $x = 2$.

If $2 < x \leq 3$, then the equation reduces to $x - 2 + 3 - x = 1$, which holds for all x in this range.

If $x > 3$, then the equation reduces to $x - 2 + x - 3 = 1$, so $x = 3$, which is out of range.

Therefore, the solutions to the equation are $2 \leq x \leq 3$.

18. The equation reduces to $||x-1|-2| = -a$, so $a \leq 0$. From $|x-1| = 2\pm a$, we solve that $x = 3 + a,\ 3 - a,\ -1 - a,\ -1 + a$. Since the equation has only three different integer solutions, the only possibility is $a = -2$, in which case the solutions are $1, 5, -3$.

Solution 2

Linear Inequalities with Absolute Values

1. Choosing specific numbers, we can see that all three are possible, so the answer is (D).
2. Choose $b = 0$, then (A) and (B) are both wrong. Choose $a = b = 1$, then (D) is wrong, so the answer is (C).
3. $a < 0, |a| > 0$. Since $p > q > 0$, we have $|p| > |q|$, and by the property of inequalities, it is easy to see that (A) is correct.
4. $|a| - |c| \leq |a - c| < |b|$, so $|a| < |b| + |c|$, and the answer is (D).
5. Since $n < m < 2n$, we have $n > 0, m > 0$, and the answer is (B).
6. It follows that $-1 < a < 0, -1 < b < 0, 01 < c < 0$, so $abc > -1$, and the answer is (C).
7. Since $a^2 x - 20 = 0$ and the solution is a prime number, where a is an integer, it can be derived that $x = 5, a = \pm 2$, and from $|ax - 7| > a^2$, we see that only $a = -2$ is valid.
8. By the geometric interpretation of absolute values, we see that the minimum of $|x + 1| + |x - 3|$ is 4, where the equality is attained when $-1 \leq x \leq 3$. Therefore, the equation has no solution when $a < 4$, and if the equation has a solution, then $a \geq 4$.
9. The inequality reduces to

$$\begin{cases} 3x + 5 \leq 10, \\ 3x + 5 \geq -10. \end{cases}$$

Therefore, $-5 \leq x \leq \frac{5}{3}$.

10. (1) If $x + 2 \geq 0$, then the inequality reduces to $x + 2 > \frac{3x+14}{5}$, which gives $x > 2$.

(2) If $x + 2 < 0$, then $x < -2$.

    If $\frac{3x+14}{5} \leq 0$, then the inequality holds automatically, in which case $x \leq -\frac{14}{3}$.

    If $\frac{3x+14}{5} > 0$, then $-(x+2) > \frac{3x+14}{5}$, in which case $-\frac{14}{3} < x < -3$.

    In summary, the solution set to the original inequality is $x > 2$ or $x < -3$.

11. The zeros of $|x+3|$ and $|2x-1|$ are $x = -3$ and $x = \frac{1}{2}$. Thus, we divide the number axis into three pieces: $x \leq -3$, $-3 < x \leq \frac{1}{2}$, $x > \frac{1}{2}$.

(1) If $x \leq -3$, then the inequality reduces to

$$-(x+3) + (2x-1) < 2.$$

    Thus, $x < 6$, and combining with the premise, we have $x \leq -3$.

(2) If $-3 < x \leq \frac{1}{2}$, then the inequality reduces to

$$x + 3 + (2x-1) < 2.$$

    Thus, $x < 0$, and combining with the premise, we have $-3 < x < 0$.

(3) If $x > \frac{1}{2}$, then the inequality reduces to

$$x + 3 - (2x-1) < 2.$$

    Thus, $x > 2$, and combining with the premise, we have $x > 2$.

    In summary, the solution set is $x < 0$ or $x > 2$.

12. Divide into three pieces: $x \leq -2$, $-2 < x \leq 3$, $x > 3$.

(1) If $x \leq -2$, then the inequality reduces to

$$-(x+2) - (x+3) > 2.$$

    Then, we get $x < -\frac{1}{2}$, and in this case, $x \leq -2$.

(2) If $-2 < x \leq 3$, then the inequality reduces to

$$x + 2 - (x+3) > 2.$$

    This is equivalent to $5 > 2$, which holds constantly, so all x such that $-2 < x \leq 3$ satisfy the inequality.

(3) If $x > 3$, then the inequality reduces to

$$x + 2 + x - 3 > 2.$$

    Thus, $x > \frac{3}{2}$, and in this case, $x > 3$.

    In summary, we conclude that the solution set of the original inequality is all real numbers.

Solution 3

Polynomial Factorization (I)

1. $k = \pm 2 \times 1 \times \frac{1}{3} = \pm \frac{2}{3}$, and the answer is (D).
2. We have $-6 = (-1) \times 6 = (-2) \times 3 = (-3) \times 2 = (-6) \times 1$, and since $a < 0$, only $a = -1$ and $a = -5$ are valid. The answer is (C).
3. Since n is odd, let $n = 2k + 1$, where k is an integer. Then,

$$\frac{1}{4}(n^2 - 1) = \frac{1}{4}(n - 1)(n + 1) = k(k + 1),$$

   which is always an even number. The answer is (B).
4. We have $n^3 - n = n(n - 1)(n + 1)$, which is always even and divisible by 3. The answer is (A).
5. $y = (x + 1)^2(x^2 + 1)$, and the answer is (C).
6. Let $a = 10^{n+1}$, then since $x = (10^{n+1})^2 \times 100 + 10^{n+1+2} + 50$, we have

$$x = 100a^2 + 100a + 50$$

$$= 25(4a^2 + 4a + 1) + 25$$

$$= 5^2(2a + 1)^2 + 25.$$

   Therefore, $x - 25$ is a perfect square, and the answer is (B).
7. Since $12345 = (111 - a)(111 + b)$, we have

$$12345 = 111^2 + 111a - 111b - ab$$

$$= 12321 + 111(a - b) - ab.$$

   and $111(a - b) = 24 + ab > 0$, so that $a - b > 0$, and $a > b$. The answer is(A).
8. Since $a + b - ab = 1$, we have $(a - 1)(b - 1) = 0$. Since a is not an integer, which means $a - 1 \neq 0$, and it follows that $b - 1 = 0$, so $b = 1$. The answer is (B).

9. The expression equals

$$2018x^2 - 2018^2x + x - 2018$$
$$= 2018x(x - 2018) + x - 2018$$
$$= (2018x + 1)(x - 2018).$$

10. The expression equals

$$(a - b)a^6 - (a - b)b^6$$
$$= (a - b)(a^6 - b^6)$$
$$= (a - b)[(a^3)^2 - (b^3)^2]$$
$$= (a - b)^2(a + b)(a^2 - ab + b^2)(a^2 + ab + b^2).$$

11. The expression equals

$$[(x + y)^2 + 2(x + y)xy + x^2y^2] - 1$$
$$= (x + y + xy)^2 - 1$$
$$= (x + y + xy - 1)(x + y + xy + 1)$$
$$= (x + y + xy - 1)(x + 1)(y + 1).$$

12. The expression equals

$$x^2 - y^2 + 4y - 4$$
$$= x^2 - (y^2 - 4y + 4)$$
$$= (x - y + 2)(x + y - 2).$$

13. The expression equals

$$x^2y - y^2z + z^2x - x^2z + y^2x + z^2y - xyz - xyz$$
$$= xy(x - z) + y^2(x - z) - xz(x - z) - yz(x - z)$$
$$= (x - z)(xy + y^2 - xz - yz)$$
$$= (x - z)(x + y)(y - z).$$

14. The expression equals $(x - m^2 + mn)(x - mn - n^2)$ by the cross method.

15. The expression equals

$$[(ax - by) + (by - cz)][(ax - by)^2 - (ax - by)(by - cz)$$
$$+ (by - cz)^2] - (ax - cz)^3$$
$$= (ax - cz)[(ax - by)^2 - (ax - by)(by - cz) + (by - cz)^2$$
$$- (ax - cz)^2]$$
$$= (ax - cz)[(ax - by)^2 - (ax - by)(by - cz)$$
$$+ (by + ax - 2cz)(by - ax)]$$
$$= (ax - cz)(ac - by)(3cz - 3by)$$
$$= 3(ax - cz)(ax - by)(cz - by).$$

16. The expression equals $\dfrac{(1-2)(1+2)+(3-4)(3+4)+\cdots+(2017-2018)(2017+2018)}{1+2+3+4+\cdots+2017+2018}$ $= -1$.

17. Let $a = 20012000$, then the original expression equals

$$\frac{a^2}{(a-1)^2 + (a+1)^2 - 2} = \frac{1}{2}.$$

18. Let $a = 2019$, then the original expression equals

$$\frac{a^3 - 2a^2 - (a-2)}{a^3 + a^2 - (a+1)} = \frac{(a-2)(a^2-1)}{(a+1)(a^2-1)} = \frac{a-2}{a+1} = \frac{2017}{2020}.$$

19. The original expression equals

$$\frac{(45.1 - 13.9)(45.1^2 + 45.1 \times 13.9 + 13.9^2)}{31.2} + 45.1 \times 13.9$$
$$= (45.1 + 13.9)^2 = 59^2 = 3481.$$

20. Since the coefficient of x is 1, we have $n = a(a+1)$, where a is an integer. Thus, n can be $1 \times 2 = 2, 2 \times 3 = 6, 3 \times 4 = 12, 4 \times 5 = 20, 5 \times 6 = 30, 6 \times 7 = 42, 7 \times 8 = 56, 8 \times 9 = 72, 9 \times 10 = 90$. There are nine different values.

21. The expression equals $3^{26}(3^2 - 3 - 1) = 3^{24} \times 45$.

22. Let the four consecutive integers be $n, n+1, n+2, n+3$, then

$$n(n+1)(n+2)(n+3) - 1$$
$$= (n^2 + 3n)(n^2 + 3n + 2) + 1$$
$$= (n^2 + 3n + 1)^2.$$

23. Let $x = m^2 + n^2$, where m, n are integers, then

$$2x = 2m^2 + 2n^2$$
$$= 2m^2 + 2n^2 + 2mn - 2mn$$
$$= (m+n)^2 + (m-n)^2.$$

24. We guess

$$\underbrace{1111\ldots1111}_{2n\ \text{copies}} - \underbrace{22\ldots22}_{2n\ \text{copies}} = \underbrace{33\ldots33}_{2n\ \text{copies}}{}^2.$$

Proof: The left-hand side equals

$$\frac{10^{2n} - 1}{9} - 2 \times \frac{10^n - 1}{9} = \frac{1}{9}(10^{2n} - 2 \times 10^n + 1) = \frac{1}{9}(10^n - 1)^2$$
$$= \left(\frac{10^n - 1}{3}\right)^2 = \underbrace{33\ldots33}^2.$$

Solution 4

Polynomial Factorization (II)

1. $m^4 + 4n^4 = m^4 + 4m^2n^2 + 4n^4 - 4m^2n^2 = (m^2 + 2n^2 + 2mn)(m^2 + 2n^2 - 2mn)$. Since m, n are both integers greater than 1, we have $m^2 + 2n^2 - 2mn = (m - n)^2 + n^2 \geq n^2 > 1$. Thus, $m^4 + 4n^4$ is the product of two integers greater than 1, and the answer is (D).

2. $m = 2006^2 - 2 \times 2006 \times 2007 + 2007^2 + 2006^2 \times 2007^2 + 2 \times 2006 \times 2007 + 1 - 1 = (2006 - 2007)^2 + (2006 \times 2007 + 1)^2 - 1 = (2006 \times 2007 + 1)^2$, which is a perfect square, and since $2006 \times 2007 + 1$ is odd, m is also an odd number. The answer is (A).

3. Assume $3x^3 - kx^2 + 4 = (3x - 1)A + 3$, and let $x = \frac{1}{3}$ on both sides. Thus, $k = 10$, and the answer is (D).

4. We can plug in $x = -2, x = 3, x = \frac{1}{2}, x = -\frac{1}{2}$ and check whether the polynomial equals 0. The answer is (C).

5. Let $x^3 + ax^2 + bx + 15 = (x^2 - 2)(x + m)$, then by the method of undetermined coefficients, we solve that $b = -2, a = m = -\frac{15}{2}$, so $a^2b^2 = 225$.

6. $a = -6$.

7. $k = 4$ or 5.

8. $m = 12$.

9. The original expression equals

$$a^4 + 16a^2b^2 + 64b^4 - 16a^2b^2$$

$$= (a^2 + 8b^2)^2 - 16a^2b^2$$

$$= (a^2 + 8b^2 - 4ab)(a^2 + 8b^2 + 4ab).$$

10. The original expression equals

$$(x^3 + x^2) + (x^2 + x) - (6x + 6)$$
$$= x^2(x + 1) + x(x + 1) - 6(x + 1)$$
$$= (x + 1)(x^2 + x - 6)$$
$$= (x + 1)(x - 2)(x + 3).$$

11. The expression equals

$$(x^3 + 2x^2) + (x^2 + 2x) + (x + 2) = (x + 2)(x^2 + x + 1).$$

12. The expression equals

$$(x^2 - x)^2 - 8(x^2 - x) + 12$$
$$= (x^2 - x - 6)(x^2 - x - 2)$$
$$= (x + 2)(x - 3)(x - 2)(x + 1).$$

13. Suppose the original expression equals $(3x - y + a)(x + 2y + b)$, then expand the right-hand side, and we have

$$3x^2 + 5xy - 2y^2 + x + 9y - 4$$
$$= 3x^2 + 5xy - 2y^2 + (a + 3b)x$$
$$+ (2a - b)y + ab.$$

Comparing the coefficients on both sides, we get

$$\begin{cases} a + 3b = 1, \\ 2a - b = 9, \\ ab = -4. \end{cases}$$

Hence, $a = 4$, $b = -1$, and the original expression equals $(3x - y + 4)$ $(x + 2y - 1)$.

14. The divisors of 6 are $1, -1, 2, -2, 3, -3, 6, -6$. Plug in these numbers, and we find that the polynomial equals 0 when $x = -1, -2, -3$. Therefore, the original expression equals $(x + 1)(x + 2)(x + 3)$.

15. Let $n = 2018$, then $2019 = n + 1$. The original expression equals

$$x^4 - nx^2 + (n + 1)x - n = (x^4 + x) - n(x^2 - x + 1)$$
$$= (x^2 - x + 1)(x^2 + x - n) = (x^2 - x + 1)(x^2 + x + 2018).$$

16. Let $u = x^2 + 5x + 6$, then the original expression equals

$$u(u + 2x) - 3x^2$$
$$= u^2 + 2ux - 3x^2$$
$$= (u + 3x)(u - x)$$
$$= (x^2 + 8x + 6)(x^2 + 4x + 6).$$

17. Suppose the original expression equals $(x - 2y + a)(x + y + b)$. Then, by expanding it, we have

$$\begin{cases} a + b = -1, \\ a - 2b = 5, \\ ab = -2. \end{cases}$$
        ①
        ②
        ③

From ①, ②, we obtain $a = 1, b = -2$, and ③ is also satisfied.

Therefore, the original expression equals $(x - 2y + 1)(x + y - 2)$.

18. The divisors of 15 are $\pm 1, \pm 3, \pm 5, \pm 15$. Plug in these values, and we see that the polynomial equals 0 when $x = 1, -3, 5$. Therefore, the original expression equals $(x - 1)(x + 3)(x - 5)$.

19. Plug in $x = -\frac{3}{2}$, and the polynomial equals

$$2 \times \left(-\frac{3}{2}\right)^4 - 5 \times \left(-\frac{3}{2}\right)^3 - 10 \times \left(-\frac{3}{2}\right)^2 + 15 \times \left(-\frac{3}{2}\right) + 18 = 0.$$

Thus, $x + \frac{3}{2}$ is a factor of $2x^4 - 5x^3 - 10x^2 + 15x + 18$, and so is $2x + 3$.

20. Let $8x^2 - 2xy - 3y^2 = A^2 - B^2$, then $(2x + y)(4x - 3y) = (A + B)(A - B)$.
Set $\begin{cases} A + B = 2x + y, \\ A - B = 4x - 3y, \end{cases}$ then $\begin{cases} A = 3x - y, \\ B = 2y - x. \end{cases}$

Therefore, $8x^2 - 2xy - 3y^2 = (3x - y)^2 - (2y - x)^2$.

21.

$$3x^2 - 8xy + 9y^2 - 4x + 6y + 13$$
$$= (2x^2 - 8xy + 8y^2) + (x^2 - 4x + 4) + (y^2 + 6y + 9)$$
$$= 2(x - 2y)^2 + (x - 2)^2 + (y + 3)^2 \geq 0.$$

Also, note that when $(x - 2)^2 = 0$, $(y + 3)^2 = 0$, we have $(x - 2y)^2 \neq 0$, so the three squares cannot be zero simultaneously. Therefore, the original expression is always positive.

22. It follows that

$$3(4a^2 - 4a|b| + b^2) + (4b^2 + 4b|c| + c^2) + 4(c^2 + 4c + 4) \leq 0,$$

or, equivalently,

$$3(2a - |b|)^2 + (2b + |c|)^2 + 4(c + 2)^2 \leq 0.$$

Hence,

$$\begin{cases} 2a - |b| = 0, \\ 2b + |c| = 0, \\ c + 2 = 0. \end{cases}$$

Solve the system of equations, and we get $a = \frac{1}{2}, b = -1, c = -2$.

Solution 5

Calculation of Rational Fractions

1. $x \neq 0$ and $x \neq \frac{1}{a}$.

2. $\frac{x^2+11}{x+1} = \frac{x^2-1+12}{x+1} = x - 1 + \frac{12}{x+1}$, so $x+1$ divides 12, and all such x are $0, 1, 2, 3, 5, 11$, whose sum is 22.

3. $m = \frac{b}{a}, \frac{a^{-2}+b^2}{a^2+b^{-2}} = \frac{b^2}{a^2} = m^2$.

4. The condition reduces to $3x = 2y$, so $\frac{x}{y} = \frac{2}{3}$.

5. $a = 20b = 200c$, so $\frac{a+b}{b+c} = \frac{210c}{11c} = \frac{210}{11}$.

6. The expression equals $6 - \frac{2}{(x+1)^2+1}$, and the minimum is 4.

7. $y_2 = \frac{1}{x}, y_3 = 2x, y_4 = \frac{1}{x}, \ldots, y_{2020} = \frac{1}{x}$, so $y_1 \cdot y_{2020} = 2x \cdot \frac{1}{x} = 2$.

8. The expression equals $\frac{(a^3+b^3)-ab(a+b)}{(|a|-|b|)^2} = \frac{(a+b)(a-b)^2}{(|a|-|b|)^2}$.

   When $ab \geq 0$, then it equals $a + b$, and when $ab < 0$, it equals $\frac{(a-b)^2}{a+b}$.

9. $\frac{ab}{a+b}$.

10. It equals $\frac{2}{x^2-1} - \frac{2}{x^2+1} - \frac{4}{x^4+1} - \frac{8}{x^8+1} = \frac{4}{x^4-1} - \frac{4}{x^4+1} - \frac{8}{x^8+1} = \frac{16}{x^{16}-1}$.

11. It equals $\frac{1}{x-1} + \frac{1}{x-2} - \frac{1}{x-1} + \frac{1}{x-3} - \frac{1}{x-2} + \cdots + \frac{1}{x-100} - \frac{1}{x-99} = \frac{1}{x-100}$.

12. It equals $\frac{(1+a)(1+x)(1-a)(1-x)}{(1+b)(1+x)(1-b)(1-x)} \cdot \frac{(1+b)(1+y)(1-b)(1-y)}{(1+a)(1+y)(1-a)(1-y)} = 1$.

13. It equals $\frac{(x-1)(x^2+x+1)}{x^2(x+1)+(x+1)^2} + \frac{(x+1)(x^2-x+1)}{x^2(x-1)-(x-1)^2} - \frac{2(x^2+1)}{x^2-1} = \frac{x-1}{x+1} + \frac{x+1}{x-1} - \frac{2(x^2+1)}{x^2-1} = 0$.

14. Let $k = \frac{ax+7}{bx+11}$, then $(a - kb)x + 7 - 11x = 0$, so $a - kb = 0, 7 - 11k = 0$. Eliminating k, we have $11a - 7b = 0$.

15. Since $\frac{a}{b} - \frac{b}{a} = \frac{a+b}{a} \cdot \frac{a-b}{b}$ and $\frac{a+b}{a} + \frac{a-b}{b} = \frac{b}{a} + \frac{a}{b}$, we see that the two desired factors are $\frac{a+b}{a}, \frac{a-b}{b}$.

16. We have $a_2 = \frac{x-1}{x}, a_3 = -\frac{1}{x-1}, a_4 = x$. This means $a_{3k+1} = a_1$. Since $2006 = 3 \times 668 + 2$, we have $a_{2006} = a_2 = \frac{x-1}{x}$.

17. It follows from the given inequality that $\frac{13}{7} < \frac{n+k}{n} < \frac{15}{8}$, so $\frac{6}{7} < \frac{k}{n} < \frac{7}{8}$, and $\frac{6n}{7} < k < \frac{7n}{8}$. Hence, there is only one integer between $\frac{6n}{7}$ and $\frac{7n}{8}$, so $\frac{7n}{8} - \frac{6n}{7} \leq 2$. This shows that $n \leq 112$, and when $n = 112$, we have $96 < k < 98$, and $k = 97$ is the only integer that satisfies the condition. Therefore, the maximum of n is 112.

Solution 6

Partial Fractions

1. It follows that $\frac{1}{n^2+3n} = \frac{(A+B)n+34}{n(n+3)}$, so $A + B = 0$, $3A = 1$. Hence, $A = \frac{1}{3}$, $B = -\frac{1}{3}$.

2. It follows that $\frac{3x-4}{x^2-3x+2} = \frac{(a+b)x-(2a+b)}{x^2-3x+2}$. Comparing the coefficients, we have

$$\begin{cases} a + b = 3, \\ 2a + b = 4. \end{cases}$$

Hence, $a = 1$, $b = 2$, and $ab = 2$.

3. $\frac{x^2-1}{x^2-5x+6} = \frac{x^2-5x+6+5x-7}{x^2-5x+6} = 1 + \frac{5x-7}{(x-2)(x-3)}$.

Let $\frac{5x-7}{(x-2)(x-3)} = \frac{a}{x-2} + \frac{b}{x-3}$, then

$$\begin{cases} a + b = 5, \\ 3a + 2b = 7. \end{cases}$$

Therefore, $M + a + b = 6$.

4. It follows easily that $m = -1$, $n = 1$. Thus,

$$\frac{ax - 1}{(x - a)(x^2 + 1)} = \frac{b(x^2 + 1) + (cx + 5)(x - 1)}{(x - 1)(x^2 + 1)},$$

so $ax - 1 = (b + c)x^2 + (5 - c)x + b - 5$. Then, we have

$$\begin{cases} b + c = 0, \\ 5 - c = a, \\ b - 5 = -1. \end{cases}$$

Therefore, $a = 9$, $b = 4$, $c = -4$.

5. Removing the denominator, we have

$$A - 17x = (7B + 4)x + (24 - B).$$

Thus,

$$\begin{cases} 7B + 4 = -17, \\ 24 - B = A. \end{cases}$$

Solving the equations, we obtain $A = 27$, $B = -3$.

6. $A = -1$, $B = -6$, $4A - 2B = 8$.

7. $A = 10$, $B = 11$, $C = -8$, so $A^2 + B^2 + C^2 = 285$.

8. It follows that $a = 2$, $b = -1$, so $c = 3$, and

$$\frac{Mx + N}{x^2 + x - 2} = \frac{2}{x + 2} - \frac{1}{x - 1}.$$

Let $x = 0$, and we obtain $N = -4$.

9. Assume that $\frac{x^2 + 2}{(x-1)^3} = \frac{A}{x-1} + \frac{B}{(x-1)^2} + \frac{C}{(x-1)^3}$. Then,

$$x^2 + 2 = Ax^2 + (B - 2A)x + (A - B + C).$$

Comparing the coefficients, we have

$$\begin{cases} A = 1, \\ B - 2A = 0, \\ A - B + C = 2. \end{cases}$$

Hence, $A = 1$, $B = 2$, $C = 3$, and $\frac{x^2 + 2}{(x-1)^3} = \frac{1}{x-1} + \frac{2}{(x-1)^2} + \frac{3}{(x-1)^3}$.

10. Since $x^3 + 1 = (x + 1)(x^2 - x + 1)$, we assume that

$$\frac{2x^2 + x + 2}{x^3 + 1} = \frac{A}{x + 1} + \frac{Bx + C}{x^2 - x + 1}.$$

Expanding, we get

$$2x^2 + x + 2 = (A + B)x^2 + (B + C - A)x + (A + C).$$

Thus,

$$\begin{cases} A + B = 2, \\ B + C - A = 1, \\ A + C = 2. \end{cases}$$

Solve the equations, and we obtain $A = 1$, $B = 1$, $C = 1$. Therefore, $\frac{2x^2 + x + 2}{x^3 + 1} = \frac{1}{x + 1} + \frac{x + 1}{x^2 - x + 1}$.

11. Let $\frac{x^2+1}{(x+1)^2(x+2)} = \frac{A}{x+1} + \frac{B}{x+2} + \frac{C}{(x+1)^2}$, then

$$x^2 + 1 = (A+B)x^2 + (3A + 2B + C)x + (2A + B + 2C).$$

Comparing the coefficients, we get

$$\begin{cases} A + B = 1, \\ 3A + 2B + C = 0, \\ 2A + B + 2C = 1. \end{cases}$$

Hence, $A = -4$, $B = 5$, $C = 2$, and $\frac{x^2+1}{(x+1)^2(x+2)} = -\frac{4}{x+1} + \frac{5}{x+2} + \frac{2}{(x+1)^2}$.

12. Since $4x^2 - 4x - 15 = (2x + 3)(2x - 5)$, we assume that

$$\frac{12x^2 + 20x - 29}{(2x - 1)(4x^2 - 4x - 15)} = \frac{A}{2x - 1} + \frac{B}{2x + 3} + \frac{C}{2x - 5}.$$

Expanding and comparing coefficients, we can solve that $A = 1$, $B = -1$, $C = 3$.

Therefore, the original expression equals $\frac{1}{2x-1} - \frac{1}{2x-3} + \frac{3}{2x-5}$.

Solution 7

Polynomial Equations and Fractional Equations with Unknown Constants

1. Note that a can be any real number, so (1), (2), (3), (4) are all wrong, and the answer is (A).
2. The equation reduces to $(3a-2b-5)x = 12-2a-3b$. If it has infinitely many solutions, then $3a - 2b - 5 = 0$, $12 - 2a - 3b = 0$. Hence, $a = 3$, $b = 2$, and the answer is (C).
3. $a - 2 = 2 - a$, so $a = 2$, and the answer is (B).
4. It follows that $x = \frac{-1-a}{2}$. Since $x > 0$ and $x \neq 1$, the answer is (C).
5. The equation reduces to $(ba - a^2)x = 2a^2 b - 2a^3$. Equivalently, $a(b - a)x = 2a^2(b - a)$.
   Since $a \neq b$, $a \neq 0$, we have $a(b - a) \neq 0$, so $x = 2a$.
6. The equation reduces to

$$\frac{z}{2(z-2)} - \frac{z+2}{3(z+1)} = \frac{z^2}{6(z-2)(z+1)}.$$

Hence, $3z(z+1) - 2(z+2)(z-2) = z^2$, so $3z = -8$. Therefore, $z = -\frac{8}{3}$, which is a valid solution.

7. Removing the denominator, we have $n(1-ax) + m(1+bx) = abx$, which reduces to $(ab - bm + an)x = m + n$.
   Since $ab + an \neq bm$, we have $ab - bm + an \neq 0$, so

$$x = \frac{m+n}{ab - bm + an}.$$

8. Removing the denominator, we have $(a-b)(x+1)+(a+b)(x-1)=2a$, which is equivalent to $ax = a+b$.

    If $a=0$, then the condition implies $a+b \neq 0$, and the equation has no solution.

    If $a \neq 0$, then $x = \frac{a+b}{a}$.

9. Since $\frac{1}{(x+n)(x+n+1)} = \frac{1}{x+n} - \frac{1}{x+n+1}$, the original equation can be reduced to $\frac{1}{x+1} = \frac{1999}{2000}$, and the solution is $x = \frac{1}{1999}$. After checking, we see that $x = \frac{1}{1999}$ is a valid solution.

10. The original equation reduces to

$$\frac{1}{x+5} + \frac{1}{x+9} = \frac{1}{x+6} + \frac{1}{x+8},$$

so $\frac{1}{x+5} - \frac{1}{x+6} = \frac{1}{x+8} - \frac{1}{x+9}$. Thus, $\frac{1}{(x+5)(x+6)} = \frac{1}{(x+8)(x+9)}$, which means $(x+5)(x+6) = (x+8)(x+9)$. Therefore, $x = -7$, which is a valid solution.

11. The solutions to the two equations are $x = \frac{27-2a}{21}$ and $x = \frac{2a}{7}$.

    Since $|\frac{27-2a}{21}| = |\frac{2a}{7}|$, we have $27 - 2a = \pm 6a$, so $a = \frac{27}{8}$ or $-\frac{27}{4}$.

    If $a = \frac{27}{8}$, then both solutions are $x = \frac{27}{28}$.

    If $a = -\frac{27}{4}$, then the first equation has a solution of $x = \frac{27}{14}$ and the second has a solution of $x = -\frac{27}{14}$.

12. The equation reduces to $(m-1)x = -10$.

    If the equation has an extraneous solution, then it must be 2 or -2. Plug into $(m-1)x = -10$, and we obtain that $m = -4$ or $m = 6$.

13. Suppose the fraction before reduction is $\frac{2k}{3k}$, where k is a positive integer. Let a be the added number, then $\frac{2k+a}{3k+a} = \frac{8}{11}$, $\frac{2k-(a+1)}{3k-(a+1)} = \frac{5}{9}$. Thus, $2k = 3a$, $3k - 4a = 4$, so $a = 8$, $k = 12$.

    Therefore, the original fraction is $\frac{24}{36}$.

14. Let $u = \frac{1}{3x+2y}$, $v = \frac{1}{3x-2y}$, then the equation system becomes

$$\begin{cases} 2u + 5v = 3, \\ 9u - 2v = \dfrac{5}{4}. \end{cases}$$

Solving the equations, we get $\begin{cases} u = \frac{1}{4}, \\ v = \frac{1}{2}. \end{cases}$ Thus, $\begin{cases} 3x + 2y = 4, \\ 3x - 2y = 2. \end{cases}$ Finally, we obtain $\begin{cases} x = 1, \\ y = \frac{1}{2}, \end{cases}$ which is a valid solution.

15. The system of equations is equivalent to

$$
\begin{cases}
\dfrac{1}{x} + \dfrac{1}{y} = 1, & \text{①} \\[2mm]
\dfrac{1}{y} + \dfrac{1}{z} = \dfrac{1}{2}, & \text{②} \\[2mm]
\dfrac{1}{x} + \dfrac{1}{z} = \dfrac{1}{3}. & \text{③}
\end{cases}
$$

From ① + ② − ③, we obtain $y = \frac{12}{7}$, so $x = \frac{12}{5}$, $z = -12$. After checking, $x = \frac{12}{5}$, $y = \frac{12}{7}$, $z = -12$ is a valid solution.

16. Since A, B have the same salinity after pouring, it is the same salinity as if we mix the salt water in the two containers in the beginning. In other words, it equals

$$
\frac{40 \times 20\% + 60 \times 4\%}{40 + 60} = 10.4\%.
$$

Suppose we move x kg from A to C and $6x$ kg from B to D, then

$$
\frac{(40 - x) \cdot 20\% + 6x \times 4\%}{40 - x + 6x} = 10.4\%.
$$

Solve the equation, and we get $x = 8$.

Therefore, we have moved 8 kg from A to C.

17. Suppose it takes x, y, z days for A, B, C, respectively, to complete the project alone. Then,

$$
\frac{1}{y} + \frac{1}{z} = \frac{a}{x}, \quad \text{so} \quad \frac{1}{z} = \frac{a}{x} - \frac{1}{y}.
$$

$$
\frac{1}{x} + \frac{1}{z} = \frac{b}{y}, \quad \text{so} \quad \frac{1}{z} = \frac{b}{y} - \frac{1}{x}.
$$

Thus, we have

$$
\frac{1}{x} = \frac{b+1}{ab-1} \cdot \frac{1}{z}, \quad \frac{1}{y} = \frac{a+1}{ab-1} \cdot \frac{1}{z}.
$$

Therefore, $\frac{1}{x} + \frac{1}{y} = \frac{a+b+2}{ab-1} \cdot \frac{1}{z}$, which means the time it takes when A and B work together equals $\frac{a+b+2}{ab-1}$ times the time it takes when C works alone.

18. Assume the front has a circumference of x m and the rear wheel has a circumference of y m. Then,

$$\begin{cases} \dfrac{12}{x} - \dfrac{12}{y} = 6, \\[4mm] \dfrac{12}{\left(1 + \dfrac{1}{4}\right)x} - \dfrac{12}{\left(1 + \dfrac{1}{5}\right)y} = 4. \end{cases}$$

Solve the system of equations by changing variables, and we obtain

$$\begin{cases} x = \frac{2}{5}, \\ y = \frac{1}{5}. \end{cases}$$

Therefore, the circumference of the front wheel is $\frac{2}{5}$ m, and the circumference of the rear wheel is $\frac{1}{2}$ m.

Solution 8

Real Numbers

1. $n = 2010^2 + (2010+1)^2 - 1 = 2 \times 2010^2 + 4020$,
$$2n + 1 = 2^2 \times 2020^2 + 2 \times 4020 + 1 = (2 \times 2010 + 1)^2,$$
   so $\sqrt{2n+1} = 4021$, and the answer is (C).
2. This number is a^2, and its adjacent integers are $a^2 + 1$, $a^2 - 1$, so the answer is (C).
3. It follows that $a \geq 2007$, and we have $\sqrt{a - 2007} = 2006$, so $a - 2006^2 = 2007$, and the answer is (C).
4. It follows that $x^2 - 4 = 0$, $2x + y = 0$, so $x = -2$, $y = 4$, or $x = 2$, $y = -4$. Hence, $x - y = 6$ or -6. The answer is (D).
5. From $ab - a - b + 1 = 0$, we get $(a-1)(b-1) = 0$, and since a is irrational, necessarily $b = 1$, and the answer is (B).
6. Let $x = \sqrt{2} - 2$, then $(x-2)(x+6) = -14$, which is rational, x^2, $(x+6)^2$ are both irrational, and $(x+2)^2$ is rational. Hence, (A), (B), (D) are all incorrect.

   On the other hand, $(x+2)(x-6) = (x-2)(x+6) - 8x$, while $(x-2)(x+6)$ is rational and $8x$ is irrational, so it follows that $(x+2)(x-6)$ is irrational. Hence, the answer is (C).
7. $\pm a$.
8. $(\sqrt{m} + 1)^2$.
9. Note that $\sqrt{203} > 14$, and $14^3 = 2744 > 2007$, so the original expression equals $\sqrt{203} \wedge \sqrt{2008} = \sqrt{203}$.
10. $-\sqrt{275} < -4\sqrt{11}$, $4 - \sqrt{2} > 1 + \sqrt{2}$, $\sqrt{61} - 1 > \sqrt{6} + 3$.
11. $m = 2$, $n = \sqrt{5} - 2$, so $mn + 4 = 2\sqrt{5}$.
12. It follows that $y = 1 - a^2 - |\sqrt{x} - \sqrt{3}|$. Plug into the second equation, and we have
$$|x - 3| = 1 - a^2 - |\sqrt{x} - \sqrt{3}| - 1 - b^2,$$

so that $|x - 3| + |\sqrt{x} - \sqrt{3}| + a^2 + b^2 = 0$. Hence, $x = 3$, $a = 0$, $b = 0$, and $y = 1$.

Therefore, $2^{x+y} + 2^{a+b} = 2^4 + 2^0 = 17$.

13. The condition reduces to $(2x + 4y - 3) + (x - y + 2)\sqrt{3} = 0$. Since x, y are rational, it follows that

$$\begin{cases} 2x + 4y - 3 = 0, \\ x - y + 2 = 0. \end{cases}$$

Thus, $\begin{cases} x = -\frac{5}{6}, \\ y = \frac{7}{6}. \end{cases}$

14. We have $(b^2 + 2) + 2b\sqrt{2} = (ac + 2) + (a + c)\sqrt{2}$, and it follows that $ac = b^2$, $a + c = 2b$. Hence,

$$(a - c)^2 = (a + c)^2 - 4ac = 4b^2 - 4b^2 = 0.$$

15. It equals

$$(1.1 - 0.14) \div \left(\frac{3}{25} - \frac{1}{12.5} \right) = 0.96 \times 25 = 24.$$

16. It equals

$$\sqrt{(82 - 1) \times 82 \times 83 \times (83 + 1) + 1}$$
$$= \sqrt{(82 \times 83 - 2) \times 82 \times 83 + 1}$$
$$= \sqrt{(82 \times 83)^2 - 2 \times 82 \times 83 + 1}$$
$$= \sqrt{(82 \times 83 - 1)^2} = 82 \times 83 - 1 = 6805.$$

17. Let $x = \frac{41}{12}$, then $x^2 + 5 = \left(\frac{49}{12} \right)^2$, $x^2 - 5 = \left(\frac{31}{12} \right)^2$.

18. The equation reduces to $(\frac{1}{2}x + \frac{1}{3}y - 4) + (\frac{1}{3}x + \frac{1}{2}y - 1)\pi = 0$. Since π is irrational, it follows that

$$\begin{cases} \dfrac{1}{2}x + \dfrac{1}{3}y - 4 = 0, \\ \dfrac{1}{3}x + \dfrac{1}{2}y - 1 = 0. \end{cases}$$

Hence, $x = 12$, $y = -6$, and $x + y = 6$.

19. Since $\sqrt{a} = \frac{7 - 5|b|}{3}$, we have

$$S = 2\sqrt{a} - 3|b| = 2 \times \frac{7 - 5|b|}{3} - 3|b| = \frac{14 - 10|b|}{3} - 3|b|.$$

Thus, $19|b| = 14 - 3S$. Since $|b| \geq 0$, we have $14 - 3S \geq 0$, so $S \leq \frac{14}{3}$.

Plug in $|b| = \frac{7-3\sqrt{a}}{5}$, and we also have

$$S = 2\sqrt{a} - 3 \times \frac{7-3\sqrt{a}}{5} = 2\sqrt{a} - \frac{21-9\sqrt{a}}{5}.$$

Thus, $19\sqrt{a} = 5S + 21$, and since $\sqrt{a} \geq 0$, we have $5S + 21 \geq 0$, or $S \geq -\frac{21}{5}$.

Therefore, the value range of S is $-\frac{21}{5} \leq S \leq \frac{14}{3}$.

20. It follows that $a \geq 3$, and the condition reduces to

$$|b+2| + \sqrt{(a-3)b^2} = 0.$$

Hence, $b+2 = 0$, $a-3 = 0$. Therefore, $b = -2$, $a = 3$, and $\sqrt{a+b} = 1$.

21. It follows from the condition that

$$(a^2 + b^2 - 25)\sqrt{3} + (a - b - 1)\sqrt{2} = 0.$$

Thus,

$$\begin{cases} a^2 + b^2 - 25 = 0, \\ a - b - 1 = 0. \end{cases}$$

Consequently,

$$ab = \frac{1}{2}[(a^2 + b^2) - (a-b)^2] = \frac{1}{2}(25 - 1) = 12.$$

Solution 9

Quadratic Radicals

1. It follows that $\frac{x^2-2}{5x-4} \geq 0$, $\frac{x^2-2}{4x-5} \geq 0$, so $\frac{x^2-2}{5x-4} = 0$, and $x^2 = 2$, $y = 2$. Thus, $x^2 + y^2 = \sqrt{6}$, and the answer is (D).

2. It follows that $a < 0$, $c - a > 0$, $a + b < 0$, $b + c < 0$. Hence,

$$\sqrt{a^2} + \sqrt{(c-a)^2} - \sqrt{(a+b)^2} + \sqrt{(b+c)^2}$$
$$= -a + c - a + a + b - b - c = -a.$$

The answer is (B).

3. It follows by condition that $x \leq 0$, so

$$\sqrt{(x-3)^2} = |x-3| = 3 - x.$$

The answer is (B).

4. If $\sqrt{-\frac{1}{1-x}}$ is defined, then $x - 1 < 0$, so

$$(x-1)\sqrt{-\frac{1}{x-1}} = -\sqrt{(x-1)^2 \cdot \left(-\frac{1}{x-1}\right)} = -\sqrt{1-x}.$$

The answer is (D).

5. It follows that $a^2 - 4\sqrt{2} = m + n - 2\sqrt{mn}$. Thus,

$$\begin{cases} m + n = a^2, \\ mn = 8. \end{cases}$$

Since m, n are positive integers, and $m > n$, the only possibilities are $(m, n) = (8, 1)$ and $(4, 2)$. Hence, $a^2 = 9$ or $a^2 = 6$, but a is a positive integer, so $a^2 = 9$ and $a = 3$. Therefore, $m = 8$, $n = 1$, and the answer is (A).

6. $4\sqrt{3+2\sqrt{2}} - \sqrt{41+24\sqrt{2}} = 4\sqrt{(\sqrt{2}+1)^2} - \sqrt{(3+4\sqrt{2})^2} = 4(\sqrt{2}+1) - (3+4\sqrt{2}) = 1$, so the answer is (B).

7. $x^2 = 6 - 2\sqrt{4} = 2$, $x < 0$, so $x = -\sqrt{2}$, and $[x] = -2$. The answer is (A).

8.

$$\left(\sqrt{x+2018} - \sqrt{x-2019}\right)\left(\sqrt{x+2018} + \sqrt{x-2019}\right)$$

$$= (x+2018) - (x-2019) = 4037.$$

Thus, $\sqrt{x+2018} + \sqrt{x-2019} = \frac{4037}{2020}$.

9. It follows that $a \geq 1999$, so $a - 1999 = 1998^2$, and $a - 1998^2 = 1999$.

10. It equals

$$\sqrt{3}\left(\sqrt{10} + \sqrt{7} - \sqrt{3}\right)\left(\sqrt{3} + \sqrt{10} - \sqrt{7}\right)$$

$$= \sqrt{3}\left[\left(\sqrt{10}\right)^2 - \left(\sqrt{7} - \sqrt{3}\right)^2\right]$$

$$= \sqrt{3} \cdot 2\sqrt{21} = 6\sqrt{7}.$$

11. It follows that $a \leq 0$, $b \leq 0$, $c \geq 0$. Thus, the expression equals

$$-b + a + b + c - a - c + b = b.$$

12. $4 - x^2 - \sqrt{1-x^2} = 3 + (1-x^2) - \sqrt{1-x^2} = (\sqrt{1-x^2} - \frac{1}{2})^2 + \frac{11}{4}$.

Thus, the expression attains minimum when $\sqrt{1-x^2} = \frac{1}{2}$, i.e. when $x = \pm\frac{\sqrt{3}}{2}$, and the minimum is $\frac{11}{4}$. When $\sqrt{1-x^2} = 0$ or 1, i.e. when $x = \pm 1$ or 0, the expression attains maximum 3.

13. $\frac{4}{\sqrt{3}+\sqrt{2}} = 4(\sqrt{3} - \sqrt{2})$, whose integer part is 1.

$\frac{4}{\sqrt{5}-\sqrt{3}} = 2(\sqrt{5} + \sqrt{3})$, whose integer part is 7.

Therefore, there are six integers that meet the requirement.

14. Since $X - Y = \sqrt{a}(\sqrt{b} - \sqrt{c}) + \sqrt{d}(\sqrt{c} - \sqrt{b}) = (\sqrt{a} - \sqrt{d})(\sqrt{b} - \sqrt{c}) > 0$ and $Y - Z = \sqrt{a}(\sqrt{c} - \sqrt{d}) + \sqrt{b}(\sqrt{d} - \sqrt{c}) = (\sqrt{a} - \sqrt{b})(\sqrt{c} - \sqrt{d}) > 0$, we conclude that $Z < Y < X$.

15. It equals $\frac{(\sqrt{3}+\sqrt{2})^2 - (\sqrt{5})^2}{\sqrt{3}+\sqrt{2}-\sqrt{5}} = \sqrt{3} + \sqrt{2} + \sqrt{5}$.

16. It equals $\frac{\sqrt{3}+\sqrt{5}}{(3+\sqrt{15})-\sqrt{2}(\sqrt{3}+\sqrt{5})} = \frac{(\sqrt{3}+\sqrt{5})}{(\sqrt{3}-\sqrt{2})(\sqrt{3}+\sqrt{5})} = \sqrt{3} + \sqrt{2}$.

17. It equals $\frac{(1+\sqrt{2}+\sqrt{3})(1-\sqrt{2}-\sqrt{3})}{(1-\sqrt{2}+\sqrt{3})(1-\sqrt{2}-\sqrt{3})} = \frac{1-(\sqrt{2}+\sqrt{3})^2}{(1-\sqrt{2})^2-3} = \frac{-4-2\sqrt{6}}{-2\sqrt{2}} = \sqrt{2} + \sqrt{3}$.

18. It equals

$$\frac{(\sqrt{7}+\sqrt{11}) + 4(\sqrt{6}+\sqrt{7})}{(\sqrt{7}+\sqrt{11})(\sqrt{6}+\sqrt{7})} = \frac{1}{\sqrt{6}+\sqrt{7}} + \frac{4}{\sqrt{7}+\sqrt{11}}$$

$$= \sqrt{7} - \sqrt{6} + \sqrt{11} - \sqrt{7} = \sqrt{11} - \sqrt{6}.$$

19. The general term is $\frac{1}{\sqrt{3n+5}+\sqrt{3n+8}} = \frac{\sqrt{3n+8}-\sqrt{3n+5}}{3} = \frac{1}{3}(\sqrt{3n+8} - \sqrt{3n+5})$. Thus, the given expression equals

$$\frac{1}{3}(\sqrt{11} - \sqrt{8} + \sqrt{14} - \sqrt{11} + \sqrt{17} - \sqrt{14} + \cdots + \sqrt{6041} - \sqrt{6038})$$

$$= \frac{1}{3}(\sqrt{6041} - \sqrt{8}) = \frac{1}{3}\sqrt{6041} - \frac{2}{3}\sqrt{2}.$$

20. It equals

$$\sqrt{10 + 8\sqrt{(\sqrt{2}+1)^2}} - (\sqrt{2}+1)$$

$$= \sqrt{4^2 + 8\sqrt{2} + (\sqrt{2})^2} - (\sqrt{2}+1)$$

$$= (4 + \sqrt{2}) - (\sqrt{2}+1) = 3.$$

21. It equals $(\sqrt{2}-1) + (\sqrt{3}-\sqrt{2}) + (\sqrt{4}-\sqrt{3}) + (\sqrt{5}-\sqrt{4}) + (\sqrt{6}-\sqrt{5}) + (\sqrt{7}-\sqrt{6}) + (\sqrt{8}-\sqrt{7}) + (\sqrt{9}-\sqrt{8}) = \sqrt{9} - 1 = 2$.

22. Let $\sqrt{7} + \sqrt{6} = a$, $\sqrt{7} - \sqrt{6} = b$, then $a + b = 2\sqrt{7}$, $ab = 1$. Thus,

$$a^2 + b^2 = (a+b)^2 - 2ab = 26,$$

$$a^6 + b^6 = (a^2 + b^2)^3 - 3a^2b^2(a^2 + b^2) = 17498.$$

Therefore, $(\sqrt{7}+\sqrt{6})^6 + (\sqrt{7}-\sqrt{6})^6 = 17498$. Note that $0 < \sqrt{7}-\sqrt{6} < 1$, so $0 < (\sqrt{7}-\sqrt{6})^6 < 1$, and the greatest integer not exceeding $(\sqrt{7}+\sqrt{6})^6$ is 17497.

23. Squaring both sides, we have

$$13 + 2\sqrt{5} + 2\sqrt{7} + 2\sqrt{35} = (x + y + 1) + 2\sqrt{x} + 2\sqrt{y} + 2\sqrt{xy}.$$

Hence, $x = 5$, $y = 7$ or $x = 7$, $y = 5$.

Solution 10

Evaluating Algebraic Expressions

1. From $(a+1)^2 = 7$, we get $a^2 + 2a - 6 = 0$, so

$$3a^3 + 12a^2 - 6a - 12 = (a^2 + 2a - 6)(3a + 6) + 24 = 24.$$

   The answer is (A).

2. $\frac{a}{b} + \frac{b}{a} - ab = \frac{a^2 + b^2}{ab} - ab = \frac{(a+b)^2 - 2ab}{ab} - ab = ab - 2 - ab = -2$. Hence, the answer is (A).

3. $\sqrt{24-a} + \sqrt{8-a} = \frac{(\sqrt{24-a})^2 - (\sqrt{8-a})^2}{\sqrt{24-a} - \sqrt{8-a}} = \frac{24-a-(8-a)}{2} = 8$. Hence, the answer is (B).

4. It follows that $\frac{a+b+c}{abc} = 0$, so $\frac{1}{ab} + \frac{1}{bc} + \frac{1}{ca} = 0$. Thus,

$$\frac{1}{a^2} + \frac{1}{b^2} + \frac{1}{c^2} = \left(\frac{1}{a} + \frac{1}{b} + \frac{1}{c}\right)^2 - 2\left(\frac{1}{ab} + \frac{1}{bc} + \frac{1}{ac}\right) = 16,$$

   and the answer is (B).

5. Since a, b, c, d are positive integers, the only possibility is $a = b = c = d = 2$. Thus,

$$\sqrt{\frac{1}{a^3} + \frac{1}{b^4} + \frac{1}{c^5} + \frac{1}{d^6}} = \sqrt{\frac{1}{8} + \frac{1}{16} + \frac{1}{32} + \frac{1}{64}} = \sqrt{\frac{15}{64}}.$$

   The answer is (C).

6. From $(a+1)^2 = 3$, we get $a^2 + 2a - 2 = 0$, so

$$a^{2014} + 2a^{2013} - 2a^{2012} = a^{2012}(a^2 + 2a - 2) = 0.$$

7. It equals $1 + \frac{1}{2+\sqrt{2}-1} = 1 + \sqrt{2} - 1 = \sqrt{2}$.

8. It follows by condition that $\frac{1}{a} + \frac{1}{b} = 3$, $\frac{1}{b} + \frac{1}{c} = 4$, $\frac{1}{c} + \frac{1}{a} = 5$. Thus,

$$a = \frac{1}{2}, \quad b = 1, \quad c = \frac{1}{3}, \quad \text{and} \quad ab + bc + ca = \frac{1}{2} + \frac{1}{3} + \frac{1}{6} = 1.$$

9. We have $x + 2 = a + \frac{1}{a}$, so

$$\sqrt{4x + x^2} = \sqrt{(x+2)^2 - 4} = \sqrt{\left(\frac{1}{a} + a\right)^2 - 4}$$

$$= \sqrt{\left(\frac{1}{a} - a\right)^2} = \left|\frac{1}{a} - a\right|.$$

Since $\sqrt{x} \geq 0$, we have $\frac{1}{\sqrt{a}} \geq \sqrt{a}$, so $\frac{1}{a} \geq a$. Therefore, $\sqrt{4x + x^2} = \frac{1}{a} - a$.

10. We have $M = \frac{\sqrt{2008} + 44}{9} = 9 + \frac{\sqrt{2008} - 37}{9}$, so

$$a = 9, \quad b = \frac{\sqrt{2008} - 37}{9},$$

and $ab = \sqrt{2008} - 37$. Further,

$$a^2 + 3\left(\sqrt{2008} + 37\right) ab + 10$$

$$= 92 + 3\left(\sqrt{2008} + 37\right)\left(\sqrt{2008} - 37\right) + 10 = 2008.$$

11. It is easy to see that $(\sqrt[3]{x} - \frac{1}{\sqrt[3]{x}})^3 = 27$, so $x - \frac{1}{x} + 3(\sqrt[3]{x} - \frac{1}{\sqrt[3]{x}}) = 27$. Thus, $x - \frac{1}{x} = 27 + 9 = 36$, and

$$x^3 - \frac{1}{x^3} = \left(x - \frac{1}{x}\right)^3 + 3x - \frac{3}{x} = 46764.$$

12. It follows that $xy = 1$, $x + y = 14$, so

$$2x^2 - 3xy + 2y^2 = 2(x+y)^2 - 7xy = 392 - 7 = 385.$$

13. It follows that $x \neq 0$, $y \neq 0$. By the property of equal proportions, we have

$$\frac{25x}{75y} = \frac{15y}{30x - 75y} = \frac{6x - 15y}{x}$$

$$= \frac{25x + 15y + 6x - 15y}{75y + 30x - 75y + x} = 1.$$

Thus, $x = 3y$, and

$$\frac{4x^2 - 5xy + 6y^2}{x^2 - 2xy + 3y^2} = \frac{36y^2 - 15y^2 + 6y^2}{9y^2 - 6y^2 + 3y^2} = \frac{9}{2}.$$

14. It follows that $b - a = 4ab$. Hence,

$$\frac{a - 2ab - b}{2a + 7ab - 2b} = \frac{-2ab - 4ab}{7ab - 8ab} = \frac{-6ab}{-ab} = 6.$$

15. $\frac{a}{1+a} + \frac{b}{1+b} + \frac{c}{1+c} = \frac{x}{x+y+z} + \frac{y}{x+y+z} + \frac{z}{x+y+z} = 1.$

16. It follows that $2(a^2 + b^2 + c^2 - ab - bc - ca) = 0$, or, equivalently, $(a - b)^2 + (b - c)^2 + (c - a)^2 = 0$. Thus, $a - b = b - c = c - a = 0$, so $a = b = c$.

Combining with $a + 2b + 3c = 12$, we have $a = b = c = 2$, and $a + b^2 + c^3 = 2 + 4 + 8 = 14$.

17. It follows that $\sqrt{a} + \sqrt{b} = m + n$, $\sqrt{a} - \sqrt{b} = (m - n) + (a - b)$, so

$$m - n = \sqrt{a} - \sqrt{b} - (a - b) = \frac{a - b}{\sqrt{a} + \sqrt{b}} - (a - b) = 0.$$

In other words, $m = n$, and since $\sqrt{a} + \sqrt{b} = m + n = 1$, we have $m = n = \frac{1}{2}$, so $m^2 + n^2 = \frac{1}{2}$.

18. It follows by condition that $y + \sqrt{y^2 + 1} = \frac{1}{x + \sqrt{x^2 + 1}} = -x + \sqrt{x^2 + 1}.$ Thus, $x + y = \sqrt{x^2 + 1} - \sqrt{y^2 + 1}$. Squaring both sides, we have

$$x^2 + y^2 + 2xy = x^2 + 1 + y^2 + 1 - 2\sqrt{x^2 + 1} \cdot \sqrt{y^2 + 1},$$

or, equivalently, $\sqrt{x^2 + 1} \cdot \sqrt{y^2 + 1} = 1 - xy.$

Then, square both sides again, and we obtain

$$x^2 y^2 + x^2 + y^2 + 1 = x^2 y^2 - 2xy + 1,$$

which reduces to $(x + y)^2 = 0$. Therefore, $x + y = 0$.

19. Let $k = 2013x^2$, then

$$\frac{1}{x} = \frac{\sqrt{2013}}{\sqrt{k}}, \quad \frac{1}{y} = \frac{\sqrt{2014}}{\sqrt{k}}, \quad \frac{1}{z} = \frac{\sqrt{2015}}{\sqrt{k}}, \quad \frac{1}{w} = \frac{\sqrt{2016}}{\sqrt{k}}.$$

Thus, $\frac{1}{\sqrt{k}}(\sqrt{2013} + \sqrt{2014} + \sqrt{2015} + \sqrt{2016}) = 1$, or, equivalently,

$$\sqrt{2013} + \sqrt{2014} + \sqrt{2015} + \sqrt{2016} = \sqrt{k}.$$

On the other hand, $2013x = 2013 \cdot \sqrt{\frac{k}{2013}} = \sqrt{2013k}$, and $2014y = \sqrt{2014k}$, $2015z = \sqrt{2015k}$, $2016w = \sqrt{2016k}$, so

$$\sqrt{2013x + 2014y + 2015z + 2016w}$$

$$= \sqrt{\sqrt{2013k} + \sqrt{2014k} + \sqrt{2015k} + \sqrt{2016k}}$$

$$= \sqrt{\sqrt{k} \cdot \sqrt{k}} = \sqrt{k}.$$

Therefore, the given expression equals $\frac{\sqrt{k}}{\sqrt{k}} = 1$.

Solution 11

Symmetric Polynomials

1. Only ④ is cyclic, and the answer is (B).
2. Let $x = y = z = 1$, then $9 = k \times 3 \times 3$, so $k = 1$. The answer is (B).
3. We check by taking specific values. Let $x_1 = 0, x_2 = 1, x_3 = -1$, then

$$\alpha = 0, \quad \beta = -1, \quad \gamma = 0, \quad x_1^3 + x_2^3 + x_3^3 = 0.$$

   Thus, (B) and (D) are incorrect. Also, let $x_1 = 0, x_2 = 1, x_3 = -1$, then

$$\alpha = 2, \quad \beta = 1, \quad \gamma = 0, \quad x_1^3 + x_2^3 + x_3^3 = 2.$$

   Thus, (C) is incorrect, and the answer is (A).
4. This is a symmetric polynomial, and when $a = -b$, it equals

$$(a + b + c)^3 - a^3 - b^3 - c^3 = 0.$$

   Hence, $a + b$ is one of its factors. Similarly, $b + c, c + a$ are also its factors.

   Let $(a+b+c)^3 - a^3 - b^3 - c^3 = k(a+b)(b+c)(c+a)$. Then, choosing $a = b = c = 1$, we have $27 - 3 = 8k$, so that $k = 3$. Therefore,

$$(a + b + c)^3 - a^3 - b^3 - c^3 = 3(a + b)(b + c)(c + a).$$

5.

$$
\begin{aligned}
x^4 + (x + y)^4 + y^4 &= (x^2 + y^2)^2 - 2x^2y^2 + (x + y)^4 \\
&= [(x + y)^2 - 2xy]^2 - 2x^2y^2 + (x + y)^4 \\
&= 2(x + y)^4 - 4xy(x + y)^2 + 2x^2y^2 \\
&= 2(x^2 + xy + y^2)^2.
\end{aligned}
$$

6. This is a cyclic polynomial, and when $a = b$, it equals 0. Hence, $a - b$ is its factor. Similarly, $b - c, c - a$ are also its factors. Thus, we may assume that

$$(a - b)^3 + (b - c)^3 + (c - a)^3 = k(a - b)(b - c)(c - a).$$

Let $a = 0, b = 1, c = 2$, and plug into the equation above. Then, we have $6 = k \times (-1) \times (-1)$, so $k = 3$. Therefore,

$$(a - b)^3 + (b - c)^3 + (c - a)^3 = 3(a - b)(b - c)(c - a).$$

7. This is a symmetric polynomial. When $a = -b$, it equals 0, so $a + b$ is its factor. Similarly, $b + c, c + a$ are also its factors. Now, using the method of undetermined coefficients, we may solve that

$$(ab + bc + ca)(a + b + c) - abc = (a + b)(b + c)(c + a).$$

8. When $x = y$, the polynomial equals 0, so $x - y$ is its factor. Similarly, $y - z, z - x$ are also its factors. Suppose it equals $k(x + y + z)(x - y)(y - z)(z - x)$, then we may compare coefficients and obtain $k = -1$. Therefore, the given polynomial equals

$$-(x + y + z)(x - y)(y - z)(z - x).$$

9. This is a cyclic polynomial. When $x = y + z$, it equals

$$x^3 + y^2(x + y + z) + z^2(y + z + y) - x^3 - y^3 - z^3 - 2yz(y + z) = 0.$$

Hence, $x - y - z$ is its factor. Similarly, $y - x - z, z - x - y$ are also its factors.

Assume that it equals $k(x - y - z)(y - x - z)(z - x - y)$. Then, by the method of undetermined coefficient, we get $k = -1$. Therefore, the given polynomial equals

$$-(x - y - z)(y - x - z)(z - x - y).$$

10. $b = x^3 + y^3 = (x + y)[(x + y)^2 - 3xy] = a(a^2 - 3xy)$, so $xy = -\frac{b}{3a} + \frac{a^2}{3}$.
Thus,

$$x^2 + y^2 = (x + y)^2 - 2xy = a^2 + \frac{2b}{3a} - \frac{2a^2}{3} = \frac{a^3 + 2b}{3a}.$$

11. When $a = 0$, the polynomial equals 0, and similarly, when $b = 0$ or $c = 0$, it also equals 0. Hence, we may assume that it equals $k = abc$, and choosing $a = b = c = 1$, we see that $k = 4$. Therefore, it equals $4abc$.

12. (1) Since $a + b = -(c + d)$, we have $(a + b)^3 = -(c + d)^3$, so $(a + b)^3 + (c + d)^3 = 0$.

(2) $a^3 + b^3 + c^3 + d^3 = (a + b)^3 - 3ab(a + b) + (c + d)^3 - 3cd(c + d)$, and since $a + b = -(c + d)$, we can obtain that

$$3 = 3ab(c + d) + 3cd(a + b).$$

Equivalently, $ab(c + d) + cd(a + b) = 1$.

13. Let $x_1 = \frac{|a|}{a}, x_2 = \frac{|b|}{b}, x_3 = \frac{|c|}{c}, x_4 = \frac{|d|}{d}$, then each of them is either 1 or -1. Thus,

$$A = x_1 + x_2 + x_3 + x_4 - x_1 x_2 - x_1 x_3 - x_1 x_4 - x_2 x_3 - x_2 x_4 - x_3 x_4$$

$$+ x_1 x_2 x_3 + x_1 x_2 x_4 + x_1 x_3 x_4 + x_2 x_3 x_4 - x_1 x_2 x_3 x_4$$

$$= 1 - (1 - x_1)(1 - x_2)(1 - x_3)(1 - x_4)$$

$$= 1 - \left(1 - \frac{|a|}{a}\right)\left(1 - \frac{|b|}{b}\right)\left(1 - \frac{|c|}{c}\right)\left(1 - \frac{|d|}{d}\right).$$

We can see that if there is at least one positive number in a, b, c, d, then $A = 1$, and if a, b, c, d are all negative, then $A = 1 - 16 = -15$.

Therefore, the value of A is either 1 or -15.

14. (1) It follows by condition that $x^2 + y^2 + z^2 + 2(xy + yz + zx) = (x + y + z)^2 = 4$, so we get $xy + yz + zx = -6$.

(2) We have $\frac{1}{xy + 2z} = \frac{1}{zy - 2(x + y - 2)} = \frac{1}{(x - 2)(y - 2)}$. Similarly,

$$\frac{1}{yz + 2x} = \frac{1}{(y - 2)(z - 2)}, \quad \frac{1}{zx + 2y} = \frac{1}{(z - 2)(x - 2)}.$$

Therefore, the given expression equals $\frac{x + y + z - 6}{(x - 2)(y - 2)(z - 2)} = -\frac{4}{13}$.

Solution 12

Proving Identities

1.

$$S_k = 1 + \left(\frac{1}{k} - \frac{1}{k+1} \right)^2 + 2 \cdot \frac{1}{k(k+1)}$$

$$= 1 + \left(\frac{1}{k(k+1)} \right)^2 + 2 \cdot \frac{1}{k(k+1)} = \left(1 + \frac{1}{k(k+1)} \right)^2$$

$$= \left(1 + \frac{1}{k} - \frac{1}{k+1} \right)^2.$$

Hence, $\sqrt{S_1} + \sqrt{S_2} + \cdots + \sqrt{S_n} = 1 + \left(\frac{1}{1} - \frac{1}{2} \right) + 1 + \left(\frac{1}{2} - \frac{1}{3} \right) + \cdots + 1 + \left(\frac{1}{n} - \frac{1}{n+1} \right) = n + 1 - \frac{1}{n+1} = \frac{n^2 + 2n}{n+1}$.

2. From $x + y = -z$ we see that $\frac{1}{x} + \frac{1}{y} = -\frac{z}{xy}$, so $\left(\frac{1}{x} + \frac{1}{y} \right) \frac{1}{z} = -\frac{1}{xy}$. Thus,

$$\sqrt{\frac{1}{x^2} + \frac{1}{y^2} + \frac{1}{z^2}} = \sqrt{\left(\frac{1}{x} + \frac{1}{y} \right)^2 - \frac{2}{xy} + \frac{1}{z^2}}$$

$$= \sqrt{\left(\frac{1}{x} + \frac{1}{y} \right)^2 + 2 \cdot \left(\frac{1}{x} + \frac{1}{y} \right) \frac{1}{z} + \frac{1}{z^2}}$$

$$= \sqrt{\left(\frac{1}{x} + \frac{1}{y} + \frac{1}{z} \right)^2} = \left| \frac{1}{x} + \frac{1}{y} + \frac{1}{z} \right|.$$

3. From $a + b = -c$, we obtain that $a^3 + 3a^2b + 3ab^2 + b^3 = -c^3$, so

$$a^3 + b^3 + c^3 = -3ab(a+b) = 3abc.$$

Therefore, $\frac{abc}{a^3+b^3+c^3} = \frac{1}{3}$.

4. It follows by condition that

$$(a^2 - b^2)^2 + (c^2 - d^2)^2 + 2(ab - cd)^2 = 0.$$

Thus, $a^2 - b^2 = c^2 = d^2 = ab - cd = 0$. Since a, b, c, d are all positive numbers, we have $a = b = c = d$.

5. The left-hand side equals $= \frac{a^8 - 2a^4b^4 + b^8 + 4a^4b^4}{(a^4 - b^4)^2} = \frac{(a^4 + b^4)^2}{(a^4 - b^4)^2}$, which is equal to the right-hand side.

6.

$$\text{LHS} - \text{RHS} = a^2 + \frac{1}{a^2} + b^2 + \frac{1}{b^2}$$

$$+ \left(ab + \frac{1}{ab} \right) \left[\left(ab + \frac{1}{ab} \right) - \left(a + \frac{1}{a} \right) \left(b + \frac{1}{b} \right) \right]$$

$$= a^2 + \frac{1}{a^2} + b^2 + \frac{1}{b^2} + \left(ab + \frac{1}{ab} \right) \left(-\frac{b}{a} - \frac{a}{b} \right) = 0.$$

Therefore, the identity holds.

7.

$$\text{LHS} = \frac{a^3(c-b) + b^3(a - b + b - c) + c^3(b-a)}{a^2(c-b) + b^2(a - b + b - c) + c^2(b-a)}$$

$$= \frac{(a^3 - b^3)(c-b) + (b^3 - c^3)(a-b)}{(a^2 - b^2)(c-b) + (b^2 - c^2)(a-b)}$$

$$= \frac{(a-b)(c-b)(a^2 + ab + b^2 - b^2 - bc - c^2)}{(a-b)(c-b)(a + b - b - c)}$$

$$= \frac{a^2 + ab - bc - c^2}{a - c} = \frac{(a^2 - c^2) + b(a - c)}{a - c} = a + b + c = \text{RHS}.$$

Therefore, the identity holds.

8.

$$\text{LHS} = \frac{1}{a} - \frac{1}{a+b} + \frac{1}{a+b} - \frac{1}{a+b+c} + \frac{1}{a+b+c} - \frac{1}{a+b+c+d}$$

$$= \frac{1}{a} - \frac{1}{a+b+c+d} = \frac{b+c+d}{a(a+b+c+d)} = \text{RHS}.$$

Therefore, the identity holds.

9.
$$\text{LHS} - \text{RHS} = \left[\frac{x}{a(x-a)} - \frac{1}{x-a}\right] + \left[\frac{y}{a(y-a)} - \frac{1}{y-a}\right]$$

$$+ \left[\frac{z}{a(z-a)} - \frac{1}{z-a}\right] - \frac{3}{a}$$

$$= \frac{x-a}{a(x-a)} + \frac{y-a}{a(y-a)} + \frac{z-a}{a(z-a)} - \frac{3}{a} = 0.$$

Therefore, the identity holds.

10.
$$\text{LHS} = \frac{(\sqrt{a})^3 + (\sqrt{b})^3}{\sqrt{a} + \sqrt{b}} - \sqrt{ab} = (\sqrt{a})^2 - \sqrt{ab} + (\sqrt{b})^2 - \sqrt{ab}$$

$$= a + b - 2\sqrt{ab}.$$

$$\text{RHS} = \left(\frac{a-b}{\sqrt{a}+\sqrt{b}}\right)^2 = (\sqrt{a} - \sqrt{b})^2 = a + b - 2\sqrt{ab}.$$

Therefore, the two sides are equal, and the identity holds.

11. Let $\frac{A}{x} = \frac{B}{y} = \frac{C}{z} = \frac{D}{t} = k$, then $A = kx, B = ky, C = kz, D = kt$. Thus,

$$\text{LHS} = \sqrt{kx^2} + \sqrt{ky^2} + \sqrt{kz^2} + \sqrt{kt^2} = \sqrt{k}(x + y + z + t),$$

$$\text{RHS} = \sqrt{(kx + ky + kz + kt)(x + y + z + t)} = \sqrt{k}(x + y + z + t).$$

The two sides are equal, so the identity holds.

12. It follows that $a = 3b, c = 3d$. Thus,

$$\text{LHS} = \frac{9b^2 + 9d^2}{3b + 3d} + \frac{b^2 + d^2}{b + d} = \frac{4(b^2 + d^2)}{b + d},$$

$$\text{RHS} = \frac{(4b)^2 + (4d)^2}{4b + 4d} = \frac{4(b^2 + d^2)}{b + d}.$$

The two sides are equal, so the identity holds.

13. It follows that $a - 1 = -b$, $b - 1 = -a$. Thus,

$$\text{LHS} = \frac{b^4 - b - a^4 + a}{(a^3 - 1)(b^3 - 1)}$$

$$= \frac{-(a - b)(a + b)(a^2 + b^2) + (a - b)}{(a - 1)(a^2 + a + 1)(b - 1)(b^2 + b + 1)}$$

$$= \frac{-(a - b)(a^2 + b^2 - 1)}{ab(a^2 + a + 1)(b^2 + b + 1)}$$

$$= \frac{-(a - b)[(a + b)^2 - 2ab - 1]}{ab[a^2b^2 + 2ab + a^2 + b^2 + 2]}$$

$$= \frac{2(a - b)}{a^2b^2 + 3} = \text{RHS}.$$

Therefore, the identity holds.

14. The given equality reduces to

$$\frac{(b + c)^2 - a^2}{bc} - 4 + \frac{(c + a)^2 - b^2}{ca} - 4 + \frac{(a + b)^2 - c^2}{ab} = 0.$$

$$\implies \frac{(b - c)^2 - a^2}{bc} + \frac{(c - a)^2 - b^2}{ca} + \frac{(a + b)^2 - c^2}{ab} = 0.$$

Factorizing the numerators, we get

$$\frac{(b - c + a)(b - c - a)}{bc} + \frac{(c - a + b)(c - a - b)}{ca}$$

$$+ \frac{(a + b - c)(a + b + c)}{ab} = 0.$$

Taking the common factor $a + b - c$, we have

$$\frac{(a + b - c)}{abc}[a(b - c - a) - b(c - a + b) + c(a + b + c)] = 0.$$

Hence, $(a + b - c)(c^2 - a^2 + 2ab - b^2) = 0$, or, equivalently,

$$(a + b - c)(c - a + b)(c + a - b) = 0.$$

Since $c > b > a > 0$, we have $c - a + b > 0$, $c + a - b > 0$, so necessarily $a + b - c = 0$, and $a + b = c$.

15. By swapping the numerators and denominators, we get

$$\frac{bz+cy}{yz} = \frac{cx+az}{xz} = \frac{ay+bx}{xy} = \frac{a^2+b^2+c^2}{x^2+y^2+z^2}, \qquad \text{①}$$

Thus, $\frac{b}{y} + \frac{c}{z} = \frac{c}{z} + \frac{a}{x} = \frac{a}{x} + \frac{b}{y}$. This implies that $\frac{a}{x} = \frac{b}{y} = \frac{c}{z}$. Let k be this ratio, then by ①, we have

$$2k = \frac{k^2 x^2 + k^2 y^2 + k^2 z^2}{x^2 + y^2 + z^2} = k^2.$$

Since each variable is nonzero, we conclude that $k = 2$, and $a+b+c = 2(x+y+z)$.

Solution 13

Linear Functions

1. We can solve that $m = 8, n = 10$ or obtain from the graph that $m < n$. The answer is (B).

2. The intersection is $\left(\frac{2m+6}{3}, \frac{2m-3}{3}\right)$, which lies in the 4th quadrant, so $\frac{2m+6}{3} > 0$ and $\frac{2m-3}{3} < 0$. Hence, $-3 < m < \frac{3}{2}$, and the answer is (D).

3. Assume $y = k_1(x + 2)$, then plug in $x = -3, y = -6$, and we have $k_1 = 6$.
   Assume $z - 4 = k_2 y$, then plug in $y = 6, z = -2$, and we have $k_2 = -1$. Hence, $z - 4 = -y = -6(x+2)$, so $z = -6x - 8$, and the answer is (D).

4. It follows that $a > 0, b > 0, -2a + b = 0$. Thus, $ax > b$ is the same as $ax > 2a$, so $x > 2$, and the answer is (C).

5. Since l corresponds to 12, the ciphertext corresponds to $\frac{12}{2} + 13 = 19$, which is s. Since o corresponds to 15, its ciphertext corresponds to $\frac{15+1}{2} = 8$, which is h. The answer is (B).

6. It follows that $a + b + c = 2(a + b + c)t$.
   If $a + b + c \neq 0$, then $t = \frac{1}{2}, y = \frac{1}{2}x + \frac{1}{4}$, which passes through quadrants I, II, and III.
   If $a + b + c = 0$, then $a + b = -c$, so $t = -1, y = -x + 1$. The graph passes through quadrants I, II, and IV. The answer is (A).

7. The intersection of the two graphs (x, y) satisfies $\begin{cases} y = bx + a, \\ y = ax + b, \end{cases}$ so the intersection is $(1, a + b)$. Thus, the x-coordinate is 1, and (A) and (C) are incorrect.
   On the other hand, the y-coordinate of (D) is a number between a and b, which cannot be $a + b$, so (D) is also wrong. Only (B) is correct.

8. The bottle contains water before the crow brings pebbles, so $y > 0$ when $x = 0$, and (A) is eliminated.

   Before the crow reaches the water for the first time, y increases as x increases, so (D) is eliminated.

   Finally, the level of water before bringing pebbles should be lower than the crow's reach, so (C) is eliminated. Only (B) is correct.

9. The conditions imply that $\triangle OA_1B_1, \triangle B_1A_2B_2, \triangle B_2A_3B_3$ are all isosceles right triangles.

   Since $y = x+1$ intersects the y-axis at A_1, we have $A_1(0,1), B_1(1,0)$, and $OB_1 = OA_1 = 1$. Thus, $S_1 = \frac{1}{2} \times 1 \times 1 = \frac{1}{2} \times 1^2$, and similarly, $S_2 = \frac{1}{2} \times 2 \times 2 = \frac{1}{2} \times 2^2, S_3 = \frac{1}{2} \times 4^2, \ldots$.

   Therefore, $S_n = \frac{1}{2} \times 2^{2n-2} = 2^{2n-3}$, and $S_{2017} = 2^{2\times 2017-3} = 2^{4031}$. The answer is (B).

10. Since $BO = \sqrt{10}$, we have $n^2 + 1 = 10$, so $n = \pm 3$. Plug in the coordinates of B, and we get $y = mx - 1$, so $n = m - 1$.

    If $n = 3$, then $n = 4$, and the line is $y = 4x - 1$. It intersects the coordinate axes at $(\frac{1}{4}, 0), (0, -1)$, and $S = \frac{1}{2} \times \frac{1}{4} \times 1 = \frac{1}{8}$.

    If $n = -3$, then $n = -2$, and the line is $y = -2x - 1$, which intersects the coordinate axes at $(-\frac{1}{2}, 0)$, $(0, -1)$. Thus, $S = \frac{1}{2} \times \frac{1}{2} \times 1 = \frac{1}{4}$.

    Therefore, $S = \frac{1}{4}$ or $\frac{1}{8}$, and the answer is (C).

11. We can solve that $A(19, 0), B\left(0, -\frac{95}{4}\right)$. The line AB has the formula $y = \frac{5}{4}(x + 1) - 25$. If a point on the line has integer coordinates, then we should have $4|x + 1$, and $0 \le x \le 19$, so $x = 3, 7, 11, 15, 19$. There are five of them.

12. It follows that $a - b = 5, c = -d = 5$. Thus,

$$a(c - d) + b(d - c) = (a - b) \times (-5) = 25.$$

13. The three intersections are $(1, a + b), \left(\frac{a-b}{a}, a\right), (0, a)$. Hence, the area of the triangle is $\frac{1}{2} \cdot \left|\frac{a-b}{a}\right| \cdot |a + b - a| = \left|\frac{b(a-b)}{2a}\right|$.

14. The formula of the line reduces to $(2x - y - 1)k - (x + 3y - 11) = 0$. Since k can be any value, this equation in k has infinitely many solutions. This means

$$\begin{cases} 2x - y - 1 = 0, \\ x + 3y - 11 = 0. \end{cases} \implies \begin{cases} x = 2, \\ y = 3. \end{cases}$$

Hence, this fixed point is $(2, 3)$.

15. It follows that $\begin{cases} -2 = 3a + b, \\ \frac{1}{4} = \frac{3}{4}a + b. \end{cases}$ Thus, $a = -1, b = 1$, and the line is $y = -x + 1$.

16. Construct $Q'(2, -1)$, which is the symmetric point of Q with respect to the x-axis. Let M be the intersection of PQ' and the x-axis. For any other point M' on the x-axis, we have $M'P + M'Q > PQ' = PM + MQ' = MP + MQ$, so this M makes $MP + MQ$ minimal. We can solve that the equation of the line PQ' is $y = 2x - 5$, so $M\left(\frac{5}{2}, 0\right)$.

17. We have $A(\sqrt{3}, 0), B(0, 1), OA = \sqrt{3}, OB = 1, AB = 2$. Thus, $S_{\triangle ABP} = S_{\triangle ABC} = 2$. Since

$$S_{\triangle AOP} = \frac{\sqrt{3}}{4}, \quad S_{\triangle BOP} = -\frac{a}{2}, \quad S_{\triangle AOB} = \frac{\sqrt{3}}{2},$$

and $S_{\triangle BOP} + S_{\triangle AOB} - S_{\triangle AOP} = 2$, we have $-\frac{a}{2} + \frac{\sqrt{3}}{2} - \frac{\sqrt{3}}{4} = 2$, and $a = \frac{\sqrt{3}-8}{2}$.

18. The intersection of the two functions (x, y) satisfies $\begin{cases} y = px - 2, \\ y = x + q. \end{cases}$ Thus,

$$\begin{cases} x = \dfrac{q + 2}{p - 1}, \\ y = \dfrac{pq + 2}{p - 1}. \end{cases}$$

Here $p - 1 \neq 0$; otherwise, they have no intersection.

Since (x, y) lies to the left of the line $x = 2$, we have $\frac{q+2}{p-1} < 2$, so $2 \leq q < 2p - 4$, and $p > 3$. Hence, $p = 4$ or 5.

If $p = 4$, then $q = 2, 3$; if $p = 5$, then $q = 2, 3, 4$. Therefore, there are five such pairs (p, q).

19. Let A, B denote the two cars. Suppose car A turns back after consuming x barrels of gasoline, preserves x barrels for the way back, and gives the remaining $(24 - 2x)$ barrels to car B. Then, B continues forward, and at this time, it has $(24 - 2x) + (24 - x) = 48 - 3x$ barrels of gasoline. By assumption, $48 - 3x \leq 24$, so $x \geq 8$.

After the two cars separate, the distance B can travel further is

$$\frac{(48 - 3x) - x}{2} \times 60 = 30(48 - 4x)\,(\text{km}),$$

and the maximal distance to the starting point is $30(48 - 4x) + 60x = 1440 - 60x$. This quantity decreases as x increases, so it is maximal when $x = 8$, and in this case, the total distance B has traveled is $2 \times (1440 - 60 \times 8) = 1920\,(\text{km})$.

20. (1) The boiler house has 96 L of water in the beginning;
   after 2 min the remaining water is 80 L;
   after 4 min, the remaining water is 72 L;
   the water flow of each tap is 8 L/min; ...

   (2) When $0 \leq x \leq 2$, assume the formula of the function is $y = k_1 x + b_1$.
   Plug in $x = 0, y = 96$ and $x = 2, y = 80$, then we have $\begin{cases} b_1 = 96, \\ 2k_1 + b_1 = 80. \end{cases}$

   Thus, $\begin{cases} k_1 = -8 \\ b_1 = 96 \end{cases}$, and $y = -8x + 96 (0 \leq x \leq 2)$.

   When $x > 2$, assume the formula of the function is $y = k_2 x + b_2$.
   Plug in $x = 2, y = 80$ and $x = 4, y = 72$, then we have $\begin{cases} 80 = 2k_2 + b_2, \\ 72 = 4k_2 + b_2. \end{cases}$

   Thus, $\begin{cases} k_2 = -4 \\ b_2 = 99 \end{cases}$, and $y = -4x + 88 (x > 2)$.

   Since the remaining water after 15 students finish collecting is $96 - 15 \times 2 = 66$ L, we have $66 = -4x + 88$, which gives $x = 5.5$.

   Therefore, it takes 5.5 min for the first 15 students to finish collecting.

   (3) Consider the following three cases:

   ① If these students start collecting from the beginning, then the time needed is $8 \times 2 \div 8 = 2$ min, which is contradictory to the assumption that they used 3 min.

   ② If these students start collecting after some students have finished. Suppose they start after t min have elapsed. If $0 < t \leq 2$, then $8(2-t) + 4[3-(2-t)] = 8 \times 2$. Thus, $16 - 8t + 4 + 4t = 16$, and $t = 1$. This is a valid solution.

   ③ If $t > 2$, then $8 \times 2 \div 4 = 4$, so it takes 4 min for them to finish collecting, which is a contradiction.

   Therefore, Mary's claim is possible, in which case the 8 students start collecting at $t = 1 \,(\text{min})$.

21. (1) Let the price of an air conditioner and an electric fan be x and y yuan, respectively. Then, the conditions imply that $\begin{cases} 8x + 20y = 17400, \\ 10x + 30y = 22500. \end{cases}$ Thus, $\begin{cases} x = 1800, \\ y = 150. \end{cases}$

   Therefore, the price of an air conditioner is 1800 yuan, and the price of an electric fan is 150 yuan.

   (2) Suppose he buys t air conditioners and $(70 - t)$ electric fans, then

$$\begin{cases} 1800t + 150(70 - t) \leq 30000, \\ 200t + 30(70 - t) \geq 3500. \end{cases}$$

Solving the inequalities, we have $8\frac{4}{17} \leq t \leq 11\frac{9}{11}$.

Since t is an integer, we have $t = 9, 10, 11$. Hence, there are three plans, namely:

Plan 1: 9 air conditioners and 61 electric fans;

Plan 2: 10 air conditioners and 60 electric fans;

Plan 3: 11 air conditioners and 59 electric fans.

Let W be the total profit after the owner sells all these appliances, then

$$W = 200t + 30(70 - t) = 170t + 2100.$$

We see that W increases as t increases, so W is maximal when $t = 11$, in which case

$$W_{\max} = 170 \times 11 + 2100 = 3970 \text{ (yuan)}$$

Therefore, plan 3 gives the highest profit, which is 3970 yuan.

Solution 14

Inversely Proportional Functions

1. $y = \frac{k}{|k|}x$ reduces to either $y = x$ or $y = -x$, and they are both axes of symmetry of $y = \frac{k}{x}$. Hence, the answer is (D).

2. It follows that $A_1 A_2$ lie in the second quadrant, A_3 lies in the fourth quadrant, and in each branch of the graph y increases as x increases. Hence, $0 < y_1 < y_2, y_3 < 0$, so $y_3 < y_1 < y_2$, and the answer is (C).

3. The answer is (B).

4. It follows that $y = k_1 x, z = \frac{k_2}{y}(k_1 \neq 0, k_2 \neq 0)$. Thus, $z = \frac{\frac{k_2}{k_1}}{x}$, where $\frac{k_2}{k_1} \neq$, so the answer is (B).

5. When $x < 0$, $y = \frac{k}{x}$ is increasing with x, so $k < 0$. Thus, $y = kx - k$ is decreasing with x, and its graph intersects the y-axis in the positive half. Therefore, the graph of $y = kx - k(k \neq 0)$ passes through quadrants I, II, and IV, and the answer is (B).

6. The answer is (D).

7. The answer is (C).

8. The answer is (C).

9. It is easy to see that $A(1,0), B(0,1)$.

   Let H be the projection of E on the y-axis, and let G be the projection of F on the x-axis.

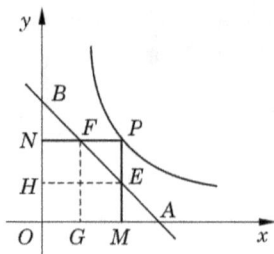

Suppose $P\left(a, \frac{1}{2a}\right)$, then $F\left(1 - \frac{1}{2a}, \frac{1}{2a}\right)$, and $E(a, 1-a)$. Thus,

$$AF = \sqrt{AG^2 + FG^2} = \sqrt{\left[1 - \left(1 - \frac{1}{2a}\right)\right]^2 + \left(\frac{1}{2a}\right)^2} = \frac{\sqrt{2}}{2a},$$

$$BE = \sqrt{BH^2 + HE^2} = \sqrt{[1-(1-a)]^2 + a^2} = \sqrt{2}a.$$

Therefore, $AF \cdot BE = \frac{\sqrt{2}}{2a} \cdot \sqrt{2}a = 1$, and the answer is (C).

10. **Solution 1.** Since A, B lie on the graph of $y = -\frac{4}{x}$, we have

$$S_{\triangle AOE} = S_{\triangle BOF} = \frac{1}{2} \times 4 = 2.$$

Connect OC. Since the graphs of $y = -\frac{4}{x}$ and $y = -\frac{1}{3}x$ are both symmetric with respect to the origin, it follows that A, B are also symmetric with respect to the origin, so we have $OA = OB$, $AE = EC$, $CF = BF$. Thus, $S_{\triangle AOC} = 2S_{\triangle AOE} = 4$, $S_{\triangle BOC} = 2S_{\triangle BOF} = 4$. Further, $S_{\triangle ABC} = S_{\triangle AOC} + S_{\triangle BOC} = 8$, and the answer is (A).

Solution 2. Construct the system of equations $\begin{cases} y = -\frac{4}{x}, \\ y = -\frac{1}{3}x. \end{cases}$ Solving the equations, we get $\begin{cases} x = -2\sqrt{3} \\ y = \frac{2}{3}\sqrt{3} \end{cases}$ or $\begin{cases} x = 2\sqrt{3} \\ y = -\frac{2}{3}\sqrt{3}. \end{cases}$ Hence, $A(-2\sqrt{3}, \frac{2}{3}\sqrt{3})$, $B(2\sqrt{3}, -\frac{2}{3}\sqrt{3})$, and

$$AC = \frac{2}{3}\sqrt{3} - \left(-\frac{2}{3}\sqrt{3}\right) = \frac{4}{3}\sqrt{3},$$

$$BC = 2\sqrt{3} - (-2\sqrt{3}) = 4\sqrt{3}.$$

Therefore, $S_{\triangle ABC} = \frac{1}{2} AC \cdot BC = \frac{1}{2} \times \frac{4}{3}\sqrt{3} \times 4\sqrt{3} = 8$, and the answer is (A).

11. The conditions imply that $A(\sqrt{2}, 2\sqrt{2}), B(-\sqrt{2}, -2\sqrt{2})$. Since $\triangle ABC$ is an isosceles right triangle with $\angle ACB = 90°$, we have

$$S_{\triangle ABC} = \frac{1}{2} \times 2\sqrt{2} \times 4\sqrt{2} = 8.$$

12. It follows that $m = -\frac{k}{2}, n = \frac{k}{5}$. Plug the coordinates of A, B into $y = ax + b$, and we get

$$\begin{cases} -2a + b = -\dfrac{k}{2}, \\ 5a + b = \dfrac{k}{5}. \end{cases}$$

Hence, $a = \frac{k}{10}$, $b = -\frac{3k}{10}$, and $3a + b = 0$.

13. The symmetric point of $A(m, n)$ with respect to the y-axis is $B(-m, n)$. Thus, $n = -\frac{2}{m}, n = -m - 4$. Equivalently, $mn = -2, m + n = -4$.

Therefore, $\frac{m}{n} + \frac{n}{m} = \frac{m^2 + n^2}{mn} = \frac{(m+n)^2 - 2mn}{mn} = -10$.

14. $S_{ABCD} = S_{\triangle ABD} + S_{\triangle BCD} = 2$.

15. According to the conditions, P_n, Q_n have the same x-coordinate ($n = 1, 2, \ldots, 2005$), and the y-coordinate of P_n is twice that of Q_n. Since the y-coordinate of P_{2005} is

$$1 + (2005 - 1) \times 2 = 4009,$$

it follows that $y_{2005} = \frac{4009}{2} = 2004.5$.

16. Let H, D be the projections of B, A on the x-axis, respectively, and construct the line through B parallel to the x-axis, which intersects AD at E (as shown in the following figure).

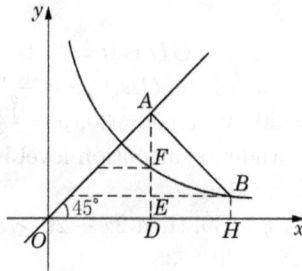

Then, $OD = AD$, $AE = BE$, and

$$OA^2 - AB^2 = 2OD^2 - 2BE^2$$

$$= 2(OD + BE)(OD - BE)$$

$$= 2(OD + DH)(AD - AE)$$

$$= 2OH \cdot BH = 8.$$

Hence, $OH \cdot BH = 4$, so $k = 4$.

17. (1) Since $A(-2,1)$ lie on the graph of $y_2 = \frac{m}{x}$, we have $m = (-2) \times 1 = -2$. Thus, the formula of the inversely proportional function is $y_2 = -\frac{2}{x}$.

On the other hand, $B(1,n)$ lies on the graph, so $n = -\frac{2}{1} = -2$, and $B(1,-2)$.

Since $A(-2,1)$ and $B(1,-2)$ both satisfy $y_1 = kx + b$, we have $\begin{cases} 1 = -2k + b \\ -2 = k + b \end{cases}$. Hence, $\begin{cases} k = -1 \\ b = -1 \end{cases}$, and the formula of the linear function is $y = -x - 1$.

(2) It follows from the graph that $y_1 > y_2$ exactly when $x < -2$ or $0 < x < 1$.

18. (1) $S_{\triangle AOE} = S_{\triangle BOF} = \frac{1}{2}k$.

(2) $E\left(-\frac{k}{3}, 3\right), F\left(-4, \frac{k}{4}\right)$.

$$S_{\triangle ECF} = \frac{1}{2}EC \cdot CF = \frac{1}{2}\left(4 - \frac{k}{3}\right)\left(3 - \frac{k}{4}\right),$$

$$S_{\triangle EOF} = S_{AOBC} - S_{\triangle AOE} - S_{\triangle BOF} - S_{\triangle ECF}$$

$$= 12 - k - S_{\triangle ECF}.$$

Hence, $S = 12 - k - 2S_{\triangle ECF} = -\frac{1}{12}k^2 + k = -\frac{1}{12}(k-6)^2 + 3$.

Therefore, S is maximal when $k = 6$.

19. (1) It follows from the graph that $A(0,20), B(10,40), C(25,40)$. Suppose the formula of AB is $y = kx + b$, then $\begin{cases} b = 20, \\ 40 = 10k + b. \end{cases}$ Thus, $k = 2, b = 20$.

Suppose the formula of CD is $y = \frac{k}{x}$, then $40 = \frac{k}{25}$, so $k = 1000$. Hence, the formula of the CD is $y = \frac{1000}{x}$ $(25 \le x \le 40)$.

(2) When $x = 5$, $y = 30$. When $x = 30$, $y = \frac{100}{3} > 30$.

Therefore, the students' attention level is higher at 30 min since the beginning of the class.

(3) For $0 \le x \le 10$, if $y \ge 36$, then $2x + 20 \ge 36$, so $8 \le x \le 10$.

For $10 < x \le 25$, $y = 40 > 36$.

For $x > 25$, if $y \ge 36$, then $\frac{1000}{x} \ge 36$, so $25 < x \le 27\frac{7}{9}$.

Thus, $y \geq 36$ for $8 \leq x \leq 27\frac{7}{9}$. Since $27\frac{7}{9} - 8 > 19$, the teacher can finish explaining the problem by starting after 8 min since the class begins.

20. (1) It follows that $A(-2, 0), B(0, 2)$. It is easy to see that the formula of AB is $y = x + 2$.

Since P lies on the graph of $y = -\frac{2}{x}$, and its y-coordinate is $\frac{5}{3}$, we have $P\left(-\frac{6}{5}, \frac{5}{3}\right)$. Plug in the coordinates, and we have $E\left(-\frac{6}{5}, \frac{4}{5}\right), F\left(-\frac{1}{3}, \frac{5}{3}\right)$.

Hence, $S_{\triangle EOF} = S_{\triangle AOF} - S_{\triangle AOE} = \frac{1}{2} \times |-2| \times \frac{5}{3} - \frac{1}{2} \times |-2| \times \frac{4}{5} = \frac{13}{15}$.

(2) The triangle with side lengths AE, BF, EF is a right triangle. The reason is explained as follows.

Given the condition, $\triangle AOB$ is an isosceles right triangle, and so are $\triangle AME, \triangle EPF, \triangle FNB$. Since $-2 < a < 0, 0 < b < 2$, we have $AM = 2 - (-a) = 2 + a$, and $AE^2 = (\sqrt{2}AM)^2 = 2a^2 + 8a + 8$.

Also, $BN = 2 - b$, so $BF^2 = (\sqrt{2}BN)^2 = 2b^2 - 8b + 8$.

Since $PE = PM - EM = PM - AM = b - (2+a) = b - a - 2$ and $ab = -2$, it follows that $EF^2 = (\sqrt{2}PE)^2 = 2a^2 + 2b^2 + 8a - 8b + 16$, and $|a| \neq |b|$ implies $AE \neq BF$.

Note that

$$(2a^2 + 8a + 8) + (2b^2 - 8b + 8) = 2a^2 + 2b^2 + 8a - 8b + 16,$$

so we conclude that $AE^2 + BF^2 = EF^2$, and the triangle with side lengths AE, BF, EF is a right triangle.

Solution 15

Statistics

1. It follows from the given condition that $a = \frac{4 \times 4 + 3 \times 5 + 3 \times 6}{10} = 49$, $b = 5$, $c = 4$. Thus, the answer is (B).

2. The formula of variance gives

$$S^2 = \frac{1}{5}[(x_1 - \bar{x})^2 + (x_2 - \bar{x})^2 + (x_3 - \bar{x})^2 + (x_4 - \bar{x})^2 + (x_5 - \bar{x})^2]$$

$$= \frac{1}{5}(x_1^2 + x_2^2 + x_3^2 + x_4^2 + x_5^2 - 5\bar{x}^2)$$

$$= \frac{1}{5}(x_1^2 + x_2^2 + x_3^2 + x_4^2 + x_5^2 - 20).$$

Hence, $\bar{x} = 2$, so the average of x_1, x_2, x_3, x_4, x_5 is 2. This implies that the average of $x_1 + 2, x_2 + 2, x_3 + 2, x_4 + 2, x_5 + 2$ is 4, so (3) is correct. The variance of these numbers is

$$\frac{1}{5}[(x_1 + 2 - 4)^2 + (x_2 + 2 - 4)^2 + (x_3 + 2 - 4)^2$$

$$+ (x_4 + 2 - 4)^2 + (x_5 + 2 - 4)^2] = S^2.$$

Therefore, (1) is correct, and the answer is (B).

3. It follows that $a = \frac{x_1 + x_2 + x_3}{3}$, $b = \frac{y_1 + y_2 + y_3}{3}$, so

$$\frac{2x_1 + 3y_1 + 2x_2 + 3y_2 + 2x_3 + 3y_3}{3} = 2a + 3b.$$

The answer is (A).

4. Suppose there are m numbers below the median (m is a positive integer).

If there are $2m + 1$ numbers in all, then $\bar{x} = \frac{70m+80+96m}{2m+1} = \frac{83(2m+1)-3}{2m+1} = 83 - \frac{3}{2m+1} \geq 82$.

If there are $2m$ numbers in all, then $\bar{x} = \frac{70m+96m}{2m} = 83$. Therefore, the answer is (D).

5. If the median of these numbers is 40, then the mode must be 40, not 39, so (A) is wrong.

   If the numbers of sizes 39 and 40 are both 5, then the median and the mode are not equal, so (B) is wrong.

   If the 10 remaining all have size 39, then the average size is 39.35; if the 10 remaining all have size 40, then the average size is 39.85. This means $39 < P < 40$ always holds, so (C) is correct.

6. It follows that the remaining $26 - 1 - 3 = 22$ people score at least

$$4.8 \times 26 - 1 \times 3 - 3 \times 4 = 124.8 = 15 \approx 110.$$

   Thus, everyone gets at least 5 points, so the number of people who get 5 points is 22.

7. Suppose there are x shooters, then

$$(x - 7 - 5 - 4) \times 6 + 7 \times 0 + 5 \times 1 + 4 \times 2$$

$$= (x - 3 - 4 - 1) \times 3 + 3 \times 8 + 4 \times 9 + 1 \times 10.$$

   This gives $x = 43$.

8. It follows that

$$S^2 = \frac{1}{n}[(x_1 - \bar{x})^2 + (x_2 - \bar{x})^2 + \cdots + (x_n - \bar{x})^2].$$

   The desired variance is

$$S_0^2 = \frac{1}{n}[(kx_1 + a - \bar{x}_0)^2 + (kx_2 + a - \bar{x}_0)^2 + \cdots + (kx_n + a - \bar{x}_0)^2].$$

   Since $\bar{x}_0 = \frac{1}{n}(kx_1 + a + kx_2 + a + \cdots + kx_n + a) = \overline{kx} + a$, we have

$$S_0^2 = \frac{1}{n}[k^2(x_1 - \bar{x})^2 + k^2(x_2 - \bar{x})^2 + \cdots + k^2(x_n - \bar{x})^2] = k^2 S^2.$$

9. The average is $\frac{4a+2b+3}{4}$, and the median is $\frac{2a+b+2}{2}$, so the difference between them is $\frac{1}{4}$.

10. (1) Before adjustment, the average ticket price is $\frac{10+10+15+20+25}{5} = 16$ (yuan).

    After adjustment, the average ticket price is $\frac{5+5+15+25+30}{5} = 16$ (yuan).

    The daily number of tourists is unchanged after the adjustment, so the average daily income is unchanged.

(2) Before adjustment, the average daily total income is $10 \times 1 + 10 \times 1 + 15 \times 2 + 20 \times 3 + 25 \times 2 = 160$ (thousand yuan), and after adjustment, it becomes $5 \times 1 + 5 \times 1 + 15 \times 2 + 25 \times 3 + 30 \times 2 = 175$ (thousand yuan), which has increased by $\frac{175-160}{160} \approx 9.4\%$.

(3) The tourists' claim better reflects the truth.

11. (1) The total number of students is $16 + 12 + 10 + 8 + 4 = 50$.

(2) The excellence rate is $\frac{16+12+5}{50} = 66\%$.

(3) Given the condition, there are $16 + 12 - 20 = 8$ students whose score may be 81 or 82. Hence, the median of their scores might be $81, 81.5, 82$.

12. (1) The average of this sample is

$$\overline{x} = \frac{0.6 + 3.7 + 2.2 + 1.5 + 2.8 + 1.7 + 1.2 + 2.1 + 3.2 + 1.0}{10}$$
$$= 2.0.$$

So, by estimation, the number of boxes of disposable chopsticks consumed by the whole town in 2016 is approximately $2 \times 600 \times 350 = 420000$.

(2) Suppose the annual growth rate is x, then according to the condition, we have

$$2(1 + x)^2 = 2.42,$$

which gives $x = 0.1 = 10\%$ or $x = -2.1$ (which is dropped).

Hence, the average annual growth rate of disposable chopsticks consumption in 2017 and 2018 is 10%.

(3) Suppose the town can produce x sets of desks and chairs with the amount of wood used for chopsticks in 2018. By (2), we know that the number of boxes consumed in 2018 is

$$2.42 \times 600 \times 350 = 508200.$$

Hence, $x = \frac{508200 \times 100 \times 5 \times 10^{-3}}{0.5 \times 10^3} \div 007 = 7260$.

Solution 16

The Sides and Angles of a Triangle

1. Since $3^2 + 4^2 = 5^2$, these lengths cannot compose a triangle, and the answer is (C).

2. Suppose the third side has a length of x, then $2003 < x < 2011$. Since 4 is even, 2017 is odd, and the perimeter is even, it follows that x is odd. Hence, $x = 2005, 2007, 2009$, and there are three such triangles. The answer is (B).

3. Let x be the third side length, then $20-15 < x < 20+15$, so $5 < x < 35$. Thus, $40 < 20 + 15 + x < 70$. This means the perimeter cannot be 72, and the answer is (D).

4. The pattern of the figures shows that starting with the second figure, the number of triangles in each figure equals 3 times the previous one plus 2. Hence, the number of triangles in (4) is $[(5 \times 3+2) \times 3+2] \times 3+2 = 161$, and the answer is (C).

5. If 9 is a leg, then the perimeter is $9 + 9 + 11 = 29$. If 11 is a leg, then the perimeter is $11 + 11 + 9 = 31$. Therefore, the perimeter is 29 or 31.

6. The side lengths compose a triangle, so $1 < 1-2x < 7$, and $-3 < x < 0$.

7. Extend DE to intersect AB at F, then $\angle BFD = \angle FAD + \angle ADF = 20° + 30° = 50°$, and $\angle BED = \angle ABE + \angle BFD = 10° + 50° = 60°$.

8. Connect GE. From the figure, we see that $\angle A + \angle B = 180° - \angle AMB, \angle C + \angle D = 180° - \angle CND$. Since $BE \parallel CE$, we have $\angle AMB + \angle CND = 180°$, $\angle MGE + \angle NEG = 180°$. Therefore, $\angle A + \angle B + \angle C + \angle D + \angle E + \angle F + \angle G = 540°$.

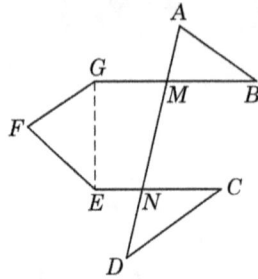

9. There are 16 in all. See the following figures.

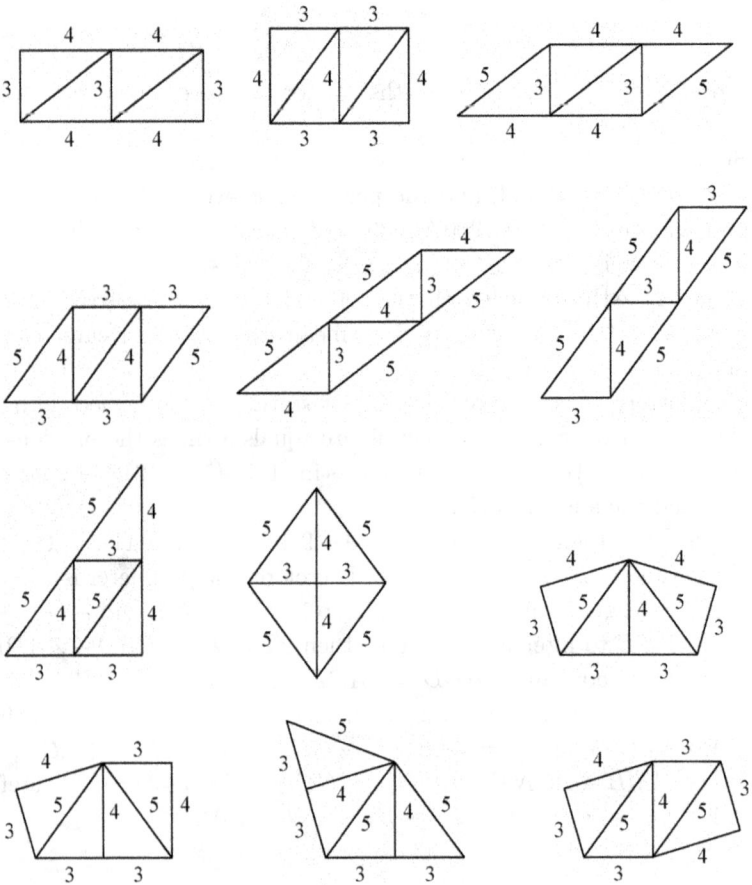

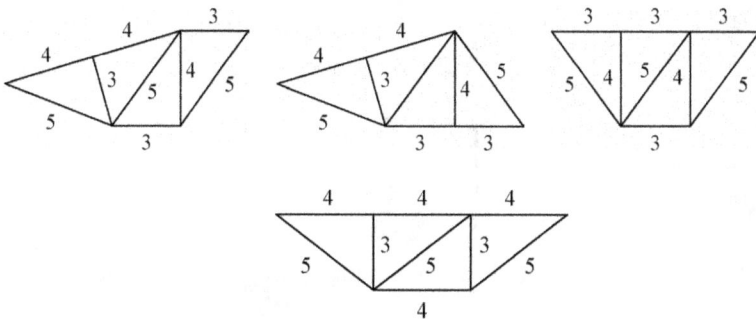

10. (1) 4 matches compose 3 sides in only one way, which is 1, 1, 2, but they cannot make a triangle.

    (2) 8 matches compose 3 sides in 12 different ways: (1, 1, 10); (1, 2, 9); (1, 3, 8); (1, 4, 7); (1, 5, 6); (2, 2, 8); (2, 3, 7); (2, 4, 6); (2, 5, 5); (3, 3, 6); (3, 4, 5); (4, 4, 4). Among them, only (2, 5, 5), (3, 4, 5), (4, 4, 4) make a triangle, where the first is an isosceles triangle, the second is a right triangle, and the third is an equilateral triangle. (Sketch is omitted.)

11. $(4a^2 + 5) - (2a + 3) = 4a^2 - 2a + 3 = 2a(2a - 1) + 2 > 0,$

$$(4a^2 + 5) - (4a^2 + 3a + 1) = -3a + 4.$$

   (1) If $-3a + 4 > 0$ and a is a nonnegative integer, then $4a^2 + 5$ is the longest side. When $a = 0$, the three lengths are $1, 3, 5$, which do not compose a triangle. When $a = 1$, the three lengths are $5, 8, 9$, and $5 + 8 > 9$, so they can compose a triangle.

   (2) If $-3a + 4 < 0$, then $4a^2 + 3a + 1$ is the longest side, and the three lengths compose a triangle if and only if the following inequality holds:

$$(4a^2 + 5) + (2a + 3) > 4a^2 + 3a + 1.$$

Thus, $a < 7$, and in this case, $\frac{4}{3} < a < 7$.

   Since a is an integer, $a = 2, 3, 4, 5, 6$, and the corresponding side lengths are (21, 7, 23), (41, 9, 46), (69, 11, 77), (105, 13, 116), (149, 15, 163).

   Therefore, there are six such triangles.

12. Let S be the area of $\triangle ABC$ and h be the third altitude. Then, the three side lengths are $\frac{2S}{5}, \frac{2S}{20}, \frac{2S}{h}$. Apparently, $\frac{2S}{5} > \frac{2S}{20}$, so

$$
\begin{cases}
\dfrac{2S}{20} + \dfrac{2S}{h} > \dfrac{2S}{5}, \\
\dfrac{2S}{20} + \dfrac{2S}{5} > \dfrac{2S}{h}.
\end{cases}
$$

Hence, $4 < h < \frac{20}{3}$, and the maximum of the third altitude is 6.

Solution 17

Congruent Triangles

1. $\triangle ABC \cong \triangle ADC, \triangle ABO \cong \triangle ADO, \triangle BCO \cong \triangle DCO$, so there are three pairs, and the answer is (C).

2. $AB = AC, BO = CO$, and OA is a common side, so $\triangle AOB \cong \triangle AOC$. Thus, $\angle BAO = \angle CAO$. Since $AO = CO$, we have $\angle CAO = \angle ACO$, so $\angle BAO = \angle ACO$. Also, $CP = AQ, \angle QAO = \angle PCO, AO = CO$, so $\triangle AQO \cong \triangle CPO$, and $\angle AQO = \angle CPO = 16°$. The answer is (A).

3. It is easy to see that $\triangle ADE \cong \triangle ADF$, so $DE = DF, AE = AF$. Hence, $\triangle AEG \cong \triangle AFG$, and $\angle AGE = \angle AGF = 90°$. Only (D) is false.

4. Since $BD = AF, \angle B = \angle CAF = 60°, AB = CA$, we have $\triangle ABD \cong \triangle CAF, \angle BAD = \angle ACF$. Thus, $\angle CED = \angle ACF + \angle CAE = \angle BAD + \angle CAE = \angle CAB = 60°$, and the answer is (B).

5. The answer is not unique. Possible answers are $AC = AE, \angle B = \angle D, \angle C = \angle E$.

6. Since $\angle 1 = \angle 2 = \angle 3$, we have $\angle ACB = \angle ECD, \angle B = \angle D$. Combining with $AC = CE$, we obtain $\triangle ABC \cong \triangle EDC$, so $AB = DE = 3$.

7. Choose a point F on AB such that $AF = AD$, and connect EF. Then, $\triangle ADE \cong \triangle AFE, \angle D = \angle AFE$. By $AD \parallel BC$, we have $\angle C + \angle D = 180°$, and since $\angle AFE + \angle BFE = 180°$, we obtain $\angle C = \angle BFE$. Thus, $\triangle BCE \cong \triangle BFE$, and $BC = BF$. Therefore, $AB = AD + BC = 6$.

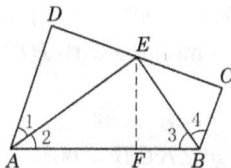

8. Extend FD to M so that $DM = DE$, and connect MQ. Since DP bisects $\angle EDF$, we have $\angle EDP = \angle FDP$. $\angle MDQ = \angle DFQ + \angle DQF = \angle DFQ + 90° - \angle DPF = 90° + \angle FDP = 90° + \angle EDP$. Also, $\angle EDQ = 90° + \angle EDP$, so $\angle MDQ = \angle EDQ$. Combining with $DM = DE, DQ = DQ$, we have $\triangle DME \cong \triangle DEQ$, which means $\angle M = \angle E$.

Since $FQ = ED + DF$, it follows that $FQ = DM + DF = MF$, so $\angle EQM = \angle M = \angle E$. In addition, $\angle MFQ = \angle E + \angle EDP = \angle E + 15°$, and $\angle MFQ = 180° - 2\angle M = 180° - 2\angle E$.

Hence, $\angle E + 15° = 180° - 2\angle E, \angle E = 55°$. Therefore, $\angle DPQ = \angle E + \angle EDP = 55° + 7.5° = 62.5°, \angle DQF = 90° - 62.5° = 27.5°$.

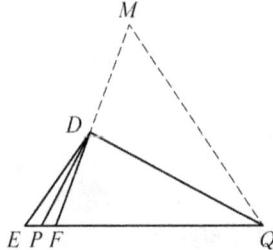

9. The answer is not unique, and the following is one option.

Conditions: $AD = BC, AC = BD$; conclusion: $\angle D = \angle C$.

Proof: Since $AD = BC, AC = BD$, and AB is common, we have $\triangle ABD \cong \triangle BAC$, so $\angle D = \angle C$.

10. Since $DE \parallel BC, \angle ACB = 90°$, we have $\angle ADE = \angle DCF = 90°$. Since D is the midpoint of AC, we have $AD = DC$. Combining with $\angle CDF = \angle A$, we have $\triangle ADE \cong \triangle DCF$, so $\angle AED = \angle F$. Since $\angle B = \angle AED = 51°$, we obtain that $\angle F = 51°$.

11. Let $\angle LBK = \angle LKB = \alpha$, then $\angle KLA = 2\alpha, \angle LKA = 90° - \angle KLA = 90° - 2\alpha$.

Let $LT \perp BC$ where T lies on BC. Since CL bisects $\angle ACB$, we have $LT = LA$.

From $LB = LK, \angle BTL = \angle KAL = 90°$, we get $\triangle BTL \cong \triangle KAL$.

Thus, $\angle LBT = \angle LKA = 90° - 2\alpha$. Further, $\angle CKB = \angle CBK = 90° - \alpha$, so $CB = CK$. Combining with $BK = CK$, we see that $\triangle BKC$ is an equilateral triangle.

Hence, $\angle CBK = 90° - \alpha = 60°, \alpha = 30°$. Therefore, $\angle ABC = 90° - 2\alpha = 30°, \angle ACB = \angle KCB = 60°$.

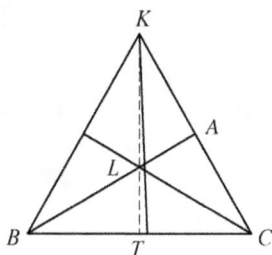

12. Extend FD to G so that $FD = DG$, and connect BG, EG.

   Since B is the midpoint of BC, we have $CD = DB$. Also, $\angle 1 = \angle 2$, so $\triangle CDF \cong \triangle BDG$, and $CF = BG$.

   Since $DE \perp DF$, we have $\angle EDF = \angle EDG = 90°$, and DE is a common side, so we obtain that $\triangle DEF \cong \triangle DEG$, and $EF = EG$.

   In $\triangle BEG$, $BE + BG > EG$, so we conclude that $BE + CF > EF$.

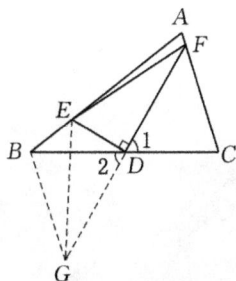

Remark: To prove inequalities among segments, it is common to construct congruent triangles to put them into a single triangle.

Solution 18

Isosceles Triangles

1. It suffices to show that $\triangle BOE \cong \triangle COD$. Among the six combinations, only ①② and ③④ do not imply this congruence. Therefore, the remaining four combinations are sufficient to prove that $\triangle ABC$ is isosceles. The answer is (B).

2. Let $\angle A = x$, then $\angle ADE = x, \angle DEB = \angle DBE = 2x, \angle BDE = 180° - 4x, \angle CDB = 180° - x - (180° - 4x) = 3x$, so $\angle CBD = 3x$. Thus, $\angle C = 180° - 6x$, and since $\angle C = \angle ABC$, we have $180° - 6x = 5x$. Therefore, $x = \frac{180°}{11}$, and the answer is (D).

3. Construct an equilateral triangle CDE, and connect AE. It follows that $AC = BC, \angle ACE = \angle BCD, EC = DC$, so $\triangle ACE \cong \triangle BCD$, and $AE + BD = 5$. Also $\angle ADE = 30° + 60° = 90°, AD = 3$, so $DE = 4$, and $CD = 4$. The answer is (B).

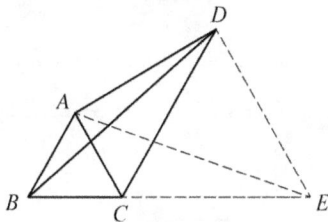

4. If the sides of $\triangle ABC$ are the bases of these isosceles triangles, then there is only one such P, which is the intersection of the perpendicular bisectors of the three sides. If at least one side of $\triangle ABC$ is a leg, then there are nine such P. They are obtained by drawing three circles centered at A, B, C, whose radius equals the side length, as well as the perpendicular bisectors of the three sides. The lines and circles intersect

at 9 points, which are $P_2 \sim P_{10}$ in the following figure. Therefore, there are 10 such points.

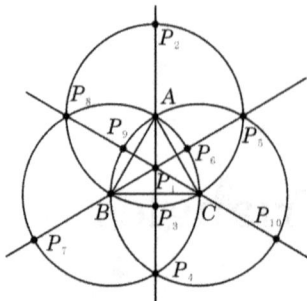

5. Since $AC = AD, \angle CAD = 75°$, we have

$$\angle ADC = \frac{180° - 75°}{2} = 52.5°.$$

Also, $AB = AD, \angle BAD = 25° + 75° = 100°$, so

$$\angle BDA = \frac{180° - 100°}{2} = 40°.$$

Hence, $\angle BDC = 52.5° - 40° = 12.5°$, and similarly, $\angle DBC = 37.5°$.

6. By assumption, $\angle B = \angle C, \angle ADE = \angle AED$. By the external angle theorem, it follows that $\angle EDC + \angle C = \angle AED$.
   We have $\angle DAE = 180° - 2\angle AED = 180° - 2\angle EDC - 2\angle C$.

$$180° = \angle B + \angle C + \angle BAD + \angle DAE$$
$$= 2\angle C + 50° + (180° - 2\angle EDC - 2\angle C)$$
$$= 180° + 50° - 2\angle EDC.$$

Thus, $2\angle EDC = 50°$, so $\angle EDC = 25°$.

7. Since BI bisects $\angle ABC$, we have $\angle DBI = \angle CBI$. Since $DE \parallel BC$, we have $\angle CBI = \angle BID$.
   Hence, $\angle DBI = \angle BID, BD = DI$. For similar reasons, $CE = EI$. Therefore, the perimeter of $\triangle ADE$ is $AB + AC = 19$.

8. It is easy to see that $AB + BN$. Choose a point D on AM such that $BD = BA$ (as shown in the following figure), then $BD = BN$, and $\angle ABD = 20°$.

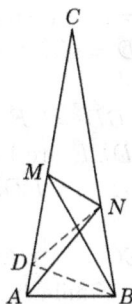

Thus, $\angle BDN = 60°, \angle MDN = 40°$. Consequently, $\angle DMN = 70°, \angle NMB = 30°$.

9. Since $ABCD$ is a square, we have $BC = DC$. $\triangle PBC, \triangle QCD$ are both equilateral triangles, so $\angle CBE = \angle CDF = 60°, \angle BCE = \angle DCF = 30°$. Hence, $\triangle BCE \cong \triangle DCF$, and $CE = EF$.

   Since the side lengths of $\triangle PBC, \triangle QCD$ are both the side length of $ABCD$, we have $CQ = CP$, so $QE = PF$.

   Now, $\angle P = \angle Q = 60°, \angle PMF = \angle QME$, so $\triangle PMF \cong \triangle QME$, which implies $PM = QM$.

10. Construct $CE \perp AD$, where E lies on AD, and connect BE.

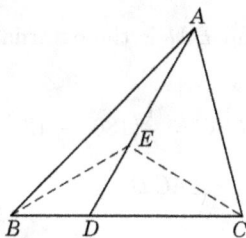

   Since $\angle ADC = 60°$, we have $\angle ECD = 30°$, and $DC = 2DE$.

   Also, $DC = 2BD$, so $DE = BD$, $\angle DBE = \angle DEB = 30°$.

   Thus, $\angle DBE = \angle ECB = 30°$, and $CE = BE$.

   Since $\angle ABD = 45°$, we have $\angle EBA = \angle EAB = 15°, AE = BE$. Hence, $AE = CE$, and $\triangle ACE$ is an isosceles right triangle with $\angle ACE = 45°$.

   Therefore, $\angle ACD = \angle DCE + \angle ACE = 75°$.

11. Any of the following five conditions work:
   (1) $BD = CE$; (2) $BE = CD$; (3) $\angle BAD = \angle CAE$;
   (4) $\angle BAE = \angle CAD$; (5) $\angle B = \angle C$.

We prove (1). Since $AD = AE$, we have $\angle ADE = \angle AED, \angle ADB = \angle AEC$. Combining with $BD = CE$, we obtain that $\triangle ABD \cong \triangle ACE$, so $AB = AC$.

12. (1) Equal segments: $AF = BD = CE, AE = BF = CD$.

    Since $\triangle ABC$ and $\triangle DEF$ are both equilateral, we have $\angle A = \angle B = \angle C = 60°, DE = EF = FD, \angle DEF = \angle EFD = \angle RDE = 60°$.

    Since $\angle CDE + \angle DEC = \angle DEC + \angle AEF = 120°$, it follows that $\angle CDE = \angle AEF$. Similarly, $\angle AEF = \angle BFD = \angle CDE$. This implies $\triangle AEF \cong \triangle BFD \cong \triangle CED$. Hence, $AF = BD = CE, AE = BF = CD$.

    (2) FD can be obtained by rotating FE clockwise around F by $60°$, and DE can be obtained by rotating FD clockwise around D by $60°$.

13. Consider three cases:

    (1) If $AO = AD$, then $\angle AOD = \angle ADO$. Since $\angle AOD = 190° - \alpha$, $\angle ADO = \alpha - 60°$, we have $190° - \alpha = \alpha - 60°, \alpha = 125°$.

    (2) If $OA = OD$, then $\angle OAD = \angle ADO$. Since $\angle OAD = 180° - (\angle AOD + \angle ADO) = 50°$, we have $\alpha - 60° = 50°, \alpha = 110°$.

    (3) If $OD = AD$, then $\angle OAD = \angle AOD$. In this case, $190° - \alpha = 50°$, $\alpha = 140°$.

14. Since $\angle ABC = 12°$, and BM is the external angle bisector of $\angle ABC$, we have

$$\angle MBC = \frac{1}{2}(180° - 12°) = 84°.$$

Also, $\angle BCM = 180° - \angle ACB = 48°$, so $\angle BMC = 180° - 84° - 48° = 48°$.

Thus, $BM = BC$. Since $\angle ACN = \frac{1}{2}(180° - \angle ACB) = 24°$, we obtain that

$$\angle BNC = 180° - \angle ABC - \angle BCN = 180° - 12° - (\angle ACB + \angle ACN)$$

$$= 168° - (132° + 24°) = 12° - \angle ABC.$$

Therefore, $CN = CB$, and $BM = CN$.

Solution 19

Right Triangles

1. Let a, b be the lengths of two legs, then $a^2 + b^2 = 2ab$, so $(a - b)^2 = 0$, and $a = b$. Thus, the hypotenuse $c = \sqrt{2}b$, and $a{:}b{:}c = 1{:}1{:}\sqrt{2}$. The answer is (D).

2. Since $\angle AHE = \angle BHD, \angle AEH = \angle BDH = 90°$, we have $\angle CAD = \angle HBD$. Combining with $BH = AC$, we obtain $\text{Rt}\triangle ACD \cong \text{Rt}\triangle BHD$, and $AD = BD, \angle ABC = 45°$. The answer is (B).

3. In right triangle ACE, $2CE = AC$, so $\angle 1 = 30°$. Hence, $CD = \frac{1}{2}AD = 2$. The answer is (A).

4. From the figure, we obtain $OA = \sqrt{17}$. Thus, the perimeter of the "sea snail" is $17 + \sqrt{17}$, and the closest integer to it is 21. The answer is (C).

5. By the Pythagorean theorem, $a^2 = (b + 1)^2 - b^2 = 2b + 1$.
   Since $b < 2011$, we see that a is an odd integer in the interval $(1, \sqrt{4023})$. Thus, a is one of $3, 5, \ldots, 63$. There are 31 such triangles, and the answer is (D).

6. Let $BC = x$, then $AB = 2x, AC = \sqrt{AB^2 - BC^2} = \sqrt{3}x$. Thus, $2x + x = 6\sqrt{3}$, so $x = 2\sqrt{3}$. Therefore, $AC = \sqrt{3} \times 2\sqrt{3} = 6$.

7. Connect BD. In right triangle ABD, $AB = 3, DA = 4$, so $BD = 5$.
   In $\triangle BCD$, we have $BD^2 + BC^2 = 5^2 + 12^2 = 13^2 = CD^2$, so it is a right triangle. Thus,

$$S_{ABCD} = S_{\triangle ABD} + S_{\triangle BCD} = \frac{1}{2} \times 3 \times 4 + \frac{1}{2} \times 5 \times 12 = 36.$$

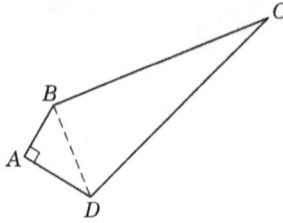

8. Since $AB = AC, \angle ABN = \angle MBC$, we have $\angle ABC = \angle C = \angle ABN + \angle MBN + \angle MBC = 2\angle MBC + \angle MBN$. Also, $BM = MN$, so $\angle MBN = \angle MNB$.

On the other hand, $\angle MBN + \angle BMN + \angle MNB = 180°$, so $\angle BMN + 2\angle MBN = 180°$, and $\angle C + \angle MBC + 2\angle MBN = 180°, 2\angle MBC + \angle MBN + \angle MBC + 2\angle MBN = 180°$. Thus, $3(\angle MBC + \angle MBN) = 180°$, and $\angle NBD = 60°$.

Construct $ND \perp BC$ with D on BC. In right triangle BND, $BN = a$, so $BD = \frac{1}{2}a$,

$$ND = \sqrt{a^2 - \left(\frac{1}{2}a\right)^2} = \frac{\sqrt{3}}{2}a.$$

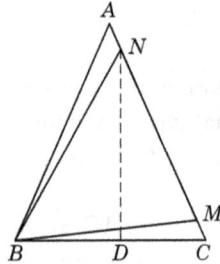

9. Since $\triangle ACD \cong \triangle AC'D$, we have $CD = BD = DC' = 2, \angle ADC = \angle ADC' = 45°$. Thus, $\angle BDC' = 90°, BC' = \sqrt{2}BD = 2\sqrt{2}$.

10. Let a be the side length of the big square, and let b be the leg of the white isosceles right triangles. Then, $a^2 = 2 \times 4 \times \frac{1}{2}b^2 = 4b^2, a = 2b$. Thus, the hypotenuse equals twice a leg, which implies that $\angle ABE = 30°, \angle BAE = 45° - 30° = 15°$.

11. The areas of squares 1 and 2 sum to $DE^2 = BC^2$, and the areas of squares 3 and 4 sum to $GF^2 = AC^2$. Thus, the sum of the four areas is $AC^2 + BC^2 = AB^2 = 13^2 = 169 \ (\text{cm}^2)$.

12. Connect AC, and construct an equilateral triangle $\triangle BCE$ as shown in the figure. Connect AE.

Since $AD = DC, \angle ADC = 60°$, $\triangle ACD$ is an equilateral triangle, and $DC = CA = AD$. From $\angle EBC = 60°, \angle ABC = 30°$, we see that $\angle ABE = 90°$. In the right triangle ABE,

$$AE^2 = AB^2 + BE^2 = AB^2 + BC^2.$$

In $\triangle ACE$ and $\triangle DCB$, we have $AC = DC, \angle ACE = \angle DCB, CE = CB$ so it follows that $\triangle ACE \cong \triangle DCB$, and $AE = BD$. Therefore, $BD^2 = AB^2 + BC^2$.

13. Extend BA to E so that $AE = AB = \sqrt{2}$, then $BE = 2\sqrt{2}$. Connect CE, then $CE \parallel AD$, and $CE = 2AD = 2\sqrt{6}$.

   In $\triangle ACE$, $AE^2 + CE^2 = 2 + 24 = 26 = AC^2$, so $\angle AEC = 90°$.

   In right triangle BCE, $CE = \sqrt{3}BC$, so $BE = \frac{1}{2}BC, \angle BCE = 30°, \angle ABC = 60°$.

14. Choose a point E on the extension of CD, such that $\angle DAE = \angle CAB$, then $\angle CAE = 90°$.

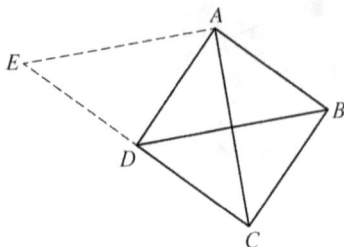

Since $\angle ACD = 45°$, we have $\angle AED = 45°, AE = AC$. Since $\angle ADB = \angle ABD = 45°$, we have $AD = AB$.

Now, $AE = AC, \angle DAE = \angle CAB, AD = AB$, so $\triangle EAD \cong \triangle CAB$, so $BC = ED$.

In the right triangle EAC, $AC = 4, \angle AEC = 45°$, so $AE = AC = 4, CE = \sqrt{AE^2 + AC^2} = 4\sqrt{2}$. Therefore, $ED = CE - CD = 4\sqrt{2}-3$, and $BC = 4\sqrt{2}-3$.

15. Let $AD = x, BE = y, CF = z$, then $BD = 5 - x, CE = 7 - y$, $AF = 6 - z$.

Connect PA, PB, PC. In right triangles PBD and PBE, the Pythagorean theorem we have $BD^2 + PD^2 = PB^2 = BE^2 + PE^2$, so that

$$(5 - x)^2 + PD^2 = y^2 + PE^2. \qquad \qquad ①$$

Similarly,

$$(7 - y)^2 + PE^2 = z^2 + PF^2, \qquad \qquad ②$$

$$(6 - z)^2 + PF^2 = x^2 + PD^2. \qquad \qquad ③$$

From ① + ② + ③, we obtain that $(5 - x)^2 + (7 - y)^2 + (6 - z)^2 = x^2 + y^2 + z^2$, so $5x + 7y + 6z = 55$. Combining with $BE - AD = 1$ (so that $y - x = 1$), we conclude that

$$AD + BE + CF = x + y + z$$

$$= \frac{1}{6}[(5x + 7y + 6z) - (y - x)]$$

$$= \frac{1}{6}(55 - 1) = 9.$$

Solution 20

Parallelograms

1. A quadrilateral satisfying ③ can be an isosceles trapezoid, so the answer is (C).
2. As shown in the figure, by central symmetry, we can construct lines through the center of the rectangle that are parallel to its sides, which divide the white region into two rectangles and four triangles. Hence, $S_{\text{shade}} = 8 \times 5 - 4 \times \frac{5}{2} \times 2 - 4 \times \frac{3}{2} \times 2 \times \frac{1}{2} - 3 \times \frac{5}{2} \times 2 \times \frac{1}{2} = 6\frac{1}{2}$. The answer is (C).

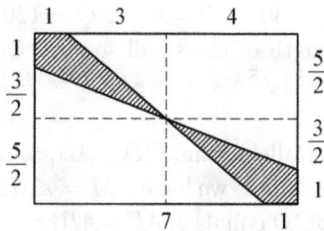

3. Let M be the midpoint of DE, and connect AM. In Rt$\triangle ADE$, we have $AM = EM = \frac{1}{2}DE = AB$. Let $\angle AED = \alpha$, then $\angle AME = 180° - 2\alpha$, $\angle ABM = \alpha - 18°$. Since $\angle ABM = \angle AMB$, it follows that $180° - 2\alpha = \alpha - 18°, \alpha = 66°$. Therefore, the answer is (A).

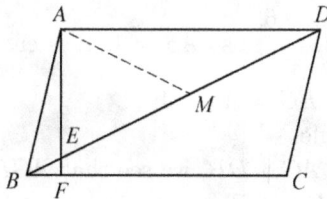

4. Let x be the side length of the small square, then $5x - 3x - \frac{1}{2} \cdot x \cdot 2x \times 2 - \frac{1}{2} \cdot x \cdot 4x \times 2 = 1$, so that $x^2 = \frac{1}{9}$. Thus, the area of $ABCD$ is $15x^2 = 15 \times \frac{1}{9} = \frac{5}{3}$. The answer is (D).

5. From $\angle BAE = 45°, \angle CAE = 15°$, we obtain $\angle BAO = 60°$. Also, $AO = BO$, so $\triangle ABO$ is equilateral, and $\angle ABO = 60°$. Thus, $\angle EBO = 30°$. It is easy to see that $AB = AE$, so $BO = BE, \angle BOE = \frac{180° - 30°}{2} = 75°$. The answer is (D).

6. Since $ABCD$ is a rhombus, $\triangle AEF$ is equilateral, and $AB = AE$, it can be proved that $\triangle ABE \cong \triangle ADF, \angle BAE = \angle DAF$.

   Now, $\angle BAE = 180° - 2\angle B = 180° - 2(180° - \angle C) = 2\angle C - 180°, \angle BAD = 60° + 2\angle BAE = 60° + 2(2\angle C - 180°) = 4\angle C - 300°$, and $\angle BAD = \angle C$, so we have $4\angle C - 300° = \angle C$, so $\angle C = 100°$, and the answer is (A).

7. Since $ABCD$ is a parallelogram, $\angle ABC + \angle A = 180°$, and since $\angle ABC = 3\angle A$, we have $\angle A = 45°$.

   Now, $\angle A = \angle C = 45°, EF \perp DC$, so $CE = EF = 1, CF = \sqrt{1^2 + 1^2} = \sqrt{2}$.

   Hence, $DE = CD - CE = CF - CE = \sqrt{2} - 1$.

8. Since $ABCD$ is a parallelogram, we have $AB = CD$, and since $BC : CD = 3:2, AB = EC$, we have $BE = \frac{1}{2}AB$. Since $\angle AEB = 90°$, it follows that $\angle BAE = 30°, \angle B = 60°, \angle C = 120°, \angle EAF = 60°$.

9. Let x be the side length of the small square. Consider the diagonal of the big square, then $2\sqrt{2}x + x = \sqrt{2}, x = \frac{4 - \sqrt{2}}{7}$. Hence, $a = 4, b = 7, a + b = 11$.

10. Since $ABCD$ is a parallelogram, $CD \parallel AB$, and $\angle NDC = \angle M$.

    Since $\angle NDC = \angle MDA$, we have $\angle M = \angle MDA, MA = AD$. Thus, the perimeter of $ABCD$ equals $2(AD + AB) = 2(MA + AB) = 2BM = 12$.

11. Since $AB \parallel CD$, we have $\angle ALD = \angle CDL$.

    Since DL bisects $\angle ADC$, we have $\angle ADL = \angle CDL, \angle ADL = \angle ALD, AD = AL = 6$.

    Similarly, we can prove that $BS = BC = 6$, so $AS = BL$.

    Now, $\begin{cases} AS + SL = 6, \\ 2AS + SL = 10, \end{cases}$ so $AS = 4$. It is easily seen that $AEGS$ is a parallelogram, so $EG = AS = 4$.

12. $AGBD$ is a rectangle.

    Since $AD \parallel BG, AG \parallel DB$, we see that $AGBD$ is a parallelogram, and since $DE = AE = EB$, we obtain that $\angle ADB = 90°$. Therefore, $AGBD$ is a rectangle.

13. Let E be the midpoint of BC, connect EA, EP, and construct $EF \perp PA$ with F on AP.

Since $ABCD$ is a square, we have $DQ = BE, \angle D = \angle B = 90°, AD = AB$. Thus, $\triangle ADQ \cong \triangle ABE, \angle DAQ = \angle BAE$. Since $\angle BAP = 2\angle QAD$, it follows that $\angle BAE = \angle EAF$.

Now, $\angle B = \angle AFE = 90°$, and AD is common, so $\triangle ABE \cong \triangle AFE$, and $AB = AF, BE = FE$.

From $CE = BE$, we see that $CE = FE$, and since $\angle C = \angle PFE$, PE is common, and we obtain $\triangle PCE \cong \triangle PFE$ and $PC = PF$. Therefore, $AP = CP + CB$.

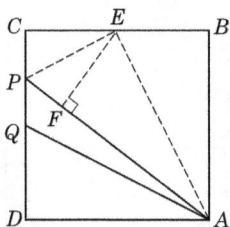

14. Connect PA, AE. Since $ABCD$ is a rhombus, it follows that $\triangle APD \cong \triangle CPD$, and $PC = PA$. Connect AC, then $\triangle ACD$ is equilateral, and $AC = 4a$. Thus, $AB = AC = 4a$.

Since E is the midpoint of BC, we have $AE \perp BC$. In right triangle ABE,

$$AE = \sqrt{AB^2 - BE^2} = \sqrt{(4a)^2 - (2a)^2} = 2\sqrt{3}a.$$

Note that $PE + PC = PE + PA \geq AE = 2\sqrt{3}a$, so the minimum of $PE + PC$ is $2\sqrt{3}a$, which is attained when P is the intersection of AE, BD.

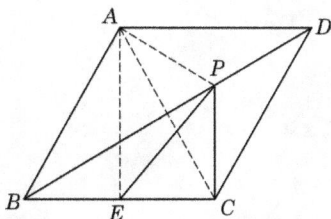

15. Assume that $AB = CD = a, AD = BC = b, AM = x, PC = y, AQ = z, NC = t$.

If $x \neq y$, without loss of generality, assume that $x > y$, then

$$a - x < a - y. \tag{1}$$

Since $S_{\triangle AQM} = S_{\triangle CNP}$, we have $xz = yt$, so $z < t$, and

$$b - t < b - z. \tag{2}$$

From ①, ②, we see that

$$\frac{1}{2}(a - x)(b - t) < \frac{1}{2}(a - y)(b - z),$$

so $S_{\triangle BMN} = S_{\triangle DPQ}$, which is contradictory to the assumption. Hence, $x = y$. Similarly, $z = t$.

Thus, $\triangle AMN \cong \triangle CPN$, and $MQ = PN$. For similar reasons, $MN = PQ$. Therefore, $MNPQ$ is a parallelogram.

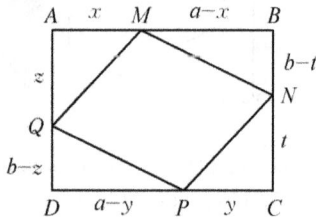

Solution 21

Trapezoids

1. The obtuse internal angles of the trapezoid compose a perigon, so each angle is 120°. Thus, $\angle A = 60°$, and the answer is (A).

2. Construct the altitudes of the trapezoid through A, D, which intersect BC at E, F, respectively. Let $BE = x$, then $FC = 3 - x$. By the Pythagorean theorem, $3^2 - x^2 = 2^2 - (3 - x)^2$, so $x = \frac{7}{3}$. Thus, the altitude equals $\sqrt{3^2 - \left(\frac{7}{3}\right)^2} = \frac{4\sqrt{2}}{3}$, and the area of the trapezoid is $\frac{1}{2} \times (1 + 4) \times \frac{4\sqrt{2}}{3} = \frac{10\sqrt{2}}{3}$. The answer is (A).

3. Construct $AE \perp BC$ with E on BC, and $CF \perp AB$ with F on AB. Connect AC.

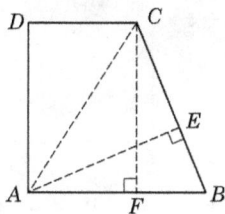

   Then, $AFCD$ is a rectangle, $CF = AD = 12, AF = CD = 8, BF = 5, BC = \sqrt{CF^2 + BF^2} = \sqrt{5^2 + 12^2} = 13$.

   Since $\frac{1}{2}AE \times BC = \frac{1}{2}CF \times AB$, we have $13 \times AE = 12 \times 13, AE = 12$. The answer is (A).

4. Construct the altitudes AE, DF, and let DG be the line through D and parallel to AC, which intersects the extension of BC at G. Then, $AE = DF = 10\sqrt{3}, BE = CF = 10$.

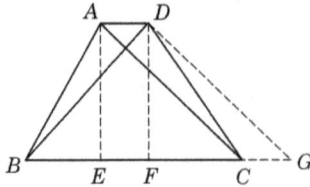

Since $ACGD$ is a parallelogram, we have $AC = DG$, $AD = CG$. Since $ABCD$ is an isosceles trapezoid, it follows that $BD = AC = DG$. Also, $BD \perp DG$ since $BD \perp AC$.

Let $AD = EF = x$, then in right triangle BDF,
$$BD^2 = BF^2 + DF^2 = (10 + x)^2 + (10\sqrt{3})^2.$$
In isosceles right triangle BDG, $BG^2 = 2BD^2$, so $(20 + 2x)^2 = 2(10 + x)^2 + 600$. Hence, $10 + x = 10\sqrt{3}$, and the perimeter of the trapezoid is $20 + 2(10 + x) = 40 + 20\sqrt{3}$. The answer is (C).

5. Let E be the symmetric point of C with respect to AD, and BE intersects AD at P, then this P makes $PB + PC$ minimal. The proof is as follows.

   Choose another point P_1 on AD, and connect BP_1, EP_1. Then, $PB + PB = EP + BP < BP_1 + EP_1 = BP_1 + CP_1$, so P is the desired point. Now, $\angle DPC = \angle DPE$, and $\angle DPE = \angle APB$, so $\angle APB = \angle DPC$, and the answer is (D).

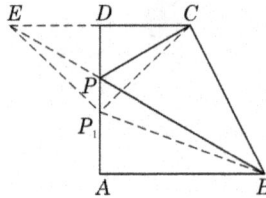

6. Construct $DF \perp AB$ with F on AB, and connect BD. Then, $\triangle ABE \cong \triangle DBE$, so $AB = DB = 25$. In right triangle BCD, $CD = \sqrt{BD^2 - BC^2} = \sqrt{25^2 - 24^2} = 7$, so $AF = 25 - 7 = 18$, $AD = \sqrt{AF^2 + DF^2} = \sqrt{18^2 + 24^2} = 30$.

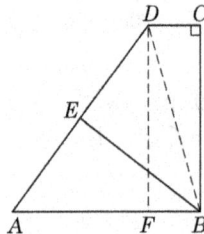

7. Let E be a point on the extension of BC such that $DE \parallel AC$. Then, $AC = DE$. It is easily seen that the area of $ABCD$ equals the area of $\triangle BDE$. Since $BD = BE = 8$, $DE = 14$, the area of $\triangle BDE$ is $4\sqrt{15}$.

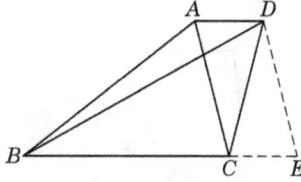

8. Two such right trapezoids can make a rectangle with a size of $800\,\text{mm} \times 400\,\text{mm}$, and the plate with a size of $2100\,\text{mm} \times 1600\,\text{mm}$ can produce 10 such rectangles. Hence, we can cut at most 20 trapezoids from the plate.

9. Construct $DE \perp BC$ with E on BC. Since $\triangle ABC$ is an isosceles right triangle, the altitude on hypotenuse BC equals half of BC, so $DE = \frac{1}{2}BC$.

   Since $BC = BD$, we have $DE = \frac{1}{2}BD, \angle DBE = 30°$ Hence, $\angle AMB = \angle MBC + \angle MCB = 30° + 45° = 75°$.

10. Construct $CF \perp AD$ with F on the extension of AD, and rotate $\triangle CDF$ counterclockwise around C by $90°$ to obtain $\triangle CGB$. Then, $ABCF$ is a square, and $\angle ECG = 45° = \angle ECD, CG = CD, CE = CE$. Thus, $\triangle ECG \cong \triangle ECD$, and $EG = ED$.

    Let $DE = x$, then $DF = BG = x - 28$, $AD = 70 - DF = 98 - x$.

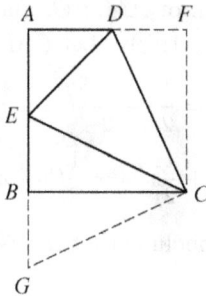

In the right triangle EAD, we have $42^2 + (98 - x)^2 = x^2$, so we solve that $x = 58$.

11. (1) Since CF bisects $\angle BCD$, we have $\angle GCF = \angle FCE$.

    In $\triangle BFC$ and $\triangle DFC$, $BC = DC, \angle GCF = \angle FCE, CF = CF$, so $\triangle BFC \cong \triangle DFC$.

(2) Extend DF to intersect BC at G. $AD \parallel BC$, $DG \parallel AB$, so $ADGB$ is a parallelogram, and $AD = BG$. Since $\triangle BFC \cong \triangle DFC$, we have $BF = DF, \angle EBC = \angle FDE$, and combining with $\angle BFG = \angle DFE$, we obtain $\triangle BFG \cong \triangle DFE$. Hence, $DE = BG = AD$.

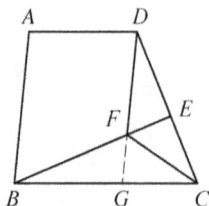

12. Construct $DE \parallel CA$, which intersects the extension of BA at E. Since $AC \perp BD$, we have $DE \perp BD$. Since $CDEA$ is a parallelogram, we have $AC = DE, CD = AE$.

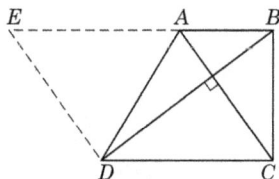

In right triangle BDE, $BD^2 + DE^2 = BE^2 = (AE + AB)^2 = (AB + AD)^2$, so

$$AC^2 + BD^2 = (AB + DC)^2.$$

13. Let E be the intersection of AM, BC, and let h_1, h_2 be the distances from E to AB, DC, respectively. Let $CM = x$, then it follows that

$$\begin{cases} \dfrac{1}{2}ah_1 = \dfrac{1}{2} \cdot \dfrac{a+b}{2}(h_1 + h_2), \\[2mm] \dfrac{1}{2}ah_1 + \dfrac{1}{2}xh_2 = \dfrac{1}{2}(b+x)(h_1 + h_2). \end{cases}$$

Eliminating h_1, h_2, we conclude that $x = \frac{a(a-b)}{a+b}$, which is the length of CM.

14. We observe that $\angle A, \angle B$ are two angles on the same base of the trapezoid. The condition that $\angle A + \angle B = 90°$ is not convenient to use in the trapezoid, so we consider putting them in a triangle, and we can use the properties of right triangles. Note that we can translate angles by constructing parallel lines, so we construct a line DP parallel to BC, which simply means translating BC to DP. Then, $\angle A + \angle APD = 90°$.

Next, we see that $AB - CD = AP$, and AP is the hypotenuse of the right triangle ADP. To prove $MN = \frac{1}{2}AP$, it suffices to show that MN equals the median of $\triangle ADP$ on the hypotenuse. Thus, we choose G as the midpoint of AP, and connect DG. Then, it suffices to show that $DGMN$ is a parallelogram. Note that $GM = AM - AG = \frac{1}{2}(AB - AP) = \frac{1}{2}BP = \frac{1}{2}CD = DN$, and $AB \parallel CD$, so the conclusion follows.

Solution 22

The Angles and Diagonals of a Polygon

1. $\angle A + \angle C + \angle E = 360° - \angle MNE, \angle B + \angle D + \angle F = 360° - \angle NMB, \angle G = 180° - (\angle GMN + \angle GNM) = \angle MNE + \angle NMB - 180°$.
   Adding the three equalities, we have

   $$\angle A + \angle B + \angle C + \angle D + \angle E + \angle F + \angle G = 540°.$$

   Hence, the answer is (C).

2. Let the extensions of FA, CB intersect at P and the extensions of GA, HB intersect at P'.

   By symmetry we have $\angle 1 = 2\angle APP', \angle 2 = 2\angle BPP'$. Since $\angle APB = 540° - \alpha$, it follows that $\angle 1 + \angle 2 = 1080° - 2\alpha$. Therefore, the answer is (B).

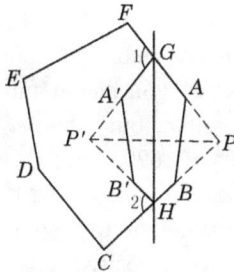

3. Every internal angle of a square equals $90°$, every internal angle of a regular pentagon equals $\frac{(5-2) \times 180°}{5} = 108°$, and every internal angle of a regular hexagon equals $120°$, so $\angle 1 = 360° - 90° - 108° - 120° = 42°$. The answer is (D).

4. The sum of external angles of a polygon is $360°$, so the sum of its internal angles is $2160° - 360° = 1800°$. Thus, $(n-2) \times 180° = 1800°$, and $n = 12$.

   The number of diagonals it has is $\frac{12 \times (12-3)}{2} = 54$, and the answer is (A).

5. The greatest internal angle of the polygon does not exceed $175°$, so its number of sides does not exceed $(175° - 95°) \div 10° + 1 = 9$. If it has 4 sides, then the sum of 4 internal angles is $95° + 105° + 115° + 125° = 440°$, which is not the sum of internal angles of a quadrilateral. Therefore, it must be a hexagon, and the answer is (A).

6. The sum of internal angles of an n-gon is $(n-2) \times 180°$, and the sum of internal angles of an $(n+x)$-gon is $(n+x-2) \times 180°$. Thus, $(n-2) \times 180° + 720° = (n+x-2) \times 180°$, and we solve that $x = 4$.

7. Suppose there are m acute angles, then $\frac{(2000-2) \times 180° - m \times 90°}{2000-m} < 180°$. Thus, $m < 4$, and since m is a natural number, the maximum of m is 3.

8. Suppose the excluded internal angle equals $m°$, then $(n-2) \times 180° = 8940° + m°$, so $n = 51 + \frac{120+m}{180}$. Apparently, when $n = 60$, we have $n = 52$, so the excluded angle equals $60°$.

9. Let the extensions of AB, DC intersect at G and the extensions of AF, DE intersect at H. Since $\angle A = \angle B = \angle C = \angle D = \angle E = \angle F$, each of them equals $120°$, so $\triangle BCG, \triangle EFH$ are both equilateral triangles, and $\angle G = \angle H = 60°$. Since $\angle D = \angle A = 120°$, it follows that $AGDH$ is a parallelogram, so $AB + BC = AB + BG = DE + EH = DE + EF = 6$.

10. Since the n-gon has exactly 4 obtuse internal angles, it has exactly 4 acute external angles. Also it has at most 3 obtuse external angles since the sum of its external angles is $360°$, so the maximum of n is 7.

11. Since an internal angle is less than $180°$, and each of them is a multiple of $15°$, we see that this polygon has at most 11 sides. The cases $n = 5, 6, 7, 8, 9, 10$ are all possible, but when $n = 11$, the sum of internal angles is $(11-2) \times 180° = 1620°$, and three of them sum to $285°$, so the remaining internal angles sum to $1620° - 285° = 1335°$. However, these angles are all multiples of $15°$, so $15 \times (k_1 + k_2 + \cdots + k_8) = 1335$, where $k_1, k_2, \ldots, k_8$ are positive integers smaller than 12. Thus, $k_1 + k_2 + \cdots + k_8 = 89$, and at least one of them is greater than 11, which is impossible. Therefore, $n = 5, 6, 7, 8, 9, 10$.

12. Since the internal angles are equal, so are the external angles, and each of them is $\frac{360°}{n}$. If the number of degrees of each angle is odd, then $\frac{360}{n}$ is an odd number. Since $360 = 2^3 \times 3^2 \times 5$, the value of n can be $2^3, 2^3 \times 3^2, 2^3 \times 5, 2^3 \times 3, 2^3 \times 3 \times 5, 2^3 \times 3^2 \times 5$. The sizes of each internal angle in these cases are (in the same order) $135°, 175°, 171°, 165°, 177°, 179°$. Therefore, there are six polygons with the desired property.

13. The nth figure is an $(n+2)$-gon, and every side has $n+2$ dots. The dots on the vertices are counted twice, so there are $(n+1)(n+2) - (n+2) = n^2 + 2n$ black dots on the nth figure.

14. Connect the diagonals of the small squares, as shown in the following figures.

   In Figure S14.1, $61° + 119° + 20° + x + 45° \times 2 = 360°$, so $20° + x = 360° - 61° - 119° - 45° \times 2 = 90°$.

   In Figure S14.2, $61° + 119° + 31° + 121° + 45° \times 4 + y = (5-2) \cdot 180°$, so $31° + 121° + y = 540° - 61° - 119° - 45° \times 4 = 180°$.

   Following the same pattern, in Figure S14.3 we have

   $$a+b+c+\cdots+d = (n+1+2-2) \cdot 180° - 45° \times 2n - 61° - 119° = n \times 90°.$$

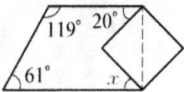

Fig. S14.1

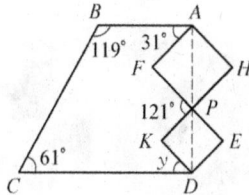

Fig. S14.2

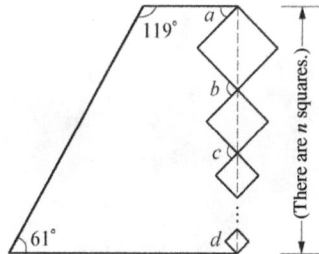

Fig. S14.3

Solution 23

Proportion of Segments

1. According to the given condition, $a = \frac{3}{4}b, c = \frac{2}{3}b, d = \frac{5}{6}b$. Thus,

$$1\frac{ac}{b^2 + d^2} = \frac{\frac{3}{4}b \cdot \frac{2}{3}b}{b^2 + \left(\frac{5}{6}b\right)^2} = \frac{18}{61}.$$

   The answer is (B).

2. Suppose $\frac{a-c}{b} = \frac{c}{a+b} = \frac{b}{a} = k > 0$, then $b = ak, c = (a+b)k = (a+ak)k, a - c = bk$. Thus, $a - (a+ak)k = ak^2$. Since $a > 0$, we have $2k^2 + k - 1 = 0$, so $k = \frac{1}{2}$ ($k = -1$ is dropped). Hence, $b = \frac{1}{2}a, a = 2b$, so $2b - c = \frac{1}{2}b, 3b = 2c$, and the answer is (A).

3. Since $BC : CD = 4 : 1$, we have $BC = 4CD, BD = 5CD$. Since $l_1 \parallel l_2$, we have $AG : BD = AG : 5CD = AF : FB = 2 : 5$, and $AG : CD = 2 : 1$. Hence, $AE : EC = 2 : 1$, and the answer is (B).

4. Since $ABCD$ is a parallelogram and AG bisects $\angle BAD$, we have $AD = DF, AB = BG$. Since $DF \parallel AB$, we have $\frac{AE}{EF} = \frac{AB}{DF}$, or equivalently $\frac{AE}{EF} = \frac{BG}{AD}$.
   Also, $AD \parallel BG$, so $\frac{BG}{AD} = \frac{EG}{AE}$. Thus, $\frac{AE}{EF} = \frac{EG}{AE}$, so $AE^2 = EG \cdot FG$, and the answer is (C).

5. Construct $EG \parallel AD$, which intersects BC at G.

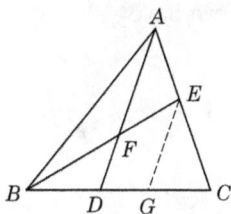

   Then, $\frac{DG}{GC} = \frac{AE}{EC} = \frac{3}{4}$, and since $\frac{BD}{DC} = \frac{2}{3}$, we obtain that $\frac{CD}{CG} = \frac{7}{4}, \frac{BD}{DG} = \frac{14}{9}, \frac{BD}{BG} = \frac{14}{23}$.

Also, $\frac{BF}{FE} = \frac{BD}{DG}$, so $\frac{BF}{FE} = \frac{14}{9}$.

Since $\frac{FD}{EG} = \frac{BD}{BG}$, we have $\frac{FD}{EG} = \frac{14}{23}$. Also, $\frac{AD}{EG} = \frac{CD}{CG}$, so $\frac{AD}{EG} = \frac{7}{4}$.

    Hence, $\frac{AD}{FD} = \frac{23}{8}$, and $\frac{AF}{FD} = \frac{15}{8}$. Therefore, $\frac{AF}{FD} \cdot \frac{BF}{FE} = \frac{15}{8} \times \frac{14}{9} = \frac{35}{12}$, and the answer is (C).

6. Let $\frac{3}{x+y} = \frac{4}{y+z} = \frac{5}{z+x} = \frac{1}{k}$, then

$$\begin{cases} x + y = 3k, \\ y + z = 4k, \\ z + x = 5k. \end{cases} \implies \begin{cases} x = 2k, \\ y = k, \\ z = 3k. \end{cases}$$

    Thus, $\frac{x^2+y^2+z^2}{xy+yz+zx} = \frac{(2k)^2+k^2+(3k)^2}{2k\cdot k+k\cdot 3k+3k\cdot 2k} = \frac{14}{11}$.

7. Since CD bisects $\angle ACB$, we have $\angle BCD = \angle ECD$. Since $DE \parallel BC$, we have $\angle ECD = \angle CDE$, so $\angle ECD = \angle CDE, DE = EC$.

    Now, $\frac{AD}{DB} = \frac{AC}{BC} = \frac{b}{a}$, so $\frac{AD}{DB} = \frac{AE}{EC} = \frac{b}{a}, \frac{AE+EC}{EC} = \frac{b+a}{a}$, which means $\frac{AC}{DE} = \frac{a+b}{a}, \frac{b}{DE} = \frac{a+b}{a}$. Hence, $DE = \frac{ab}{a+b}$.

8. $\frac{1}{AB} + \frac{1}{CD} = 2$.

    Proof: Since $MO \parallel AB$, we have $\frac{MO}{AB} = \frac{DM}{AD}$.   ①

    Also, $MO \parallel CD$, so $\frac{MO}{CD} = \frac{AM}{AD}$.   ②

    From ① + ②, we obtain that $\frac{MO}{AB} + \frac{MO}{CD} = \frac{DM}{AD} + \frac{AM}{AD}$. Hence, $\frac{MO}{AB} + \frac{MO}{CD} = 1$, or, equivalently, $\frac{1}{AB} + \frac{1}{CD} = \frac{1}{MO}$.

    Next, we show that $OM = ON$.

    Since $\frac{MO}{AB} = \frac{DM}{AD}, \frac{ON}{AB} = \frac{CN}{BC}, \frac{DM}{AD} = \frac{CN}{BC}$, we see that $\frac{MO}{AB} = \frac{ON}{AB}$, and $MO = ON = \frac{MN}{2}$. Therefore, $\frac{1}{AB} + \frac{1}{CD} = \frac{1}{MO}, \frac{1}{AB} + \frac{1}{CD} = \frac{2}{MN}$, and since $MN = 1$, we conclude that $\frac{1}{AB} + \frac{1}{CD} = 2$.

9. Since $AB \parallel EF \parallel CD$, we have

$$\frac{CE}{CF} = \frac{AC}{BC}, \qquad \frac{DE}{CF} = \frac{BD}{BC}.$$

    Adding the two equalities, we have $\frac{EC+ED}{CF} = \frac{AC+BD}{BC}$, so $\frac{192}{CF} = \frac{240}{100}$, and $CF = 80$.

10. Construct $FG \parallel AB$, which intersects BC at G.

    Since $\frac{AB}{DB} = \frac{CF}{FA} = \frac{2}{3}$, we have $\frac{AB}{DB} = \frac{5}{3}, \frac{CF}{CA} = \frac{2}{5}, \frac{FG}{AB} = \frac{CF}{CA} = \frac{2}{5}$, and $FG = \frac{2}{5}AB$.

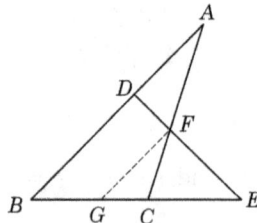

Hence, $\frac{EF}{ED} = \frac{FG}{DB} = \frac{\frac{2}{5}AB}{DB} = \frac{2}{5} \times \frac{5}{3} = \frac{2}{3}$, and $EF : DF = 2 : 1$.

11. Since $l \parallel BD$, it follows that

$$\frac{PN}{OD} = \frac{CP}{CO} = \frac{PR}{OB} \implies \frac{PN}{PR} = \frac{OD}{OB}; \qquad \text{①}$$

$$\frac{PM}{OB} = \frac{AP}{AO} = \frac{PS}{OD} \implies \frac{PS}{PM} = \frac{OD}{OB}. \qquad \text{②}$$

From ①, ②, we obtain that $\frac{PN}{PR} = \frac{PS}{PM}$, so $PM \cdot PN = PR \cdot PS$.

12. Construct $PK \parallel OQ, QL \parallel OR, RM \parallel OP$, which intersect BC, CA, AB at K, L, M, respectively.

It is easy to see that $PK = KQ = QL = OR = x$. Let $BK = y, QC = z$, then $x + y + z = BC = a$.

Since $\frac{PK}{AC} = \frac{BK}{BC}, \frac{QL}{AB} = \frac{QC}{BC}$, it follows that $\frac{x}{b} = \frac{y}{a}, \frac{x}{c} = \frac{z}{a}$, so we have $\frac{a}{a}x + \frac{a}{b}x + \frac{a}{c}x = a$. Hence, $x = \frac{1}{\frac{1}{a}+\frac{1}{b}+\frac{1}{c}}$.

13. The desired equality is equivalent to $\frac{1}{AD} - \frac{1}{EF} = \frac{2}{GH} - \frac{2}{BC}$, i.e. $\frac{GH \cdot BC}{AD \cdot EF} = \frac{2(BC-GH)}{EF-AD}$.

Since $\frac{BC-GH}{EF-AD} = \frac{BG}{AE}, \frac{BC}{AD} = \frac{BE}{AE}$, it also suffices to show that $\frac{GH}{EF} \cdot \frac{BE}{AE} = \frac{2BG}{AE}$, or, equivalently, $\frac{BE}{BG} = \frac{2EF}{GH}$.

Since $\frac{GH}{EF} = \frac{GN+NH}{EF} = \frac{BG}{BE} + \frac{CH}{CF}$ and $\frac{BG}{BE} = \frac{CH}{CF}$, we have $\frac{GH}{EF} = \frac{2BG}{BE}$, which means exactly $\frac{BE}{BG} = \frac{2EF}{GH}$. Therefore, the equality $\frac{1}{AD} + \frac{2}{BC} = \frac{1}{EF} + \frac{2}{GH}$ holds.

Solution 24

Similar Triangles

1. Since $ABCD$ is a parallelogram, we have $\triangle ADE \sim \triangle GBE, \triangle DEF \sim \triangle BEA$, so $\frac{AE}{EG} = \frac{DE}{EB}, \frac{DE}{EB} = \frac{EF}{AE}$. Hence, $\frac{AE}{EG} = \frac{EF}{AE}$, and $AE^2 = EF \cdot EG$. The answer is (B).

2. $BC = \sqrt{AB^2 - AC^2} = \sqrt{20^2 - 12^2} = 16$, and $BD = \frac{1}{2}AB = 10$. It follows easily that $\triangle ABC \sim \triangle EBD$, so $\frac{AC}{DE} = \frac{BC}{BD}$. Thus, $\frac{12}{DE} = \frac{16}{10}$, and $DE = 7.5$. The answer is (D).

3. Let $AP = x$, then $BP = 6 - x$.

   If $\frac{AP}{BP} = \frac{AD}{BC}$, then $\triangle PAD \sim \triangle PBC$, and $\frac{x}{6-x} = \frac{\sqrt{3}}{3\sqrt{3}}, x = 15$.

   If $\frac{AP}{BC} = \frac{AD}{BP}$, then $\triangle PAD \sim \triangle CBP$, and $\frac{x}{3\sqrt{3}} = \frac{\sqrt{3}}{6-x}$. Thus, $(x-3)^2 = 0$, and $x = 3$.

   Therefore, there are two such points P, and the answer is (B).

4. Since $\angle BAC = 90°$, and AD is the median on the hypotenuse, let E be the projection of D on AB, then $ED = \frac{1}{2}AC, AE = \frac{1}{2}AB, ED \parallel AC$.

   Thus, $S_{\triangle ADE} = \frac{120}{4} = 30, S_{\triangle ACE} = \frac{120}{2} = 60, \triangle DEF \sim \triangle ACF$.

   Hence, $\frac{S_{\triangle ADE}}{S_{\triangle ACE}} = \frac{1}{2}, \frac{DE}{AC} = \frac{1}{2}$ and $\frac{S_{DEF}}{S_{ACF}} = \frac{1}{4}$.

   Let $S_{\triangle DEF} = x, S_{\triangle AEF} = xt$, then $S_{\triangle ACF} = 4t$.

   Since $S_{\triangle ADE} = S_{\triangle DEF} + S_{\triangle AEF}, S_{\triangle ACE} = S_{\triangle ACF} + S_{\triangle AEF}$ and $\frac{S_{ADE}}{S_{ACE}} = \frac{1}{2}$, it follows that $2(t + xt) = xt + 4t$. Solve the equation, and we have $x = 2$, so $2S_{\triangle DEF} = S_{\triangle AEF}$. Since $S_{\triangle ADE} = 30$, we have $S_{\triangle ACF} = 30 \times \frac{2}{3} = 20$. The answer is (B).

5. Since $\frac{a}{b} = \frac{a+b}{a+b+c}$, we have $a(a + b + c) = b(a + b)$. Equivalently, $a(a + c) + ab = b^2 + ab$, so $a(a + c) = b^2$, and $\frac{a}{b} = \frac{b}{a+c}$.

   Extend CB to D so that $BD = AB$, and connect AD. Then, $CD = a + c$.

In $\triangle ABC$ and $\triangle DAC$, $\angle C$ is common, and $\frac{BC}{AC} = \frac{AC}{DC}$, so $\triangle ABC \sim \triangle DAC$, and $\angle BAC = \angle D$. Since $\angle BAD = \angle D$, we obtain that

$$\angle ABC = \angle D + \angle BAD = 2\angle D = 2\angle BAC.$$

Therefore, the answer is (B).

6. Since $\angle BAE + \angle DAE = \angle DAE + \angle AED = 90°$, we have $\angle BAE = \angle AED$, $\text{Rt}\triangle ABF \backsim \text{Rt}\triangle EAD$. Thus, $\frac{BF}{AD} = \frac{AF}{BE}$.
   Let x be the side length of the square, then $x\sqrt{x^2 - 9} = 2 \times 3$, which reduces to

$$x^4 - 9x^2 - 36 = 0, \quad (x^2 - 12)(x^2 + 3) = 0, \quad x^2 + 3 > 0,$$

so $x^2 - 12 = 0$, and $x = 2\sqrt{3}$.

7. Let AD be the angle bisector of $\angle A$ which intersects BC at D. Since $\angle A = 2\angle B$, we have $\angle CAD = \angle B = \angle BAD, AD = BD$. Since $\angle C$ is common, it follows that $\triangle ACD \sim \triangle BCA$, and $\frac{AC}{BC} = \frac{CD}{AC} = \frac{AD}{AB}$.
   Let $BC = a$, $CD = x$, then $\frac{4}{a} = \frac{x}{4} = \frac{a-x}{5}$, so $a = 6$.

8. Construct $GA \perp AC$ such that $AG = CF$, and connect EG, FG. Since $AC = BC, BE = ED = CF$, we have $CE = AF$, so $\triangle ECF \cong \triangle FAG$. Thus, $EF = FG, \angle CEF = \angle AFG$.

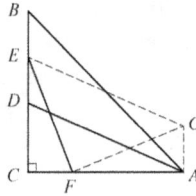

Since $\angle CEF + \angle EFC = 90°$, we have $\angle AFG + \angle EFC = 90°$, so $\angle EFG = 90°$, which means $\triangle EFG$ is an isosceles right triangle, and $\angle FEG = 45°$.

$GA \perp AC, BC \perp AC$, so $BC \parallel AG$, and $ED = CF = AG$, so $EDAG$ is a parallelogram, and $EG \parallel DA$. Thus, $\angle ADC = \angle CEG = \angle CEF + \angle FEG, \angle CEF = \angle ADC - \angle FEG$, and further, $\angle CEF + \angle CAD = \angle ADC - \angle FEG + \angle CAD = 90° - 45° = 45°$.

9. Construct $GM \perp EC$, which intersects the extension of EC at M. Since $AD \parallel EC$, we have $\triangle ADF \sim \triangle BEF$. This means $\frac{AD}{EB} = \frac{FD}{EF} = \frac{1}{2}$. Since $AD = 5$, we have $EB = 10$, $EC = 15$.

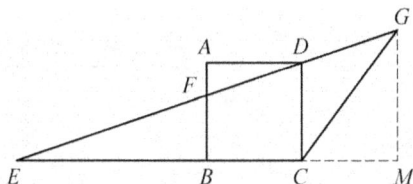

Since $DC \parallel GM$, it follows that $\triangle ECD \sim \triangle EMG$, and $\frac{DC}{GM} = \frac{ED}{EG} = \frac{3}{4}$. Thus, $GM = \frac{20}{3}$.

Therefore, $S_{\triangle ECG} = \frac{1}{2} EC \cdot GM = \frac{1}{2} \times 15 \times \frac{20}{3} = 50 (\text{cm}^2)$.

10. Since $\angle MBC = 90°$, $BP \perp CM$ we have $\triangle BMP \sim \triangle CBP$, and $\frac{BM}{BC} = \frac{BP}{CP}$.

Since $BC = CD$, $BM = BN$, we have $\frac{BN}{CD} = \frac{BP}{CP}$.

On the other hand, $\angle DCP + \angle PCB = 90°$, $\angle PBC + \angle PCB = 90°$, so $\angle DCP = \angle PBC$, which implies $\triangle BPN \sim \triangle CPD$, and $\angle NPB = \angle DPC$.

Also, $\angle NPB + \angle NPC = 90°$, so $\angle NPC + \angle CPD = 90°$, and $PD \perp PN$.

11. Since $\angle 2 = \angle 3$, we have $DE \parallel CA$, $\triangle BED \sim \triangle BAC$.

Since $BD = 6$, $DC = 2$, we have $\frac{m_1}{m} = \frac{BD}{BC} = \frac{3}{4}$.

Now, $\angle C = \angle C$, $\angle 2 = \angle 1$, so $\triangle CAD \sim \triangle CBA$, $\frac{AC}{CD} = \frac{CB}{AC}$. Equivalently, $AC^2 = CD \times CB = 2 \times 8 = 16$, so $AC = 4$. Hence,

$$\frac{m_2}{m} = \frac{AC}{BC} = \frac{4}{8} = \frac{1}{2}.$$

Therefore, $\frac{m_1 + m_2}{m} = \frac{5}{4}$.

12. It is easy to prove that $\triangle AP'B \sim \triangle CQB \sim \triangle CQ'D \sim \triangle ERD \sim \triangle ER'F \sim \triangle APF$, so their areas are proportional to the square of their corresponding side lengths. Let k be this ratio, then

$$S_{\triangle AP'B} = AB^2 k = a_1^2 k, \quad S_{\triangle CQB} = CB^2 k = b_1^2 k,$$

$$S_{\triangle CQ'D} = CD^2 k = a_2^2 k, \quad S_{\triangle ERD} = ED^2 k = b_2^2 k,$$

$$S_{\triangle ER'F} = EF^2 k = a_3^2 k, \quad S_{\triangle APF} = FA^2 k = b_3^2 k.$$

Since the non-overlapping regions of two equilateral triangles have the same area, we have

$$(a_1^2 + a_2^2 + a_3^2)k = (b_1^2 + b_2^2 + b_3^2)k.$$

Equivalently, $a_1^2 + a_2^2 + a_3^2 = b_1^2 + b_2^2 + b_3^2$.

Solution 25

The Midsegment

1. It is easy to prove that $EFGH$ is a rhombus, and its side length is 10 cm, so its perimeter is 40 cm, and the answer is (B).
2. Let G be the midpoint of AD, and connect EG, FG.

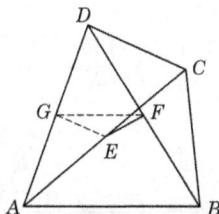

Since GE, EF are midsegments of $\triangle ACD, \triangle ABD$, respectively, we have $EG = \frac{1}{2}CD, GF = \frac{1}{2}AB$, so $GE + GF = \frac{1}{2}(AB + CD)$.
In $\triangle EFG$, $EF < GE + GF$, so we obtain $EF < \frac{1}{2}(AB + CD)$. The answer is (C).

3. Construct $MQ \parallel BP$ with Q lying on AC. Since $AM = BM$, we have $AQ = PQ$. Since $AP = 2PC$, it follows that $AQ = PQ = CP$. Now, $PN = \frac{1}{2}MQ, MQ = \frac{1}{2}BP$, so $BP = 4PN = 4$. The answer is (C).

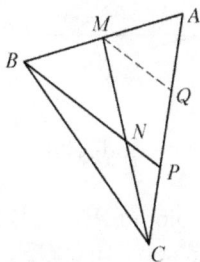

4. Let G be the midpoint of CD, and connect EG. Since E is the midpoint of AC, we have $EG \parallel FD$. Also, $CD = 2BD$, and it follows that $BD = DG$.

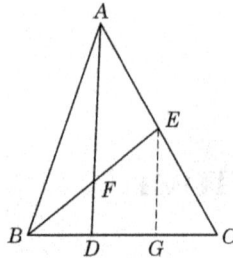

Since $\triangle BDF \sim \triangle BGE$, we have $\frac{S_{BDF}}{S_{BGE}} = \left(\frac{BD}{BG}\right)^2 = \frac{1}{4}$. Given the condition, $S_{\triangle BDF} = 1$, so $S_{\triangle BGE} = 4$. Hence, $S_{\triangle CGE} = 2$, $S_{\triangle ABC} = 2(4 + 2) = 12$. The answer is (B).

5. Extend BA, CD, and let them intersect at P. Since $\angle B = 30°, \angle C = 60°$, we have $\angle BPC = 90°$. In right triangle BPC, M, N, P are collinear, $PM = 3.5, PN = 0.5$, so $AN = 0.5, AD = 1$, and $EF = \frac{1}{2}(AD + BC) = 4$. The answer is (A).

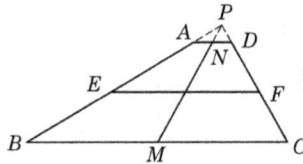

6. Construct $MD \parallel BP$, which intersects AP at D. Since M is the midpoint of AB, D is also the midpoint of AP.

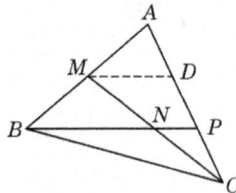

Since $AP = 2CP$, P is the midpoint of CD. Thus, $DM = 2PN = 2, BP = 2DM = 4$.

7. Apparently G is the midpoint of AC, so $AB = 2GF$. Since $EG + GF = 16, EG - GF = 4$, we can solve that $GF = 6$. Hence, $AB = 12\,(\text{cm})$.

8. Since AD bisects $\angle BAC$, we have $\angle EAD = \angle CAD$.

Also, $DE \parallel AC$, so $\angle CAD = \angle ADE$, and this implies $\angle EAD = \angle ADE, AE = ED$.

By $BD \perp AD$, we have $\angle ADE + \angle BDA = 90°$, and $\angle EAD + \angle EBD = 90°$, so $\angle BDE = \angle EBD$, and $BE = DE$.

Thus, $DE = \frac{1}{2}AB = 2.5$.

9. Let M be the midpoint of BC, and let N be the intersection of AM and BP.

   Since E is the midpoint of CD and F is the midpoint of AD, we can derive that $CF \perp BE, AM \perp BE$, so $AM \parallel CF$, and N is the midpoint of BP. Hence, $AB = AP$.

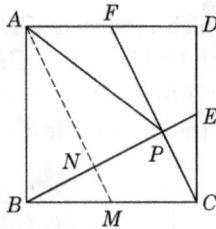

Since $\angle CBP = 27°$, we have $\angle ABP = 63°$, $\angle BAP = 180° - 2\angle ABP = 54°$.

10. Construct $PG \perp l$, which intersects l at G. Since $BE \perp l, PG \perp l$, $CF \perp l$, we have $BE \parallel PG \parallel CF$. Also, P is the midpoint of BC, so PG is the midsegment of the trapezoid $BCFE$.

    Since $S_{\triangle PEF} = \frac{1}{2}PG \cdot EF = \sqrt{2}$, we have $PG \cdot EF = 2\sqrt{2}$.

    Hence, the area of $BCFE$ is $\frac{1}{2}(BE + CF) \cdot EF = PG \cdot EF = 2\sqrt{2}$.

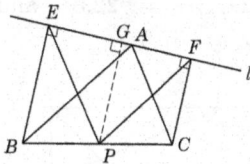

11. Let H be the midpoint of BC, and connect MH, NH.

    Since M, H are the midpoints of BE, BC, respectively, we have $MH \parallel EC$, and $MH = \frac{1}{2}EC$.

    Since N, H are the midpoints of CD, BC, respectively, we have $NH \parallel BD$, and $NH = \frac{1}{2}BD$.

    Now, $BD = CE$, so $MH = NH, \angle HMN = \angle HNM$.

Since $MH \parallel EC$, we have $\angle HMN = \angle PQA$, and similarly, $\angle HNM = \angle QPA$, so $\angle PQA = \angle QPA$. Therefore, $\triangle APQ$ is an isosceles triangle, where $AP = AQ$.

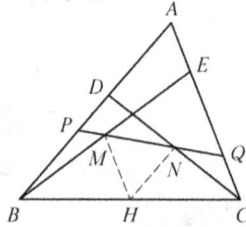

12. Construct $EF \parallel AD$, which intersects CD at F. Since E is the midpoint of AB, F is also the midpoint of CD, and $EF = \frac{1}{2}CD = DF$. Thus, $\angle DEF = \angle EDF$. Also, $\angle DEF = \angle ADE$, so $\angle ADE = \angle EDF$. Finally, $EA \perp AD$, so the distance from E to CD equals EA.

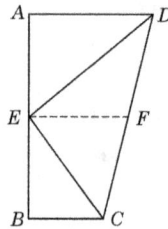

13. Let E be the midpoint of CG, and connect OE. Then, OE is a midsegment of $\triangle ACG$, $OE \parallel AG$, and $\angle CAG = \angle COE$.

Since AG bisects $\angle CAB$, we have $\angle CAG = \angle BAG$. Since $\angle CAB = 45°$, it follows that $\angle CAG = 22.5°$, and $\angle COE = 22.5°$. Here, $\angle COB = 90°$, so $\angle BOE = 67.5°$.

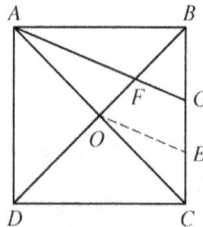

Note that $\angle CBD = 45°$, so $\angle BEO = 67.5°$. Therefore, $OEGF$ is an isosceles trapezoid. Thus, $OF = GE$, and $CG = 2OF$.

14. Let the extensions of BE, AC intersect at G, then it follows that $\triangle ABE \cong \triangle ACE$. Thus, $BE = GE, AB = AG$. Since $BM = CM$, we have

$$ME = \frac{1}{2}CG = \frac{1}{2}(AG - AC) = \frac{1}{2}(AB - AC).$$

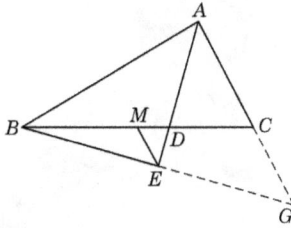

Solution 26

Translation and Symmetry

1. By the definition of translation, we have $BC \parallel EF$, so $\triangle DBG \sim \triangle DFE$. Thus, $\dfrac{\text{perimeter of}\triangle DBG}{\text{perimeter of}\triangle DEF} = \dfrac{DB}{DE} = \dfrac{1}{2}$. In right triangle ABC, it is easy to get $AC = 6\,\text{cm}$, so the perimeters of $\triangle ABC$ and $\triangle DEF$ equal $10 + 8 + 6 = 24\,\text{cm}$, and the perimeter of $\triangle DBG$ is $12\,\text{cm}$.

2. $2\angle A' = \angle 1 + \angle 2$. Connect AA', then $\angle 1 = \angle DAA' + \angle DA'A$, $\angle 2 = \angle EAA' + \angle EA'A$, and $DA = DA', EA = EA'$. Therefore, $\angle 1 = 2\angle DA'A, \angle 2 = 2\angle EA'A$, and $2\angle A' = \angle 1 + \angle 2$.

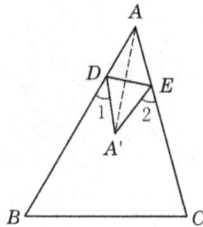

3. As shown in the figure, by the property of axial symmetry, we know that the hypotenuse of $\triangle IJK$ equals the perimeter of $EFGH$. Since the two legs of $\triangle IJK$ are exactly twice the side lengths of $ABCD$, it follows that the hypotenuse equals twice the diagonal of the rectangle, which is $2 \times 2 = 4$.

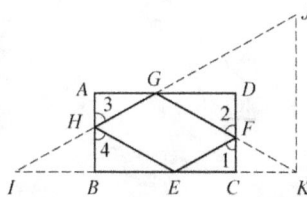

4. (1) When P is the midpoint of AB, P_3 coincides with P, and in this case, the distance the squirrel travels is the sum of the three midsegments of the triangle, which is $\frac{1999}{2}$ (cm).

   (2) When P is not the midpoint of AB, we observe that:

$$P_1P_2 \parallel AB, \quad P_2P_3 \parallel CA, \quad P_3P_4 \parallel BC, \quad P_4P_5 \parallel AB,$$

$$P_5P_6 \parallel CA, \quad P_6P_7 \parallel BC.$$

Hence, $\triangle P_3P_2B$ can be obtained by translating $\triangle AP_1P$, $\triangle P_4CP_5$ can be obtained by translating $\triangle P_3P_2B$, and $\triangle AP_7P_6$ can be obtained by translating $\triangle P_4CP_5$. This implies $AP = AP_6, AP_1 = AP_7$, so P_6 coincides with P, and P_7 coincides with P_1. Therefore, the squirrel travels back to P after six turnings, and the distance it travels is

$$PP_1 + P_1P_2 + P_2P_3 + P_3P_4 + P_4P_5 + P_5P$$

$$= BP_2 + AP_3 + AP_1 + CP_2 + BP_3 + CP_1$$

$$= AB + BC + CA = 1999 \, (\text{cm}).$$

5. Since $ABCD$ is a square, we have $AB = BC, \angle CBD = 45°$. By the definition of folding, we have $A'B = AB$. Thus, $\triangle A'BC$ is isosceles with $A'B = BC$, and $\angle CBD = 45°$. Since $\angle BA'C = \angle BCA'$, it follows that $\angle BA'C = \frac{1}{2}(180° - \angle CBD) = \frac{180° - 45°}{2} = 67.5°$.

6. Construct $\triangle AED$, which is in axial symmetry with $\triangle ABD$ with respect to AD. Then, $\angle EAD = 21°, AE = AB$, so $DE = BD$.

It is easy to see that $\angle ADC = 21° + 46° = 67°$, so $\angle ADE = \angle ADB = 180° - 67° = 113°, \angle CDE = 113° - 67° = 46°$.

Connect CE. Since $DC = AB$, it follows that $\triangle CDE \cong \triangle ABD \cong \triangle AED$.

Let O be the intersection of AE, CD, then $\angle AOC = \angle ADC + \angle DAE = 67° + 21° = 88°$. Since $\angle ODE = \angle OED = 46°$, we obtain $OD = OE$. Since $DC = AE$, we have $AO = CO$, so $\angle OCA = \angle OAC = 46°$. Therefore, $\angle DAC = \angle DAE + \angle EAC = 67°$.

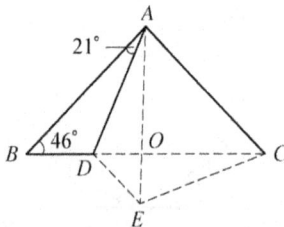

7. It follows that $\angle BAF = \angle CEF = 45°$, so $\triangle CEF$ is an isosceles right triangle. Since $EC = 8 - 6 = 2$, we have $S_{\triangle CEF} = \frac{1}{2} \times 2^2 = 2$.

8. (1) $DE = EF, NE = BF$.

   Since BF bisects $\angle CBM$, we have $\angle CBF = \angle MBF = 45°$, so $\angle EBF = 135°$.

   In square $ABCD$, $AN = AE, \angle A = 90°$, so $\angle ANE = \angle AEN = 45°$, and $\angle DNE = 135° = \angle EBF$. Since $\angle NDE + \angle DEA = 90°, \angle BEF + \angle DEA = 90°$, we obtain that $\angle NDE = \angle BEF$. It is easy to prove that $DN = EB$, so we can derive that $\triangle DNE \cong \triangle EBF$, which implies $DE = EF, NE = BF$.

   (2) Choose a point N on AD such that $DN = EB$, then N satisfies the condition. In this case, $DE = EF$, and the proof is similar to (1).

9. Translate $\triangle A'OC$ in direction $A'A$ by distance $|A'A|$ to obtain $\triangle AQR$, and translate $\triangle BOC'$ in direction BB' by distance $|BB'|$ to obtain $\triangle B'PR$.

Since $OP = OQ = 2, \angle POQ = 60°$, we have $PQ = 2$. However, $QR + RP = OC + OC' = 2$, so P, R, Q are collinear. Therefore, in equilateral triangle POQ, we have

$$S_{\triangle AOB'} + S_{\triangle BOC'} + S_{\triangle COA'} = S_{\triangle AOB'} + S_{\triangle B'PR} + S_{\triangle AQR}$$

$$< S_{\triangle OPQ} = \frac{\sqrt{3}}{4} \cdot 2^2 = \sqrt{3}.$$

10. (1) $PB = PQ$.

   Construct $MN \parallel BC$, which passes through P and intersects AB, DC at M, N, respectively. Then, $AMND$ and $BCNM$ are both rectangles, and $\triangle AMP, \triangle CNP$ and both isosceles right triangles. Thus, $NP = NC = MB$.

   Since $\angle BPQ = 90°$, we have $\angle QPN + \angle BPM = 90°$. Also, $\angle PBM + \angle BPM = 90°$, so $\angle QPN = \angle PBM$ Combining with $\angle QNP = \angle PMB = 90°$, we conclude that $\triangle QNP \cong \triangle PMB$, and $PQ = PQ$.

(2) It is possible.

    Case 1: When P coincides with A, and Q coincides with D, we have $PQ = QC$, and $\triangle PQC$ is isosceles. In this case, $x = 0$.

    Case 2: When Q lies on the extension of DC, and $CP = CQ$, $\triangle PCQ$ is also isosceles.

Solution   In this case, $QN = PM = \frac{\sqrt{2}}{2}x, CP = \sqrt{2} - x, CN = \frac{\sqrt{2}}{2}CP = 1 - \frac{\sqrt{2}}{2}x$. Thus, $CQ = QN - CN = \frac{\sqrt{2}}{2}x - \left(1 - \frac{\sqrt{2}}{2}x\right) = \sqrt{2}x - 1$. By $\sqrt{2} - x = \sqrt{2}x - 1$, we obtain $x = 1$.

Solution   In this case, $\angle CPQ = \frac{1}{2}\angle PCN = 22.5°$, $\angle APB = 90° - 22.5° = 67.5°$, so $\angle ABP = 180° - (45° + 67.5°) = 67.5°$. Thus, $\angle APB = \angle ABP$, and $AP = AB = 1$, which means $x = 1$.

11. Since $ABCD$ is a rhombus, we have $AD \parallel BC$, and $\angle A = 120°$, so $\angle B = 60°$. Let P' be the symmetric point of P with respective to BD, and connect $PP', P'Q, PC$. Then, the length of $P'Q$ is exactly the minimum of $PK + QK$. From the figure, we see that when Q coincides with C and $CP' \perp AB$, $PK + QK$ is minimal.

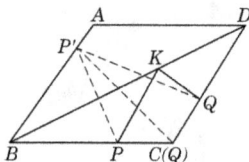

12. In $\triangle PMN$, since $PM = PN$, $\angle P = 90°$, we have $\angle PMN = \angle PNM = 45°$. Let the extension of AD intersect PM, PN at G, H, respectively, and let F, T be the projections of G, H on MN, respectively.

    Since $DC = 2$ cm, we have $MF = GF = 2$ cm, $TN = HT = 2$ cm. Since $MN = 8$ cm, we have $MT = 6$ cm. Thus, during the process that $ABCD$ moves until it stops, there are three steps:

(1) C is moving from M to F ($0 \le x \le 2$). As shown in Figure S12.1, let E be the intersection of CD, PM. Then, the overlapping is $\triangle MCE$, where $MC = EC = x$.

    Thus, $y = \frac{1}{2}MC \cdot EC = \frac{1}{2}x^2 (0 \le x \le 2)$.

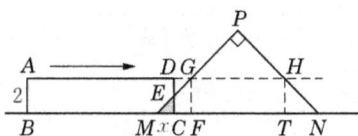

Fig. S12.1

(2) C is moving from F to T $(2 < x \le 6)$. As shown in Figure S12.2, the overlapping is a right trapezoid $MCDG$. Since $MC = x, MF = 2$, we have $FC = DG = x - 2, DC = 2$.

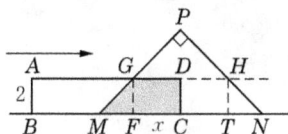

Fig. S12.2

Thus, $y = \frac{1}{2}(MC + GD) \cdot DC = 2x - 2(2 < x \le 6)$.

(3) C is moving from T to N $(6 < x \le 8)$. As shown in Figure S12.3, let Q be the intersection of CD, PN. Then, the overlapping is the pentagon $MCQHG$.

Now, $MC = x$, so $CN = CQ = 8 - x$, and $DC = 2$. Hence,

$$y = \frac{1}{2}(MN + GH) \cdot DC - \frac{1}{2}CN \cdot CQ$$

$$= -\frac{1}{2}(x - 8)^2 + 12 = -\frac{1}{2}x^2 + 8x - 20(6 < x \le 8).$$

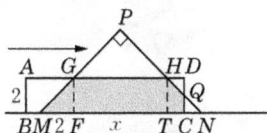

Fig. S12.3

Solution 27

The Area

1. Let a be the side length of $EFGB$, then by $\triangle CBM \sim \triangle CGF$, we have $\frac{2}{2+a} = \frac{BM}{a}$, so
$$2a = 2BM + a \cdot BM.$$
$$S = \frac{1}{2}AM \cdot (EF + BC) = \frac{1}{2}(2 - BM)(2 + a)$$
$$= \frac{1}{2}(4 - 2BM + 2a - a \cdot BM) = 2.$$
   The answer is (A).

2. Let E, H, Q be the projections of A, M, B on DC, respectively, then $AE \parallel MH \parallel BQ$. Since M is the midpoint of AB, H is also the midpoint of EQ, and MH is the midsegment of the trapezoid $AEQB$, $2MH = AE + BQ$.

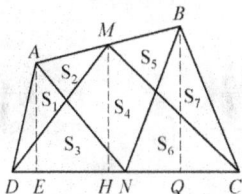

   Now, we have $S_3 + S_4 + S_6 = S_{\triangle MDC} = \frac{1}{2}DC \cdot MH$, $S_6 + S_7 = S_{\triangle BNC} = \frac{1}{2}NC \times BQ$, $S_1 + S_3 = S_{\triangle ADN} = \frac{1}{2}DN \times AE$. Since $DN = CN$, it follows that
$$S_6 + S_7 + S_1 + S_3 = \frac{1}{2}NC \times AE + \frac{1}{2}DN \times AE = \frac{1}{2}CD \times MH.$$
   Hence, $S_6 + S_7 + S_1 + S_3 = S_3 + S_4 + S_6$, and $S_1 + S_7 = S_4$. The answer is (B).

3. Since the four squares have areas of 25, 1144, 48, 121, their side lengths are $5, 12, 4\sqrt{3}, 11$, respectively. Since $PR = 13, PS = 12, RS = 5$, we have $PS \perp SR$. Since $PQ = 4\sqrt{3}, QR = 11$, we also have $PQ \perp QR$. Hence,

$$S_{PQRS} = \frac{1}{2}(PS \cdot SR + PQ \cdot QR) = 30 + 22\sqrt{3}.$$

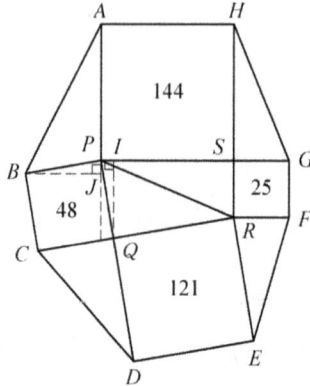

Apparently, $S_{\triangle HSG} + S_{\triangle CDQ} = S_{PQRS}$. Construct $QI \perp PS$ with I on PS, and $BJ \perp AP$ with J on the extension of AP. Since $BP = PQ, \angle BJP = \angle QIP = 90°$, and $\angle APB + \angle QPS = 360° - 90° - 90° = 180°$, which implies $\angle QPS = \angle BPJ$, we obtain that $\triangle PQI \cong \triangle PBJ$. Hence, $QI = BJ$, and $S_{\triangle APB} = S_{\triangle PSQ}$. Similarly, $S_{\triangle EFR} = S_{\triangle QSR}$, and $S_{\triangle APB} + S_{\triangle EFR} = S_{PQRS}$. Therefore, the area of the octagon is

$$3(30 + 22\sqrt{3}) + 144 + 48 + 121 + 25 = 428 + 66\sqrt{3}.$$

The answer is (A).

4. Connect ED. Since $\frac{AG}{GE} = \frac{S_{\triangle ABD}}{S_{\triangle BDE}} = \frac{S_{\triangle BDC}}{S_{\triangle BDE}} = \frac{BC}{BE} = \frac{2}{1}$, we have $\frac{S_5}{S_2} = \frac{AG}{GE} = \frac{2}{1}$. In other words, $S_5 = 2S_2$. Similarly, we have $S_4 = 2S_3$, and since $S_{\triangle ABE} = S_{\triangle ADF} = \frac{1}{4}$, we have

$$S_2 = S_3 = \frac{1}{12}S, \quad S_4 = S_5 = \frac{1}{6}S.$$

Also, $S_1 + S_4 + S_5 = S_{\triangle ABD} = \frac{1}{2}S$, so $S_1 = \frac{1}{6}S$, and $BG:GH:HD = S_5:S_1:S_4 = 1:1:1$.

$$S_6 = S_{ABCD} - S_{\triangle ABD} - S_2 - S_3 = \left(1 - \frac{1}{2} - 2 \times \frac{1}{12}\right)S = \frac{1}{3}S.$$

Therefore, $S_2:S_4:S_6 = \frac{1}{12}:\frac{1}{6}:\frac{1}{3} = 1:2:4$, and the answer is (D).

5. $S_{\triangle ABC} = \frac{1}{2}a \cdot h_a = \frac{1}{2}b \cdot h_b = \frac{1}{2}c \cdot h_c$. Since $a = 3, b = 4, c = 6$, we have $3h_a = 4h_b = 6h_c$. Hence, assume $\frac{h_a}{4} = \frac{h_b}{3} = \frac{h_c}{2} = k$, then $h_a = 4k, h_b = 3k, h_c = 2k$. Thus,

$$(h_a + h_b + h_c)\left(\frac{1}{h_a} + \frac{1}{h_b} + \frac{1}{h_c}\right) = (4k + 3k + 2k)\left(\frac{1}{4k} + \frac{1}{3k} + \frac{1}{2k}\right)$$

$$= 9k \times \frac{13}{12k} = \frac{39}{4}.$$

The answer is (B).

6. Construct the perpendicular line from G to AB, which intersects AB, CD at P, H, respectively. Then, $\frac{GH}{DF} = \frac{CG}{CF} = \frac{2}{5}$, and $GH = \frac{2}{5}, GP = \frac{8}{5}, S_{\triangle BEG} = \frac{4}{5}$.

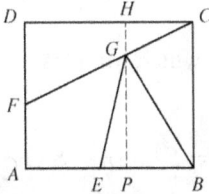

7. Let E, F be the projections of B, D on AC, respectively. Then, $\triangle BEO \cong \triangle DFO$, and $\frac{BE}{DF} = \frac{BO}{DO}$. Thus,

$$\frac{S_{\triangle ABC}}{S_{\triangle ADC}} = \frac{\frac{1}{2}AC \cdot BE}{\frac{1}{2}AC \cdot DF} = \frac{BE}{DF} = \frac{BO}{DO}.$$

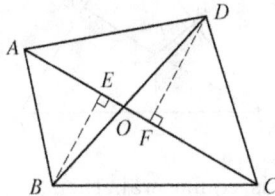

Given the condition, we have $\frac{BO}{DO} = \frac{S_{\triangle ABC}}{S_{\triangle ACD}} = \frac{S_1}{S_2} = \frac{5}{10} = \frac{1}{2}$, and $\frac{S_{\triangle ABO}}{S_{\triangle ABD}} = \frac{S_{\triangle ABO}}{S_3} = \frac{BO}{BD} = \frac{1}{3}$. Therefore, $S_{\triangle ABO} = \frac{1}{3} \times S_3 = \frac{1}{3} \times 6 = 2$.

8. Connect BD, and let S_1, S_2, S_3 denote the areas of $\triangle BFC, \triangle DEF, \triangle FDB$, respectively. Since $S_{ABCD} = 30$, we have $S_3 + S_2 = 15$. Also, $S_2 = S_1 + 9$, so $S_3 + S_2 = S_1 + 9 + S_3$, which means $S_1 + S_3 = 15 - 9 = 6$.

Since $BC = AD = 5$, the altitude of the parallelogram $ABCD$ is $h = \frac{30}{5} = 6$.

Thus, $S_1 + S_3 = \frac{1}{2} \times DE \times h = \frac{1}{2} \times DE \times 6 = 6$. Therefore, $DE = 2\,\mathrm{cm}$.

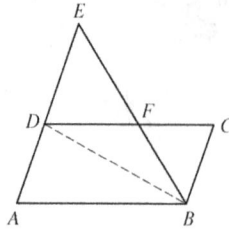

9. We see from the figure that triangles X, T have the same altitude, and Y, T have the same altitude. Thus, $\frac{S_X}{S_T} = \frac{BE}{ED} = \frac{80}{40} = 2$, so $S_X = 2S_T$.
   Similarly, $\frac{S_Y}{S_T} = \frac{CE}{EA} = \frac{60}{30} = 2$, so $S_Y = 2S_T$.
   Also, $\frac{S_Z}{S_X} = \frac{CE}{EA} = \frac{60}{30} = 2$, so $S_Z = 2S_X = 4S_T$. Therefore, $\frac{S_Z + S_T}{S_X + S_Y} = \frac{4S_T + S_T}{2S_T + 2S_T} = \frac{5}{4}$.

10. Connect PA, PB, PC, which divides $\triangle ABC$ into three triangles. Then,

$$S_{\triangle ABC} = S_{\triangle PAB} + S_{\triangle PCB} + S_{\triangle PCA}$$

$$= \frac{1}{2}cz + \frac{1}{2}az + \frac{1}{2}by.$$

Hence, $ax + by + cz = 2S_{\triangle ABC}$, which means $ax + by + cz$ is a constant.

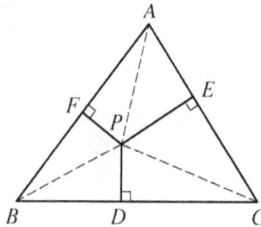

Remark: If $\triangle ABC$ is equilateral, then $x + y + z = \frac{2S_{\triangle ABC}}{a} = h$. Equivalently, the distances from any point inside the equilateral triangle to the three sides sum to a constant, which is the altitude of the equilateral triangle.

11. Construct a line through A parallel to BC, a line through B parallel to AC, and a line through C parallel to AB. The three intersections of these lines all satisfy the condition, and the area of $\triangle PAB$ is 1. The center of $\triangle ABC$ also satisfies the condition, and in this case, the area of $\triangle PAB$ is $\frac{1}{3}$.

12. Consider three cases:

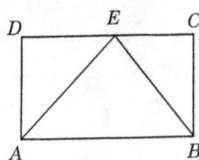
Fig. S12.1

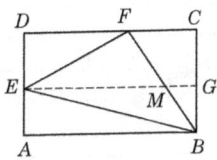
Fig. S12.2

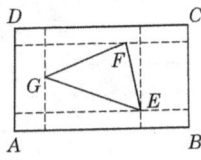
Fig. S12.3

(1) As shown in Figure S12.1, $S_{\triangle ABE} = \frac{1}{2}S$.

(2) As shown in Figure S12.2, let G be the projection of E on BC and M is the intersection of EG, FB. Apparently, $S_{\triangle EFM} < \frac{1}{2}S_{CDEG}$, $S_{\triangle BEM} < \frac{1}{2}S_{ABGE}$, so $S_{\triangle BEF} < \frac{1}{2}S$.

(3) As shown in Figure S12.3, construct lines parallel to the sides of $ABCD$ that pass through E, F, G, to obtain a rectangle that encloses $\triangle EFG$. From (1) and (2), we see that $S_{\triangle EFG}$ does not exceed half the area of this rectangle. Hence, $S_{\triangle EFG} < \frac{1}{2}S$.

In summary, the desired assertion is proven.

13. Connect AE, DC, FB.

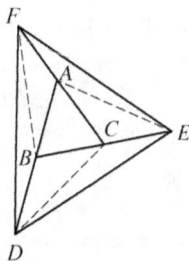

Since $\triangle EFA, \triangle ACE$ have the same base and altitude, we have $S_{\triangle FAE} = S_{\triangle ACE}$.

Since $\triangle ABC, \triangle ACE$ have the same base and altitude, we have $S_{\triangle ABC} = S_{\triangle ACE} = 1$.

Similarly,

$$S_{\triangle ABF} = S_{\triangle FBD} = S_{\triangle BDC} = S_{\triangle DCE} = 1.$$

Therefore, $S_{\triangle DEF} = 7$.

14. (1) As shown in Figure S14.1, the area of $\triangle MON$ is minimal when $?N$, so the answer is (D). The reason is explained in (3).

the area of $\triangle MON$ is minimal when $PM = PN$. Since is a right triangle, we have $OP = \frac{1}{2}MN$, so $MN = 2a$.

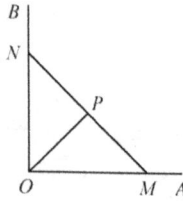

Fig. S14.1

Since $\angle POM = 30°$, we have $\angle PMO = 30°$, and $NO = a, MO = \sqrt{3}a$. Thus, $S_{\triangle MON} = \frac{1}{2}NO \cdot MO = \frac{\sqrt{3}}{2}a^2$.

(3) The construction method is explained as follows and shown in Figure S14.2.

① Construct $PC \parallel OA$, which intersects OB at C.
② Choose C on OB such that $CN = OC$.
③ Connect NP, and extend to intersect OA at M.

Then, MN is the desired segment. In this case, $PC \parallel OM, OC = CN$, so $PM = PN$, and the area of $\triangle MON$ is minimal.

Proof of Minimality: If there is another line EF through P which intersects OA, OB at E, F, respectively, we construct $NG \parallel OA$, which intersects EF at G, then $\triangle PEM \cong \triangle PGN$ (ASA). Thus, $S_{\triangle PEM} = S_{\triangle PNG} < S_{\triangle PNF}$. Consequently,

$$S_{\triangle MON} = S_{OEPN} + S_{\triangle PEM} < S_{OEPN} + S_{\triangle PNF} = S_{\triangle OEF}.$$

On the other hand, if $E'F'$ passes through P and intersects OA, OB at E', F', respectively (as shown in Figure S14.3), then we construct $MG' \parallel OB$, which intersects $E'F'$ at G'. For similar reasons, we have $S_{\triangle MON} < S_{\triangle OE'F'}$.

Therefore, $\triangle MON$ has the minimal area.

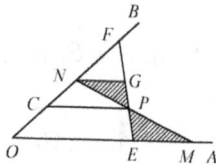

Fig. S14.2

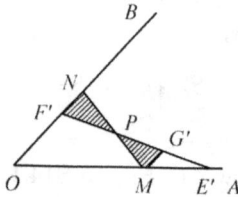

Fig. S14.3